Automotive Mathematics

Jason C. Rouvel
Western Wisconsin Technical College
La Crosse, Wisconsin

Upper Saddle River, New Jersey
Columbus, Ohio

Library of Congress Cataloging-in-Publication Data

Rouvel, Jason C.
 Automotive mathematics / Jason C. Rouvel.
   p. cm.
 Includes index.
 ISBN 0-13-114873-7 (alk. paper)
  1. Automobiles—Design and construction—Mathematics. 2. Engineering mathematics.
I. Title.

 TL154.R686 2007
 629.2'310151—dc22   2005058666

Editor in Chief: Vernon R. Anthony
Editor: Gary Bauer
Assistant Editor: Linda Cupp
Editorial Assistant: Jacqueline Knapke
Production Editor: Kevin Happell
Design Coordinator: Diane Ernsberger
Cover Designer: Candace Rowley
Cover art: Jason Rouvel
Production Manager: Pat Tonneman
Marketing Manager: Leigh Ann Sims

This book was set in Sabon by Techbooks. It was printed and bound by Bind Rite Graphics. The cover was printed by Phoenix Color Corp.

Copyright © 2007 by Pearson Education, Inc., Upper Saddle River, New Jersey 07458.
Pearson Prentice Hall. All rights reserved. Printed in the United States of America. This publication is protected by Copyright and permission should be obtained from the publisher prior to any prohibited reproduction, storage in a retrieval system, or transmission in any form or by any means, electronic, mechanical, photocopying, recording, or likewise. For information regarding permission(s), write to: Rights and Permissions Department.

Pearson Prentice Hall™ is a trademark of Pearson Education, Inc.
Pearson® is a registered trademark of Pearson plc
Prentice Hall® is a registered trademark of Pearson Education, Inc.

Pearson Education Ltd.   Pearson Education Australia Pty. Limited
Pearson Education Singapore Pte. Ltd.   Pearson Education North Asia Ltd.
Pearson Education Canada, Ltd.   Pearson Educación de Mexico, S.A. de C.V.
Pearson Education—Japan   Pearson Education Malaysia Pte. Ltd.

10 9 8 7 6 5 4 3 2 1
ISBN: 0-13-114873-7

preface

Most industrial trades rely heavily on a workforce with a sound knowledge of basic mathematical skills. The automotive trades are no exception to this, and workers who not only possess basic mathematical skills but also can apply those skills will have a distinct advantage in the workplace. This automotive mathematics text is intended to be a versatile tool for teaching basic mathematics skills alongside realistic applications.

Chapters 1–8 focus primarily on the development of basic mathematical skills. Applications are integrated in both the presentation of topics and the practice problems. Chapters 9–16 assume a working knowledge of the basic skills and focus on more involved applications. These later chapters develop all concepts and all formulas for greater understanding, as well as cover many topics not fully explored by other texts. These topics include engine balancing, camshaft event timing, modifying compression ratio, planetary gear ratios, and additional hydraulic topics.

Versatility, one of the primary goals of this text, is accomplished in two ways. First, there are more topics included than can likely be covered in the number of hours typically allotted for an automotive mathematics course. The instructor can therefore choose the most appropriate topics for a given curriculum. Second, the practice problems include problems at a broad range of skill levels. Students who are developing basic skills can practice basic automotive applications, and more proficient students can be challenged by the more involved application problems. All problems are categorized in basic (●), moderate (●●), or high (●●●) skill levels. In all chapters, attention has been paid to include the most realistic and practical applications possible.

Additional features include line drawings and photographs to more clearly illustrate concepts. The answers to odd-numbered problems are included in the text, and all solutions are available in an online instructor's solutions manual. Most even-numbered problems are similar to the odd-numbered problems so that similar assignments can be made with or without access to solutions. A computerized test bank (TestGen) is also available as a support resource.

To access supplementary materials online, instructors need to request an instructor access code. Go to **www.prenhall.com**, click the **Instructor Resource Center** link, and then click **Register Today** for an instructor access code. Within 48 hours after registering you will receive a confirming e-mail including an instructor access code. Once you have received your code, go the site and log on for full instructions on downloading the materials you wish to use.

ABOUT THE AUTHOR

Jason C. Rouvel is an applied mathematics instructor at Western Wisconsin Technical College in La Crosse, Wisconsin. His educational background includes a bachelor of science degree in mathematics with dual minors in physics and automotive engineering technology, and a master of arts degree in mathematics, all from Minnesota State University, Mankato. Mr. Rouvel has a strong interest in automotive technology and has worked closely with the Automotive Technology instructors at WWTC to develop a curriculum that is relevant for students entering this trade.

ACKNOWLEDGMENTS

I would like to thank the entire automotive technology department at Western Wisconsin Technical College for their input on the content of this text. Thanks also to Andrew Rybolt for his valuable feedback about many aspects of the project. In addition, sincere thanks are offered to the following reviewers who provided thoughtful and meaningful feedback about the early drafts: John Eichelberger, St. Philip's College; John Jacobs, Massachusetts Bay Community College; Michael Longrich, Cuyahoga Community College; Virginia E. Olson, Chippewa Valley Technical College; Victoria Seals, Athens Technical College; and Scott R. States, University of Northwestern Ohio.

I thank my family for their unconditional support with this and every other challenge I've ever taken on. Finally, I thank Rachel for her help, patience, and support on this and every new project.

contents

CHAPTER 1 *Whole Number Operations with General Applications* 1

Our Number System 1
Addition of Whole Numbers 2
Subtraction of Whole Numbers 2
Multiplication of Whole Numbers 4
Division of Whole Numbers 7
Exponents 9
Order of Operations 9
Practice Problems 11

CHAPTER 2 *Decimal Operations with Applications to Measurement and Finance* 20

Reading and Writing Decimal Numbers 20
Rounding Decimal Numbers 21
Addition and Subtraction of Decimal Numbers 22
Reading a Micrometer 24
Multiplication of Decimal Numbers 26
Division of Decimal Numbers 28
Square Roots 30
Practice Problems 31

CHAPTER 3 *Fractional Operations with Applications to Measurement* 43

Equivalent Fractions 43
Common Denominators 47
Addition of Fractions 49
Subtraction of Fractions 51
Multiplication of Fractions 54
Division of Fractions 56
Converting Fractions to Decimals 58
Practice Problems 59

CHAPTER 4 Ratio and Proportion 68

Equivalent Ratios 68
Writing and Solving Proportions 71
Finding Nearest Fractional Part 74
Practice Problems 75

CHAPTER 5 Percent and Percent Applications 79

Converting Fractions, Decimals, and Percents 79
Base, Rate, and Amount 82
Discount, Markup, and Sales Tax 84
Technical Applications 87
Practice Problems 89

CHAPTER 6 The Metric System and Unit Conversion 98

Metric Units of Measure 98
Metric Prefixes and Conversions 100
Conversion of Units Using Ratios 103
Conversion of Units Using Tables 105
Temperature Conversions 107
Practice Problems 108

CHAPTER 7 Geometry 113

Measurement 113
Perimeter 114
Pythagorean Theorem 117
Area 119
Volume 123
Angle Measure and Applications 126
Practice Problems 130

CHAPTER 8 Signed Numbers 152

Signed Numbers 152
Addition and Subtraction of Signed Numbers 154
Multiplication and Division of Signed Numbers 157
Practice Problems 158

CHAPTER 9 Engine 163

General Engine Measurements 163
Displacement 165
Compression Ratio 167
Changing Compression Ratio 175
Diesel Applications 176
Practice Problems 178

CHAPTER 10 Crankshafts and Camshafts 186

Engine Balancing 186
Torque and Horsepower 194
Camshaft Event Timing 197
Practice Problems 202

CHAPTER 11 *Thermodynamics* 209

Air/Fuel Ratios and Volumetric Efficiency   209
Induction System and Carburetor Sizing   211
Indicated Horsepower and Torque   213
Mechanical Efficiency   215
Altitude Compensation   216
Practice Problems   218

CHAPTER 12 *Transmission and Gear Ratios* 222

Gear Ratios   222
Overall Drive Ratios   226
Planetary Gear Ratios   228
Tire Sizing   231
Speedometer Calibration   233
Practice Problems   233

CHAPTER 13 *Hydraulic Systems* 247

Force, Pressure, and Area   247
Braking Systems   249
Hydraulic Pumps and Flow Rates   253
Hydraulic Cylinders, Force, and Actuation Speed   255
Practice Problems   259

CHAPTER 14 *Electrical Systems* 265

Current, Voltage, Resistance, and Ohm's Law   265
Electrical Power   267
Resistor Circuits   269
Practice Problems   274

CHAPTER 15 *Motion* 280

Distance, Speed, and Time Relationships   280
Velocity, Acceleration, and Time Relationships   285
Practice Problems   287

CHAPTER 16 *Repair Orders* 293

Filling Out a Repair Order   293
Practice Problems   297

APPENDIX   303

INDEX   325

CHAPTER 1

Whole Number Operations with General Applications

OUR NUMBER SYSTEM

The most common numbering system used around the world today is called the **decimal** system. As we will see later, the prefix "deci" means "ten," which means our numbering system is based on the number 10.

To keep track of numbers when counting, we use a typical placeholder system. For example, take the number 32,409,587. Each digit is holding a position in this number. Each position has a name, based on its value.

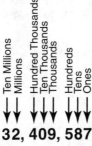

FIGURE 1.1 The place values of each digit in the number 32,409,587.

The number in Figure 1.1, then, is said to have seven "ones," eight "tens," five "hundreds," nine "thousands," and so on. Notice that a comma is used after the third and sixth digit to the left of the decimal. Commas or spaces are used every three digits from the right to make numbers easier to read.

Numbers can also be written longhand, as they often are in writing checks. When written longhand, the numbers twenty-one through ninety-nine use a hyphen, while others usually do not.

Example 1-1 Write the numbers 312, 86, and 29,422 in longhand.

312	would be written as	Three hundred twelve
86	would be written as	Eighty-six
29,422	would be written as	Twenty-nine thousand, four hundred twenty-two

Example 1-2 Write the values one hundred forty-eight, six hundred thirteen, and two thousand seventy using numbers.

One hundred forty-eight	would be written as	148
Six hundred thirteen	would be written as	613
Two thousand seventy	would be written as	2,070

ADDITION OF WHOLE NUMBERS

To add whole numbers, begin by writing all the numbers to be added, called **addends**, in a column. Make sure that all of the digits in the ones place align in the column on the right. Now, beginning with the ones place, add up all the digits in this position. If the total, or **sum**, of all the digits in any column is ten or more, a digit must be carried to the next place-holder column to the left.

Example 1-3   Find the sum of 6,835 and 724.

$$\begin{array}{r} \boxed{1} \\ 6{,}835 \\ +\ \ 724 \\ \hline 7{,}559 \end{array}$$

← Addends

← Sum

Example 1-4   Find the sum of these four values: 39, 417, 1,008, and 520.

$$\begin{array}{r} \boxed{2} \\ 39 \\ 417 \\ 1{,}008 \\ +\ \ 520 \\ \hline 1{,}984 \end{array}$$

A quick way to double-check your work is to perform the addition again, this time adding from bottom to top. This checking method will often catch any mental errors you may have made. Try to get into the practice of double-checking your work.

$$\begin{array}{r} 1{,}984 \\ 39 \\ 417 \\ 1{,}008 \\ +\ \ 520 \\ \boxed{2} \end{array}$$

Example 1-5   Installing a wiring harness, you'll need the following lengths of wire: 48 in., 32 in., 126 in., and two pieces measuring 58 in. each. How many total inches of wire will you need?

$$\begin{array}{r} \boxed{2}\boxed{3} \\ 48\ \text{in.} \\ 32\ \text{in.} \\ 126\ \text{in.} \\ 58\ \text{in.} \\ +\ \ 58\ \text{in.} \\ \hline 322\ \text{in.} \end{array}$$

You'll need 322 in. of wire. Double-check this by adding from bottom to top.

SUBTRACTION OF WHOLE NUMBERS

Subtraction is used to find the **difference** of two numbers. When subtracting, the numbers must be aligned in the same way as addition, making sure the ones place for each number align in a column. In subtraction, the order of the numbers is important. The number we

WHOLE NUMBER OPERATIONS WITH GENERAL APPLICATIONS 3

are subtracting *from* is called the **minuend.** The number we are taking away is called the **subtrahend,** and is usually the smaller number. The minuend is always on the top, and the subtrahend is always on the bottom.

Example 1-6 Find the difference of 1,397 and 251.

$$
\begin{array}{r}
1{,}397 \\
- \ \ 251 \\
\hline
1{,}146
\end{array}
\begin{array}{l}
\leftarrow \text{Minuend} \\
\leftarrow \text{Subtrahend} \\
\leftarrow \text{Difference}
\end{array}
$$

Sometimes, a digit in the minuend is smaller than the digit below. In this case, **borrow** from the digit in next column to the left. This increases the digit that was too small by ten, and decreases the digit to its left by one.

Example 1-7 Find the difference of 5,652 and 3,248.

$$
\begin{array}{r}
\overset{4\ 12}{5{,}6\cancel{5}\cancel{2}} \\
-3{,}248 \\
\hline
2{,}404
\end{array}
$$

Example 1-8 Find the difference of 13,236 and 6,149.

$$
\begin{array}{r}
\overset{0\ 13\ 1\ \overset{12}{2}\ 16}{\cancel{13{,}236}} \\
-\ \ 6{,}149 \\
\hline
7{,}087
\end{array}
$$

An easy way to double-check your answer when subtracting is to add your answer to the subtrahend. These two should add up to the minuend. Again, this quick check will probably catch any error you made in subtracting.

$$
\begin{array}{r}
\boxed{1}\boxed{1} \\
7{,}087 \\
+\ 6{,}149 \\
\hline
13{,}236
\end{array}
$$

The sum here is the same as the minuend. Our subtraction, then, was done correctly.

Example 1-9 Subtract: 1,500 − 748.

In this example, we'll need to borrow twice.

$$
\begin{array}{r}
\overset{\ \ \ \ 9\ 10}{4\ \cancel{10}} \\
1{,}\cancel{5}\cancel{0}0 \\
-\ \ 748 \\
\hline
752
\end{array}
$$

Example 1-10 The minimum compression for any cylinder in a certain engine is 125 psi. This means that every cylinder must have at least 125 psi of compression when checked. The measurement for a certain cylinder is 142 psi. How far above the minimum is this measurement?

$$
\begin{array}{r}
142\ \text{psi} \\
-125\ \text{psi} \\
\hline
17\ \text{psi}
\end{array}
$$

Example 1-11 A customer brings in a vehicle for new tires. The odometer currently reads 85,718 mi. Your records indicate that the odometer reading when the tires were purchased was 39,599 mi. How many miles are on the old tires?

$$\begin{array}{r}{\scriptsize 10}\\ {7\,15\,6\,\cancel{0}\,18}\\ {8\cancel{5},\cancel{7}\cancel{1}8\text{ miles}}\\ {-39{,}599\text{ miles}}\\ \hline 46{,}119\text{ miles}\end{array}$$

Double-check this answer by using addition.

$$\begin{array}{r}{\scriptsize 1\quad 11}\\ {46{,}119\text{ miles}}\\ {+39{,}599\text{ miles}}\\ \hline 85{,}718\text{ miles}\end{array}$$

MULTIPLICATION OF WHOLE NUMBERS

Multiplication is used when a certain number must be added repeatedly. The numbers being multiplied are called **factors**, and the result is called the **product**. To multiply whole numbers, begin by aligning the two factors as before, so that the ones place of each number align in a column. Then, multiply the rightmost digit in the bottom factor by each digit in the top factor, carrying to the next placeholder as needed.

If the bottom factor has more than one digit, begin each new row by placing another zero on the right. Then, multiply the next digit in the bottom factor by each digit in the top factor, again carrying as needed.

Example 1-12 Multiply 251 by 4.

$$\begin{array}{r}{\boxed{2}}\\ {251}\\ {\times\quad 4}\\ \hline 1{,}004\end{array}\begin{array}{l}\longleftarrow\text{Factors}\\ \longleftarrow\text{Product}\end{array}$$

Example 1-13 Multiply 342 by 126.

Begin by multiplying each digit in the top factor by 6.

$$\begin{array}{r}{\boxed{2}\boxed{1}}\\ {342}\\ {\times\ 126}\\ \hline 2{,}052\end{array}$$

After placing a zero on the right in the next row, multiply every digit in the top factor by 2. (We do this because the 2 is in the tens place. That means we're actually multiplying by 20.)

$$\begin{array}{r}{342}\\ {\times\ 126}\\ \hline 2{,}052\\ 6{,}840\end{array}$$

Now, place two zeros on the right in the next row, and multiply every digit in the top factor by 1. (We add two zeros because the 1 is in the hundreds place, so we're actually multiplying by 100.)

$$\begin{array}{r}{342}\\ {\times\ 126}\\ \hline 2{,}052\\ 6{,}840\\ 34{,}200\end{array}$$

WHOLE NUMBER OPERATIONS WITH GENERAL APPLICATIONS

Finally, add the digits in these three rows to find the product.

$$
\begin{array}{r}
342 \\
\times\ 126 \\
\hline
2{,}052 \\
6{,}840 \\
34{,}200 \\
\hline
43{,}092
\end{array}
$$

Example 1-14 Multiply 2,041 by 73.

First, multiply each digit in the top factor by 3.

$$
\begin{array}{r}
\overset{1}{} \\
2{,}041 \\
\times\ \ 73 \\
\hline
6{,}123
\end{array}
$$

Now, place a zero on the right in the next row, and multiply each digit in the top factor by 7.

$$
\begin{array}{r}
\boxed{2} \\
2{,}041 \\
\times\ \ 73 \\
\hline
6{,}123 \\
142{,}870
\end{array}
$$

Then, add these two rows together to find the product.

$$
\begin{array}{r}
2{,}041 \\
\times\ \ 73 \\
\hline
6{,}123 \\
142{,}870 \\
\hline
148{,}993
\end{array}
$$

Example 1-15 Multiply 5,932 by 207.

$$
\begin{array}{r}
5{,}932 \\
\times\ 207 \\
\hline
41{,}524 \\
0 \\
1{,}186{,}400 \\
\hline
1{,}227{,}924
\end{array}
$$

Like addition, the order of multiplication does not matter. This means that we can check our result by performing the multiplication again, but switching the placement of our two factors. We should get the same product in either case. This is a useful check because rearranging the numbers makes it difficult to make the same mistake twice.

Let's check Example 4.

$$
\begin{array}{r}
207 \\
\times 5{,}932 \\
\hline
414 \\
6{,}210 \\
186{,}300 \\
1{,}035{,}000 \\
\hline
1{,}227{,}924
\end{array}
$$

Example 1-16 A dealership does an average of 23 oil changes per week. How many oil changes will be done in a year? (There are about 52 weeks in a year.)

$$\begin{array}{r} 23 \text{ oil changes per week} \\ \times\ 52 \text{ weeks} \\ \hline 46 \\ 1{,}150 \\ \hline 1{,}196 \text{ oil changes in a year} \end{array}$$

Example 1-17 A parts manager earns $21 per hour. If there are 40 hours in a workweek, and she gets paid for 50 weeks of the year, what is her annual income?

$$\begin{array}{ll}
(1)\quad \begin{array}{r} \$21 \text{ per hour} \\ \times\ 40 \text{ hours per week} \\ \hline 0 \\ 840 \\ \hline \$840 \text{ per week} \end{array}
&
(2)\quad \begin{array}{r} \$840 \text{ per week} \\ \times\ 50 \text{ weeks per year} \\ \hline 0 \\ 42{,}000 \\ \hline \$42{,}000 \text{ per year} \end{array}
\end{array}$$

Example 1-18 A new spark plug wire should have no more than 900 ohms (Ω) per in. of resistance. What is the maximum allowable resistance for a 27 in. spark plug wire?

$$\begin{array}{r} 900 \text{ ohms per in.} \\ \times\ 27 \text{ in.} \\ \hline 6{,}300 \\ 18{,}000 \\ \hline 24{,}300 \text{ ohms} \end{array}$$

Notice that we can shorten this process. Ignore the zeros in 900, and multiply 27 by 9. Then, add the zeros back on after multiplying to find the product.

$$\begin{array}{r} 27 \\ \times\ 9 \\ \hline 243 \end{array}$$

$27 \times 9\underline{00} = 24{,}3\underline{00}$ ohms

In Example 1-18, we calculated the maximum acceptable resistance for a 27 in. plug wire, allowing for 900 Ω/in. Figure 1.2 shows a multimeter being used to measure the actual resistance.

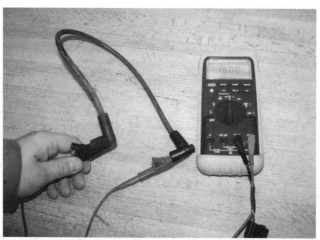

FIGURE 1.2 The allowable resistance for a spark plug wire depends on the length of the wire.

Source: James D. Halderman and Chase D. Mitchell, Jr., *Automotive Technology: Principles, Diagnosis, and Service,* Second Edition © 2003. Reprinted by permission of Pearson Education, Inc., Upper Saddle River, NJ.

WHOLE NUMBER OPERATIONS WITH GENERAL APPLICATIONS

DIVISION OF WHOLE NUMBERS

Division is used when we need to know how many times one number fits into the other. The number being divided is called the **dividend,** while the **divisor** is the number by which the dividend is being divided. The result is called a **quotient.** To divide whole numbers, begin by writing the dividend on the inside of the division sign, and the divisor to the left. The process for division is given in the following example.

Example 1-19   Divide 96 by 4.

Here, 96 is our dividend, and 4 is the divisor. One way to write this is

$$96 \div 4$$

but we'll use another division symbol to solve this problem.

$$\text{Divisor} \longrightarrow 4\overline{)96} \longleftarrow \text{Dividend}$$

Begin by finding how many times 4 fits into 9. 4 fits into 9 two times. Place a 2 over the 9.

$$\begin{array}{r} 2 \\ 4\overline{)96} \end{array}$$

Now, multiply 2 by the 4, and place the product below the 9. Subtract 8 from 9, and bring down the 6.

$$\begin{array}{r} 2 \\ 4\overline{)96} \\ -8 \\ \hline 16 \end{array}$$

Begin the cycle over again. How many times will 4 fit into 16? The answer is 4, so write that result above the 6. Then, multiply by the divisor again, and subtract.

$$\begin{array}{r} 24 \\ 4\overline{)96} \\ -8 \\ \hline 16 \\ -16 \\ \hline 0 \end{array}$$

After subtracting, we see the difference is 0. That means $96 \div 4 = 24$ exactly. We say that 24 is our quotient.

Example 1-20   Divide 139 by 7.

Writing this another way, we're looking for the quotient of $139 \div 7$.

$$\begin{array}{r} 19 \\ 7\overline{)139} \\ -7 \\ \hline 69 \\ -63 \\ \hline 6 \end{array}$$

In this case, 7 does not divide evenly into 139. After doing the final subtraction, we are left with 6. There are three ways to write the answer, and each way has its place. We'll discuss this more later. For now, we'll say our answer is 19 with a remainder of 6.

Example 1-21 Divide 1,059 by 17.

This is the same as writing 1,059 ÷ 17. Again, 1,059 is the dividend and 17 is the divisor.

$$17\overline{)1059}$$

Here, the divisor has more than one digit. The process is the same, but this time we'll begin by finding how many times 17 fits into 10. The answer of course is 0, so we'll try again with another decimal place. How many times does 17 fit into 105? With a little practice, we can see that the answer is 6. Now, write a 6 above the 5, and multiply by 17 and subtract as before.

$$\begin{array}{r} 6 \\ 17\overline{)1059} \\ -102 \\ \hline 3 \end{array}$$

Next, bring down the 9 and start the cycle again.

$$\begin{array}{r} 62 \\ 17\overline{)1059} \\ -102 \\ \hline 39 \\ -34 \\ \hline 5 \end{array}$$

Now we can see that 1059 ÷ 17 = 62 with a remainder of 5.

Example 1-22 Solve 2,547 ÷ 21.

$$\begin{array}{r} 121 \\ 21\overline{)2547} \\ -21 \\ \hline 44 \\ -42 \\ \hline 27 \\ -21 \\ \hline 6 \end{array}$$

By dividing, we see that 21 fits into 2,547 a full 121 times, with 6 left over.

Example 1-23 Fuel economy refers to the average number of miles a vehicle is able to drive on a gallon of gasoline. Another term for fuel economy is gas mileage, which is found by dividing the distance a vehicle travels by the amount of gasoline used. A truck is able to drive 247 miles on 13 gallons of gasoline. What is the gas mileage of the truck?

We're looking for the number of miles per gallon. The word "per" means we need to divide, so we'll divide 247 miles by 13 gallons.

$$\begin{array}{r} 19 \\ 13\overline{)247} \\ -13 \\ \hline 117 \\ -117 \\ \hline 0 \end{array}$$

This truck gets 19 miles per gallon, or 19 MPG.

WHOLE NUMBER OPERATIONS WITH GENERAL APPLICATIONS

Example 1-24 A transmission shop has an annual electricity budget of $6,900. How much, on average, can the shop spend per month on electricity to be within the budget?

We'll divide $6,900 by 12 months.

$$
\begin{array}{r}
575 \\
12\overline{)6900} \\
\underline{60} \\
90 \\
\underline{84} \\
60 \\
\underline{60} \\
0
\end{array}
$$

The shop can spend $575 per month for electricity.

EXPONENTS

An **exponent** is used when a number is multiplied by itself repeatedly. Exponents often occur in automotive formulas, including many we'll use later. An exponent or **power** indicates how many times a number, called the **base**, should be multiplied by itself.

Example 1-25 Find the value of 4^3.

Here, the base is 4 and the exponent is 3. That means we should multiply by 4 three times.

Base $\longrightarrow 4^3$ = 4 × 4 × 4 = 64
          $\nearrow$Exponent

Example 1-26 Find the value of 2^5.

The base is 2, and the exponent is 5. Multiply by 2 five times.

$$2^5 = 2 \times 2 \times 2 \times 2 \times 2 = 32$$

Example 1-27 Write 8 × 8 × 8 × 8 × 8 × 8 × 8 using exponential form.

Since 8 is being multiplied seven times, we can write this product in exponential form as 8^7.

Example 1-28 Find the value of $3^3 \times 4^2$.

$$3^3 \times 4^2 = 3 \times 3 \times 3 \times 4 \times 4 = 27 \times 16 = 432$$

ORDER OF OPERATIONS

Many formulas used in the automotive industry rely on the idea that everyone in the world does the computation in the same way. When a formula involves more than one operation, everyone must agree on the order the operations are done in. A standard **order of operations** has been established so that everyone using a certain formula will get the same result.

Consider the following expression.

$$6 + 4^2 \div 2 - 4 \times 3 + (10 \div 5)$$

If there were no established way to find the result, three different people would probably get three different answers. To avoid that problem, mathematical operations must always be done in the following order:

1. **Parentheses** Do any operations enclosed by parentheses ( ), brackets [ ], or braces { } first. If there is more than one set, such as brackets within parentheses, work from the inside out.

2. **Exponents** Find the value of any exponential values.
3. **Multiplication and Division** If multiplication and division occur next to each other, work from left to right. If they do not occur next to each other, do them at the same time.
4. **Addition and Subtraction** These should be the last operations you do. Simply work from left to right.

Example 1-29 Find the value of the following expression.

$$6 + 4^2 \div 2 - 4 \times 3 + (10 \div 5) =$$

First, simplify inside the parentheses. The simplified value is underlined.

$$6 + 4^2 \div 2 - 4 \times 3 + \underline{2} =$$

Now, evaluate the exponent.

$$6 + \underline{16} \div 2 - 4 \times 3 + 2 =$$

Do the multiplication and division.

$$6 + \underline{8} - \underline{12} + 2 =$$

Finally, add and subtract.

$$\underline{14} - 12 + 2 =$$
$$\underline{2} + 2 = 4$$

Example 1-30 Simplify the following expression.

$$2 + (7)(8) - (24 \div 2) - (7 - 2)^2 =$$

Notice that there is no operation indicated between the (7) and (8). When no other operation is listed, multiplication should be used. With that in mind, begin by simplifying inside the parentheses.

$$2 + (7)(8) - \underline{12} - \underline{5}^2 =$$

Evaluate the exponents, then multiply.

$$2 + (7)(8) - 12 - \underline{25} =$$
$$2 + \underline{56} - 12 - 25 =$$

Finish by doing the addition and subtraction, from left to right.

$$2 + 56 - 12 - 25 = 21$$

Example 1-31 Find the value of the following expression.

$$(3 \times 6 - 9)^2 + 4 \times (11 - 3^2)^3 =$$

First, do the operations inside parentheses.

$$(\underline{18} - 9)^2 + 4 \times (11 - \underline{9})^3 =$$
$$\underline{9}^2 + 4 \times \underline{2}^3 =$$

Now, evaluate exponents.

$$\underline{81} + 4 \times \underline{8} =$$

Multiply, then add.

$$81 + \underline{32} = 113$$

Example 1-32 Find the value of the following expression.

$$24 \div 4 \times 2 \div (3 - 5 \times 0) =$$

Working inside the parentheses, do the multiplication first.

$$24 \div 4 \times 2 \div \underline{3} =$$

WHOLE NUMBER OPERATIONS WITH GENERAL APPLICATIONS

Multiplication and division are next to each other, so work left to right.

$$\underline{6 \times 2} \div 3 =$$
$$\underline{12 \div 3} = 4$$

Example 1-33 Simplify the following fraction.

$$\frac{5 \times 7 - 2^3}{15 + 34 - 3 \times 6} =$$

Even though there are no parentheses here, we must simplify the top and the bottom of the fraction as if they were enclosed in parentheses.

Simplify the top of the fraction. This is called the **numerator**.

$$\frac{\underline{35 - 8}}{15 + 34 - 3 \times 6} =$$

$$\frac{27}{15 + 34 - \underline{3 \times 6}} =$$

Now, simplify the bottom of the fraction. It is called the **denominator**.

$$\frac{27}{15 + 34 - \underline{18}} = \frac{27}{31}$$

We'll leave the answer as a fraction. We'll do more work with fractions later.

CHAPTER 1 *Practice Problems*

Our Number System

Write the numbers below using words.

1. 49 _____
2. 103 _____
3. 561 _____
4. 3,920 _____
5. 10,634 _____
6. 348,211 _____

Write the following values using numbers.

7. Eighty-eight   _____
8. Three hundred forty   _____
9. Nine hundred sixty-one   _____
10. Ten thousand two hundred eleven   _____
11. Seven thousand four   _____
12. Two hundred twelve thousand, six hundred twenty _____

Addition of Whole Numbers

Find the sum for each problem below.

●1. 37
    +12

●2. 98
    +45

●3. 136
    + 84

●4. 2,302
    + 599

●●5. 51,352
    +12,691

●6. 42
     15
    +38

●7. 528
     316
    + 64

●8. 983
      15
    + 6

●●9. 4,825
      6,320
      1,555
    +   708

●●10. 1,999
      2,999
      3,999
     +4,999

●11. 18 + 371 =

●12. 581 + 327 =

●13. 12,630 + 6,722 =

●14. 321 + 1,450 + 99 =

●●15. 13,648 + 6,733 + 12,904 + 350 =

●16. A lighting circuit on a truck supplies current to six light bulbs and the horn. Four of the bulbs draw 2 A each, two of the bulbs draw 3 A each, and the horn draws 5 A. Find the total current needed for this circuit.

●17. Three pieces of fuel line are needed. One must be 12 in. long, another must be 49 in. long, and the third must be 96 in. long. How much fuel line is needed?

●●18. The following lengths of wire are needed to install a power-braking system: three pieces 137 in. long, two pieces 56 in. long, two pieces 36 in. long, and one piece 44 in. long. Find the total length of the wire necessary, in inches.

●19. On a trip, a customer fills the tank of the family van several times. They purchase the following amounts of gasoline: 14 gallons, 16 gallons, 14 gallons, 11 gallons, 17 gallons, 9 gallons, and 15 gallons. How much fuel was used on this trip?

●20. On the same trip, the family drives the following distances on each tank: 251 miles, 293 miles, 244 miles, 202 miles, 310 miles, 165 miles, and 266 miles. What was the total distance traveled?

●21. The following components make up what is called the reciprocating weight of a piston assembly: Piston, 522 g; wrist pin, 88 g; compression and oil rings, 52 g; small end of the connecting rod, 149 g. To balance an engine, the reciprocating weight must be known. Find the reciprocating weight of this assembly.

WHOLE NUMBER OPERATIONS WITH GENERAL APPLICATIONS

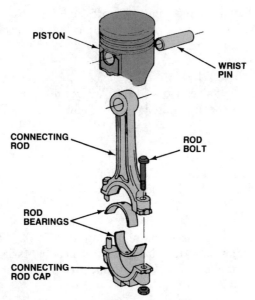

The piston, wrist pin, rings, and the small end of the connecting rod make up the reciprocating weight of a piston and connecting rod assembly.

Source: James D. Halderman and Chase D. Mitchell, Jr., *Automotive Technology: Principles, Diagnosis, and Service*, Second Edition © 2003. Reprinted by permission of Pearson Education, Inc., Upper Saddle River, NJ.

●22. Find the reciprocating weight (see problem 21) of the following assembly: Piston, 435 g; wrist pin, 67 g; compression and oil rings, 48 g; small end of the connecting rod, 137 g.

●23. The base price of a new vehicle is $18,742. The following options can be added: power windows/locks, $425; AM/FM/CD/MP3 player, $360; antitheft system, $330; keyless entry system, $290. If a buyer would like to add power windows/ locks and the AM/FM/CD/MP3 player to the base model, what will be the total price?

●24. If a buyer would like all the options listed in problem 23, what will be the total price?

Subtraction of Whole Numbers

Find the difference for each of the problems below.

●1. 57 − 16

●2. 63 − 37

●3. 264 − 72

●4. 1,409 − 534

●●5. 12,583 − 6,724

●6. 400 − 17

●7. 4,327 − 30

●8. 2,015 − 430

●●9. 127,038 − 16,878

●●10. 15,012 − 13,967

●11. 68 − 47 =

●12. 872 − 655 =

●13. 13,461 − 2,540 =

●14. The sidewall of a tire recommends a pressure of 44 psi for the tire. After checking the actual pressure with a gauge, you find that the pressure in the tire is 36 psi. How many pounds must be added?

●15. The retail price of a vehicle is $18,467. Right now, the factory is offering a $1,750 rebate. What is the discounted price of the vehicle?

●16. A local dealership is selling a new half-ton pickup for $19,437. A dealership a few miles away is selling the same truck for $20,273. How much can you save by buying the truck at the local dealership?

●17. A certain alternator has a maximum output of 75 A. While the vehicle is running at night, the ignition system draws 7 A, the headlights draw 24 A, the heater draws 6 A, and the radio draws 2 A. How much remaining current is there to charge the battery?

●18. A sports car weighs 3,150 lb with a full tank of gas. The driver wants to do a little drag racing, so she removes the following: passenger seat, 40 lb; spare tire, 30 lb; and 11 gallons of gas, 70 lb. What is the new weight of the vehicle?

●19. To keep track of gas mileage, the driver reads the odometer each time the fuel tank is completely filled. On June 4, the driver fills the tank, and the odometer displays 13,829 miles. On June 12, the driver fills the tank again, and the odometer displays 14,152 miles. How far did the driver travel between fill-ups?

●20. On August 27, a driver fills his tank, and the odometer reads 16,780. On September 4, the driver fills the tank again, and the odometer reads 17,052.
   a. How many days passed between fill-ups?
   b. How many miles were driven between fill-ups?

●21. A driver is able to average 268 miles on a tank of gas with typical in-town driving. On a long trip, the driver can average 311 miles on a tank. How many more miles per tank can the driver get on long trips?

●22. A driver buys a used car with 41,396 miles. Three years later, she sells it with 90,408 miles. How many miles did she put on the car?

●23. Your dealership installs a new transmission with a 36,000 mile warranty. When installed, the vehicle's odometer read 137,435. The car is returned to the dealership with an odometer reading of 170,903 miles. Is the transmission still under warranty?

●●24. A compression check is done on an eight-cylinder engine. Compression on the lowest cylinder must be within 35 psi of the highest cylinder. Determine if each of the cylinders below is within specifications.

WHOLE NUMBER OPERATIONS WITH GENERAL APPLICATIONS   15

Cylinder number	Compression reading (psi)	Difference from highest cylinder	Within specifications?
1	138		
2	122		
3	141		
4	104		
5	115		
6	139		
7	129		
8	107		

Multiplication of Whole Numbers

Find the product for each of the problems below.

•1. 13 × 3

•2. 25 × 3

•3. 63 × 9

•4. 79 ×16

•5. 342 × 62

••6. 975 × 79

••7. 1,467 × 28

••8. 27,521 × 34

••9. 687 ×379

••10. 1,409 ×2,635

••11. 18 × 307 =

•12. 307 × 18 =

••13. 37,000 × 9 =

••14. 5,000 × 70 =

••15. 1,500 × 1,500 =

••16. 12,000 × 700 =

•17. A P215/60R16 tire is on sale for $136 each. How much will four tires cost?

•18. On a long trip, a driver averages about 55 MPH. How many miles can he expect to travel, on average, in a nine-hour period?

•19. Due to poor weather conditions, a driver can only average about 45 MPH on a long trip. How far can she expect to travel, on average, in a 12-hour period?

•20. The average speed on a long trip is about 80 kilometers per hour (km/hr). How many kilometers can be traveled in a 10-hour period?

•21. An apprentice makes $12 per hour. How much money can she make in a 40-hour workweek?

••22. How much can the apprentice in problem 21 make in a year if she works 51 weeks of the year?

●●23. After taking a new job, the technician in problems 21 and 22 receives a $2 per hour pay increase. How much can she now make in a year, if she works 51 weeks?

●●●24. A technician has two job offers to choose from. He can work at a local garage, make $16 per hour and spend $2 a day on gasoline, or he can work at a dealership in a nearby city, make $18 per hour and spend $9 per day on gasoline. Both jobs are based on five eight-hour days.
   a. What is his weekly pay at the local shop?
   b. How much will he spend per week on gasoline commuting to the local shop?
   c. What is his net income per week, after subtracting gasoline expenses?
   d. What is his weekly pay at the dealership?
   e. How much will he spend on gasoline commuting to the city, per week?
   f. What is his net income per week at the dealership, after subtracting gasoline expenses?

●25. A new car advertises 32 miles per gallon highway mileage. If the vehicle has a 17-gallon tank, what is the maximum distance the car could travel, on average, using one tank?

●26. Suppose the car in problem 25 is rated at 26 miles per gallon in the city. How far can the vehicle drive in the city on 15 gallons of gasoline?

●●27. A certain spark plug wire should have no more than 1200 ohms per in. of resistance. What is the maximum allowable resistance of a 19-in. plug wire?

●●28. A certain spark plug wire should have a maximum of 400 ohms per cm of resistance. What is the allowable resistance for a 35-cm plug wire?

●●29. A used truck is sold for $12,500. The buyer gets a loan and pays $309 per month for four years.
   a. How much does the buyer pay over four years?
   b. How much interest did the buyer pay?

Division of Whole Numbers

Find the quotient for each of the problems below.

●1. $7\overline{)84}$ ●2. $3\overline{)96}$ ●3. $6\overline{)162}$ ●4. $5\overline{)87}$ ●5. $9\overline{)678}$

●6. $12\overline{)132}$ ●7. $23\overline{)276}$ ●●8. $13\overline{)91}$ ●●9. $12\overline{)160}$ ●●10. $18\overline{)384}$

●●11. $33\overline{)528}$ ●●12. $14\overline{)1358}$ ●●13. $25\overline{)7250}$ ●●14. $50\overline{)8463}$ ●●●15. $72\overline{)10896}$

●16. $56 \div 7 =$ ●17. $560 \div 7 =$ ●18. $56{,}000 \div 7 =$ ●●19. $5{,}600 \div 70 =$

WHOLE NUMBER OPERATIONS WITH GENERAL APPLICATIONS

●20. Fuel-line antifreeze is added at a rate of 1 can for every 8 gallons of fuel. How many cans are needed for a 24 gallon fuel tank?

●●21. A driver is able to drive 338 miles on 13 gallons of fuel. What is her average gas mileage?

●●22. A driver is able to drive 518 km on 37 L of fuel. What is his fuel economy, in kilometers per liter (km/L)?

●●●23. A truck driver fills the tank on his truck while the odometer reading is 115,974. Later, when the odometer reads 116,208, it takes 18 gallons to fill his tank. What is his fuel economy, in miles per gallon (MPG)?

●●24. A technician's weekly paycheck is $520 after taxes. What is his hourly pay after taxes if he works a 40-hour week?

●●●25. An oil change shop used 21,216 qt of oil last year.
   a. If there are 52 weeks per year, how many quarts did the shop use each week, on average?
   b. If the shop was open six days per week, how many quarts did it use, on average, each day?
   c. Assuming each oil change uses 4 qt, how many oil changes were done on average, each day?

●26. A body shop allows $2,040 per year for garbage disposal. How much can the shop afford to pay per month?

●●27. You can rent a prime 2,300 sq. ft shop space for $6,900 per month. What is the price per square foot?

●28. With the headlights on, two 12-V bulbs draw 36 A of current. What is the current used by each bulb?

●●●29. A low-risk driver pays $1,044 per year for insurance. He can pay either once, every three months, or monthly.
   a. If he pays once every three months, there is a $2 surcharge per payment. How much would each bill be if he pays every three months?
   b. If he pays monthly, there is a $3 surcharge per payment. How much would each bill be if he pays monthly?

Exponents

Write each of the expressions below using exponential form.

●1. $5 \times 5 \times 5 =$ ●2. $7 \times 7 \times 7 \times 7 \times 7 \times 7 \times 7 \times 7 \times 7 =$

●3. $62 \times 62 =$ ●4. $100 \times 100 \times 100 \times 100 =$

Write each of the exponential values below using multiplication.

●5. $8^2 =$ ●6. $1^9 =$

●7. $14^4 =$ ●8. $333^3 =$

●9. $10^5 =$ ●10. $10^7 =$

Find the value of each of the expressions below.

●11. $6^2 =$ ●12. $4^3 =$ ●13. $3^5 =$

●14. $10^5 =$ ●●15. $23^2 =$ ●●16. $17^3 =$

●17. $3^2 \times 2^4 =$ ●●18. $6^2 \times 4^3 =$ ●●19. $7^1 \times 2^4 \times 5^2 =$

●●●20. $2^5 \times 3^4 \times 1^7 \times 6^4 =$ ●●●21. $10^4 \times 10^1 \times 10^2 =$

●22. A square shop space is 12 yd wide and long. What is the area of the shop, in square yards (sq. yd)?

●●23. A square shop measures 42 ft by 42 ft. What is the area, in sq. ft?

Order of Operations

Use proper order of operations to find the value of each expression below.

●1. $3 \times 8 - 6 =$ ●2. $30 \div 5 + 5 =$ ●3. $6 + 4^2 =$

●4. $12 \div 2 \times 3 =$ ●5. $12 \div (2 \times 3) =$ ●6. $(6 + 4)^2 =$

●7. $5^2 - 3^2 =$ ●8. $(5 - 3)^2 =$ ●9. $3 + 2(8 - 3) =$

●●10. $(3 + 3 \times 3)^2 =$ ●●11. $(5)(4^2) - 16 \div 4 =$ ●●12. $(16 - 9)^2 \div 7 + 1 =$

●●●13. $(4^2 - 12 - 2^2) + 25 \div (5)(5) =$ ●●●14. $[3 \times (10 - 6) + 2^3] \div 5 =$

●●●15. $\{7 - [3 + (8 - 4)^2 - 15]\}^2 =$

●16. $\dfrac{4^2 - 11}{6 \times 5 - 21} =$

●●17. $\dfrac{(6 + 21) \div 9 + 10}{(36 + 6) \div 3} =$

●●●18. $\dfrac{(6)(2)^2 - [13 + (8 - 5)]}{(3 \times 2)^2 - (15)(2) + 3} =$

CHAPTER 2

Decimal Operations with Applications to Measurement and Finance

READING AND WRITING DECIMAL NUMBERS

Most of the measurements a technician deals with involve partial units. For example, a battery will not produce exactly 12 V, nor will a manifold weigh exactly 15.0 lb. A cooling system may call for 11.5 qt of coolant, or a bearing journal diameter may need to be 2.467 in. These are only a few examples of places you'll encounter decimal numbers, and being able to communicate decimal values is important.

To read a decimal number, we first need to know the value of the placeholders. Since the places to the left of the decimal count ones, tens, hundreds, and so on, it is the places to the right of the decimal that we must focus on.

Example 2-1 Consider the number 3.578 264.

This number has three whole units, five tenths, seven hundredths, eight thousandths, and so on. Notice that a space is used after every third decimal place to the right of the decimal. A comma on the right of a decimal should never be used, but spaces make the number much easier to read.

Example 2-2 The charge of a battery is given as 12.75 V. How is this number written, in words?

The whole number is written first, then the word *and* is used to show where the decimal is. Finally, the last decimal digit is located in the hundredths spot. Then we can write:

Twelve and seventy-five hundredths

Notice that again, the numbers twenty-one through ninety-nine should be hyphenated, even if they are in the decimal part of the number.

Example 2-3 A camshaft journal measures 1.749 in. How is this measurement written in words?

The whole number is one, and the last decimal digit is located in the thousandths spot. We write:

One and seven hundred forty-nine thousandths

Example 2-4 The tolerance of a certain clearance in a turbocharger is given as 0.000 15 in. How is this written, in words?

There is no whole number, and the last digit is in the hundred-thousandths spot. We write:

Fifteen hundred-thousandths

Example 2-5 A spark plug should have the gap set at 0.040 in. How should this value be written?

The last digit is in the thousandths spot. That means we should write:

Forty thousandths of an inch

Notice that in Example 2-5, 0.040 = 0.04 (four hundredths). However, the zero in the thousandths position *should* be maintained to show how precisely the gap should be set. The **precision** of a measurement shows how exact the measurement is.

Example 2-6 A shop manual explains that the bearing clearance for a certain part "should be no more than six thousandths of an inch." Write this number using a decimal.

There is no whole number given, only a decimal part. We're measuring thousandths, so we'll need three decimal places. We then write:

0.006 in.

Example 2-7 Write "eight and nine hundred thirty-seven ten-thousandths" as a decimal.

The whole number is 8, and our decimal part is ten thousandths; this means we need four decimal places. This number should be written:

8.093 7

ROUNDING DECIMAL NUMBERS

Often times, more decimal places are included in a number than are useful or needed. For example, when calculating fuel economy, a driver would not likely say that her truck averages 20.069 93 MPG. It is more likely she would say that she gets 20.1 or even 20 MPG. Often, it is appropriate to round decimal values.

First determine the digit that occupies the rounding place. Either the desired rounding place will be stated (tenth, thousandth, etc.) or good judgment must be used.

Next, look at the digit to the right of the rounding place. If it is 0 to 4, the digit in the rounding place stays the same, and all remaining digits to the right of the rounding place are dropped. If it is 5 to 9, the digit in the rounding place increases by 1, and all remaining digits to the right of the rounding place are dropped.

Example 2-8 Round 14.152 739 to the nearest hundredth.

From the previous section, we know that the hundredths spot is the place where the 5 is. Rounding to the hundredth means our rounded number will have two decimal places. Looking to the right of the 5, the next digit is a 2. This means the 5 will stay the same. Our number then is rounded to 14.15.

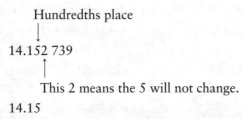

14.15

22   CHAPTER 2

Example 2-9   Round 795.183 69 to the nearest thousandth.

The thousandths place is occupied by the 3. The digit to the right of that is a 6. This means the three must be increased to 4 before dropping the other digits. Our number, then, is rounded to 795.184.

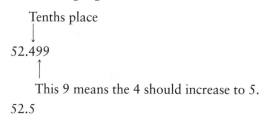

795.184

Example 2-10   Round 52.499 to the nearest tenth.

The digit in the tenths place is a 4. To the right of that is a 9, so we'll increase the 4 to a 5, and drop the remaining digits. Our rounded value is 52.5.

```
    Tenths place
        ↓
      52.499
        ↑
This 9 means the 4 should increase to 5.
```

52.5

Example 2-11   Round 2.8967 to the nearest hundredth.

The digit in the hundredths place is 9, and the digit one position to the right is a 6. Then the 9 increases to 10, but we'll carry that extra digit to the tenths place. The 8 then becomes a 9.

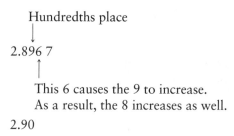

2.90

Example 2-12   Round $83.528 to the nearest cent.

Rounding to the nearest cent is the same as rounding to the hundredth. Here, a 2 occupies the hundredths place, and there's an 8 to the right. We'll round to $83.53.

ADDITION AND SUBTRACTION OF DECIMAL NUMBERS

When adding and subtracting decimal numbers, we used the same process that we did when working with whole numbers. When adding and subtracting whole numbers, we needed to make sure the ones columns for all the numbers were aligned. An easier way to say the same thing is to note that *when adding or subtracting decimal numbers, the decimal point in each number must align with all the others*. Finally, we can add zeros on the right of a decimal number to act as placeholders without changing the value of the number.

Example 2-13   Add the following measurements: 4.3 cm, 3.862 cm, and 11.41 cm.

Make sure the decimals align, then add zeros to fill empty placeholders.

$$\begin{array}{r} \overset{1}{} \\ 4.300 \text{ cm} \\ 3.862 \text{ cm} \\ +11.410 \text{ cm} \\ \hline 19.572 \text{ cm} \end{array}$$

Example 2-14 The following parts were needed to fix a customer vehicle: brake caliper, $137.54; brake rotor, $98.29; inner bearing, $13.90; outer bearing, $15.31; and brake pads, $29.06. What is the total bill for these parts?

$$\begin{array}{r} \overset{1\ 3\ 2\ 2}{137.54} \\ 98.29 \\ 13.90 \\ 15.31 \\ +\ \ 29.06 \\ \hline \$294.10 \end{array}$$

Example 2-15 Find the difference between the following measurements: 4.852 in. and 3.6 in.

Since we want to find the difference of two numbers, we'll need to subtract the smaller from the larger. Make sure the decimal point is aligned when writing these numbers, and put zeros in the empty placeholders.

$$\begin{array}{r} 4.852 \text{ in.} \\ -3.600 \text{ in.} \\ \hline 1.252 \text{ in.} \end{array}$$

Example 2-16 The price of a starter, including tax, is $156.76. If the tax on the starter was $7.47, find the price before tax.

To find the retail price, or price before tax, we'll simply subtract the amount of tax from the total price.

$$\begin{array}{r} \overset{4\ 16\ 6\ 16}{\$1\cancel{5}\cancel{6}.\cancel{7}\cancel{6}} \\ -\ \ \$7.47 \\ \hline \$149.29 \end{array}$$

Example 2-17 The maximum diameter of a brake drum is 9.587 in. A used brake drum has a diameter of 9.559 in., but must be machined to remove scoring. How much metal can be removed and still stay within the maximum diameter?

FIGURE 2.1 If a brake drum must be machined beyond its maximum inside diameter, it must be replaced.

Source: James D. Halderman and Chase D. Mitchell, Jr., *Automotive Technology: Principles, Diagnosis, and Service*, Second Edition © 2003. Reprinted by permission of Pearson Education, Inc., Upper Saddle River, NJ.

24 CHAPTER 2

An example of a brake drum is shown in Figure 2.1. The current measurement is below the allowable limit. We need to find the difference between the measurements.

$$\begin{array}{r} \overset{7\,17}{9.58\rlap{/}{8}} \text{ in.} \\ -9.559 \text{ in.} \\ \hline 0.028 \text{ in.} \end{array}$$

The drum's diameter can be increased by up to twenty-eight thousandths of an inch.

READING A MICROMETER

English micrometers are used to measure distances with a precision of .001 in. (to the nearest thousandth) or .000 1 in. (to the nearest ten-thousandth). Metric micrometers can usually measure with a precision of .01 mm (nearest hundredth). Both micrometers work on the same principle, but their scales differ. The micrometer is opened and closed by turning the thimble. Readings are taken in three parts, by taking readings from the barrel and thimble. See Figure 2.2.

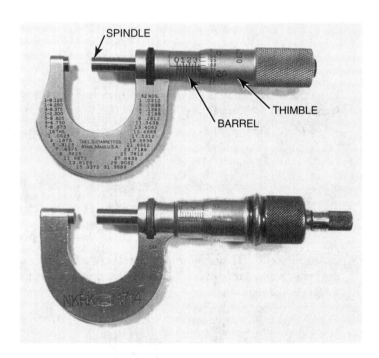

FIGURE 2.2 Micrometers are the most common measuring tool used in the automotive industry.

Source: James D. Halderman and Chase D. Mitchell, Jr., *Automotive Technology: Principles, Diagnosis, and Service*, Second Edition © 2003. Reprinted by permission of Pearson Education, Inc., Upper Saddle River, NJ.

English Micrometers

Example 2-18 Read the 0–1″ English micrometer below.

Step 1: As the thimble is turned to open the micrometer, numbered and non-numbered lines are exposed on the barrel. The first value is taken by reading the largest numbered line on the barrel that is not covered by the thimble. In this case, that number is 9. Since there are 10 numbered lines per inch, this 9 means 0.9 in.

Step 2: The next value is taken by viewing the largest unmarked line that is not covered by the thimble. There are four unmarked lines for each marked line. Each unmarked line has a value of 0.100 in. ÷ 4 = 0.025 in. In this diagram, there are two exposed lines past the 9, so we add 0.025 in. + 0.025 in. = 0.050 in.

Step 3: The final value is taken by reading the number on the thimble that best lines up with the scale on the barrel. Here, it looks like that number is 18, which actually means 0.018 in.

To find the overall measurement, we simply add the values from parts 1, 2, and 3 as shown in Figure 2.3.

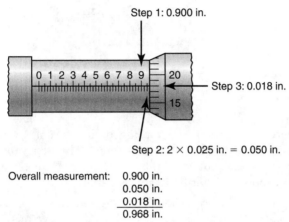

FIGURE 2.3 Read the measurement of a micrometer by following steps 1, 2, and 3 from the text.

The overall reading then is 0.9 in. + 0.050 in. + 0.018 in. = 0.968 in.

Example 2-19 Read the 1–2 in. English micrometer below.

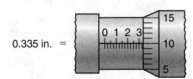

Step 1: The largest numbered mark that is exposed here is 3. Since this is a 1–2 in. micrometer, the first value is 1.3 in.

Step 2: Here, there is only one unmarked line exposed beyond the 3, so the next value is 0.025 in.

Step 3: The 10 on the thimble lines up with the scale on the micrometer, so our final value is 0.010 in.

Our overall reading then is 1.3 in. + 0.025 in. + 0.010 in. = 1.335 in.

Example 2-20 Read the 3–4 in. English micrometer below.

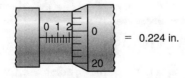

Step 1: In this case, the largest exposed number is a 2. Because we are told this is a 3–4 in. micrometer, our first value is 3.2 in.

Step 2: Be careful here. The first unmarked line becomes exposed when the thimble is fully rotated, or reaches 0. The thimble reading is 24, so the first unmarked line must not be fully exposed. Therefore, there are no exposed lines, and our second value is 0.000 in. or simply 0.

Step 3: As we just noted, the thimble marking which lines up with the scale on the barrel is 24, so our final value is 0.024 in.

The overall measurement is 3.2 in. + 0.000 in. + 0.024 in. = 3.224 in.

Metric Micrometers

Example 2-21 Read the metric micrometer below.

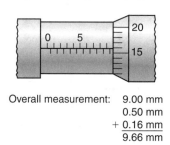

Overall measurement: 9.00 mm
0.50 mm
+ 0.16 mm
9.66 mm

Step 1: On metric micrometers, the top row of lines, usually with 0, 5, 10, 15, and so on, marked, represent whole millimeters (mm). The first value is taken by noting which of the top marks is not covered by the thimble. Here, the ninth mark on the top row of lines is exposed, so our first value is 9 mm.

Step 2: The next value is taken by looking at the bottom row of lines on the barrel. Each mark here represents 0.5 mm. Unlike the English micrometer, there will be at most one of these lines exposed before another whole number is reached on the barrel. In this example, there is one line on the bottom exposed beyond the 9, so our second value is 0.5 mm.

Step 3: As before, the final value is taken by reading the number on the thimble that best lines up with the scale on the barrel. Here, that number is 16, which actually represents 0.16 mm.

Our final reading is 9 mm + 0.5 mm + 0.16 mm = 9.66 mm.

Example 2-22 Determine the measurement on the 50–75 mm metric micrometer below.

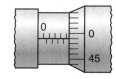

Step 1: On the top row, the fourth mark is fully exposed. Since we're told this is a 50–75 mm micrometer, our first value is 54 mm.

Step 2: There is no line on the bottom exposed beyond the 4, so our second value is 0.0 mm.

Step 3: Again, care must be taken as the thimble nears zero. The reading on the thimble is 49, so the final value is 0.49 mm.

Our final reading is 54 mm + 0.0 mm + 0.49 mm = 54.49 mm.

MULTIPLICATION OF DECIMAL NUMBERS

To multiply decimal numbers, use the same process used when multiplying whole numbers. When done, care must be taken to put the decimal point in the correct position of the product. Counting the total number of decimal places in each of the two factors determines the number of decimal places in the product.

Example 2-23 Multiply 4.57 × 8.2.

Since the order of multiplication does not matter, we can save a bit of work by placing the factor with fewer digits on the bottom. Multiply every digit in the bottom

DECIMAL OPERATIONS WITH APPLICATIONS TO MEASUREMENT AND FINANCE

number by every digit in the top number, as we did for whole numbers. Don't forget to use an additional zero each time a new row is begun.

$$
\begin{array}{r}
4.57 \\
\times\ 8.2 \\
\hline
914 \\
36560 \\
\hline
37474
\end{array}
$$

To place the decimal in the correct position, count the total number of decimal places in the factors. The factor 4.57 has two decimal places while 8.2 has one. Since there are a total of three decimal places, our answer will have three decimal places. Finally, put the decimal in the correct position.

$$
\begin{array}{r}
4.57 \longleftarrow \text{Two decimal places} \\
\times\ 8.2 \longleftarrow \text{One decimal place} \\
\hline
914 \\
36560 \\
\hline
37.474 \longleftarrow \text{Three decimal places}
\end{array}
$$

Example 2-24 Multiply 7.952×134.6.

Begin by multiplying as usual, then add the rows together. Notice that 7.952 has three decimal places, while 134.6 has one. Our product, then, must have four decimal places.

$$
\begin{array}{r}
7.952 \\
\times\ 134.6 \\
\hline
47712 \\
318080 \\
2385600 \\
7952000 \\
\hline
1070.3392
\end{array}
$$

Example 2-25 Gross income refers to the amount of income before payroll taxes and other deductions are removed from an employee's paycheck. What is the gross income for a tech who works 42.3 hours at $16.75 per hour?

Simply multiply 42.3 hours by $16.75. Again, to save a bit of writing, we'll put the number with fewer digits on the bottom.

$$
\begin{array}{r}
16.75 \\
\times\ 42.3 \\
\hline
5025 \\
33500 \\
670000 \\
\hline
708.525
\end{array}
$$

The product here will have three decimal places. Since the solution is an amount of money, though, we'll round to the nearest penny. Our solution is $708.53.

Example 2-26 The twisting force, or torque, available at the rear wheels is found by multiplying the engine's torque by the overall drive ratio. How much torque is available at the rear wheels of a vehicle if the engine produces 281 ft·lb and the overall drive ratio is 8.42:1?

We need to multiply 281 × 8.42 to find the torque at the rear wheels.

$$\begin{array}{r} 281 \\ \times\ 8.42 \\ \hline 562 \\ 11240 \\ 224800 \\ \hline 2{,}366.02 \end{array}$$

The rear wheels have 2,366.02, or about 2,366 ft·lb of torque.

DIVISION OF DECIMAL NUMBERS

When decimal numbers are divided, any decimal point in the divisor must be moved all the way to the right of the last digit. The decimal in the dividend must then be moved to the right the same number of places. Now, the division can be performed in almost the same way that was done with whole numbers.

Example 2-27 Set up the division for 82.63 ÷ 4.3.

$$4.3 \overline{)82.63}$$

$$43 \overline{)826.3}$$

Note that the decimal in the divisor was moved one place to the right, so the decimal in the dividend was moved one place to the right as well.

Example 2-28 Set up the division for 3.863 42 ÷ 12.83.

$$12.83 \overline{)3.863\ 42}$$

$$1283 \overline{)386.342}$$

In this case, the decimal in 12.83 had to be moved two places to the right, so the decimal point in 3.863 42 was moved two places also.

Example 2-29 Divide 16.92 by 4.7.

Our division process does not work if the divisor contains a decimal. We avoid that issue by moving both decimal places one place to the right.

$$4.7 \overline{)16.92}$$

$$47 \overline{)169.2}$$

Now, bring the decimal point directly up above the division symbol, and divide.

$$\begin{array}{r} 3.6 \\ 47 \overline{)169.2} \\ \underline{141} \\ 282 \\ \underline{282} \\ 0 \end{array}$$

The quotient is 3.6.

Example 2-30 Divide 9.1 by 2.6.

Here, 2.6 is the divisor, so the decimal places must each be moved one place.

$$2.6 \overline{)9.1}$$

$$26 \overline{)91}$$

Now divide as usual. Notice that more zeros must be added to the right of the 1 in the dividend. Adding zeros to the right of a decimal doesn't change the value of the number.

$$\begin{array}{r} 3.5 \\ 26\overline{)91.0} \\ \underline{78} \\ 130 \\ \underline{130} \\ 0 \end{array}$$

The quotient is 3.5.

Example 2-31   Divide 64.8 by 8.

In this case, 8 is the divisor. There is no decimal to move.

$$\begin{array}{r} 8.1 \\ 8\overline{)64.8} \\ \underline{64} \\ 0\,8 \\ \underline{8} \\ 0 \end{array}$$

The quotient is 8.1.

Example 2-32   Divide 72 by 2.4.

The divisor here is 2.4, so the decimal must be moved one place to the right. Since the dividend has no decimal, we must place it at the end of the number, then move it while adding necessary zeros.

$$2.4\overline{)72.0}$$
$$24\overline{)720}$$

$$\begin{array}{r} 30 \\ 24\overline{)720} \\ \underline{72} \\ 00 \end{array}$$

The quotient is 30.0 or simply 30.

Example 2-33   Divide 15.2 ÷ 6.7. **Round to the nearest hundredth.**

Move both decimal points the necessary one place. Then divide. Add zeros after the decimal in the dividend to get three decimal places, so we can correctly round to two decimal places.

$$\begin{array}{r} 2.268 \\ 67\overline{)152.000} \\ \underline{134} \\ 180 \\ \underline{134} \\ 460 \\ \underline{402} \\ 580 \end{array}$$

Our quotient continues, but we can now round this result to 2.27.

Example 2-34 Your shop purchases a case of 24 oil filters for $57.12. How much was each filter?

Divide the case price by the number of filters.

$$\begin{array}{r}2.38\\24\overline{)57.12}\\\underline{48}\\91\\\underline{72}\\192\\\underline{192}\\0\end{array}$$

The price is $2.38 per filter.

Example 2-35 A small car is able to drive 362.9 miles while using 10.3 gallons of gasoline. What was the average gas mileage for this tank?

Divide the number of miles by the number of gallons used. Again, the decimals must be moved one place to the right. We'll approximate our mileage by rounding to the tenth, so the division must be carried out to two decimal places.

$$\begin{array}{r}35.23\\10.3\overline{)362.900}\\\underline{309}\\539\\\underline{515}\\240\\\underline{206}\\340\end{array}$$

While the division could go on, we can now safely round to 35.2 MPG.

SQUARE ROOTS

Many formulas use the square root function. The square root of a value is that number, which if multiplied by itself will give the original value. The symbol for square root is $\sqrt{}$. Although most numbers have both a positive and negative square root, we'll consider only the positive root for the purpose of this text.

Example 2-36 Find the square root of 25.

Since $5 \times 5 = 25$, it must be true that $\sqrt{25} = 5$.

Example 2-37 Find the square root of 121.

Again, we can see that $11 \times 11 = 121$, so $\sqrt{121} = 11$.

Example 2-38 Find the square root of 17.

There is no whole number that, when multiplied by itself, yields 17. Use a calculator with a square root ($\sqrt{}$) button to evaluate $\sqrt{17}$.
Then $\sqrt{17} = 4.123105626\ldots$
Rounding to the nearest thousandth, we would say $\sqrt{17} \approx 4.123$.
We can check this result by multiplying $4.123 \times 4.123 = 16.999$.
Our approximate square root is very close.

CHAPTER 2 *Practice Problems*

Reading and Writing Decimal Numbers

Write the numbers below using words.

●1. 3.7 _____

●2. 12.79 _____

●3. 0.125 _____

●4. 9.371 _____

●5. 16.020 _____

●●6. 8.430 6 _____

●●7. 10.030 0 _____

●●8. 21.627 3 _____

Write the following values using numbers.

●9. Nine and two tenths _____

●10. Fifteen and ninety-four hundredths _____

●11. Fifteen and ninety-four thousandths _____

●12. Six and three hundred forty-four thousandths _____

●13. Three hundred seventy-five thousandths _____

●14. One and twelve ten-thousandths _____

●●15. Four and six hundred two ten-thousandths _____

●●16. Two thousand, five hundred ten-thousandths _____

Rounding Decimal Numbers

●1. Round 4.529 to the nearest tenth.

●2. Round 13.672 to the nearest hundredth.

●3. Round 4.986 to the nearest hundredth.

●4. Round 138.6734 to the nearest thousandth.

●5. Round 6.783 651 to the nearest thousandth.

●6. Round 46.449 949 to the nearest hundredth.

●7. Round 3.001 356 to the nearest ten-thousandth.

●8. Round 0.000 465 to the nearest thousandth.

CHAPTER 2

Addition and Subtraction of Decimal Numbers

Find the sum or difference, as appropriate for the problems below.

●1. 2.8
    +6.3

●2. 64.791
    +13.7

●3. 136
    + 84.63

●4. 16.992
    + 8.998

●5. 16.783 + 16.2 =

●6. 6.931 1 + 4.38 + 2.640 + 7 =

●7. 2300 + 0.006 63 =

●8. 32.7 + 18.63 + 99 + 13.72 =

●●9. 551.638 + 121.4 + 8.627 + 2.220 2 =

●10. 42.7
    − 9.5

●11. 15.63
    − 4.85

●12. 120.5
    − 53.371

●●13. 100
    − 62.854 7

●14. 63.872 − 17.4 =

●15. 81.234 53 − 17.110 =

●16. 22.1 − 21.923 =

●●17. 1,328.568 237 − 542.473 158 =

●●18. 1.010 701 − 0.909 39 =

●19. 17.2 + 16.3 − 12.4 =

●20. 371.6 − 122.8 + 13.9 =

●21. 8.2 + 9.7 + 1.2 + 6.3 − 13.4 =

●22. 27.602 + 43.44 − 8.7 =

●●23. 62.3 + 14 + 2.998 − 6.78 − 2.7 =

●●24. 18.9 − 17.81 + 6.95 − 5.853 + 4 =

●25. On a vacation, a driver filled the tank of his van four times. He put in 13.7 gallons, 14.9 gallons, 12.6 gallons, and 12.2 gallons. How much fuel was used for the entire trip?

●26. Find the total price of the following parts: alternator, $184.32; voltage regulator, $21.35; battery cables, $6.83 and $12.

DECIMAL OPERATIONS WITH APPLICATIONS TO MEASUREMENT AND FINANCE

●27. An engine has an original bore of 3.680 in. If it is bored 0.030 in. over, what is the new bore?

●28. Voltage drops across three consecutive components are 1.2 volts, 1.7 volts, and 2 volts. What is the total voltage drop?

●29. During the week, a technician bills the following hours: 4.2 hours, 3.5 hours, 8 hours, 3.9 hours, 7.2 hours, 4.4 hours, 6.7 hours, and 0.2 hours. How many total hours does he bill?

●30. The retail price of a muffler is $57.23. The tax on this part is $3.43. What is the total bill?

●31. Find the reciprocating weight (see problem 21 in Chapter 1, Addition) for a piston assembly, given the following measurements: piston, 542.3 g; wrist pin, 97.3 g; rings, 63.2 g; small end of connecting rod, 151.0 g.

●●32. Piston-to-bore clearance is the difference between the diameter of a piston and the bore it fits into. A piston has a diameter of 4.125 5 in., and requires a clearance of 0.001 5 in. in the cylinder. What is the minimum bore?

●33. A camshaft journal measures 35.20 mm, and must have a clearance of 0.04 mm in the bearing. What is the minimum bearing bore?

●34. The gross pay for a technician's week of work is $663.30. Taxes and insurance totaling $272.41 must be deducted from that pay. What is the net pay for the week?

●35. The total voltage drop for a solenoid and cable is 2.7 volts. The cable alone has a voltage drop of 0.2 volts. What is the voltage drop across the solenoid?

●36. The odometer reading on a truck on January 1, 2003 was 98,372.8. On December 31, 2003, the odometer read 113,147.2. How many miles were driven in 2003?

●37. Wrist pin height is the distance from the center of the wrist pin to the top of the piston. Determine the wrist pin height of the piston below.

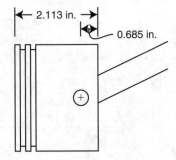

●●38. The bore of a cylinder is measured to be 3.941 in. The piston measures 3.937 inches. What is the clearance for this piston?

●●39. Cylinders wear more at the top of the cylinder than they do at the bottom. This uneven wearing is called **taper**. To determine the taper of a cylinder, the bore is measured both at the bottom and below the ring ridge on the top. If the bore on the bottom of the cylinder is 4.137 in. while the bore just before the ring ridge is 4.151 in., what is the taper of this cylinder?

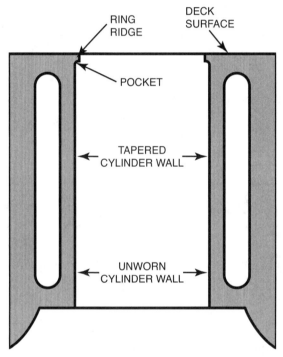

Source: James D. Halderman and Chase D. Mitchell, Jr., *Automotive Technology: Principles, Diagnosis, and Service,* Second Edition © 2003. Reprinted by permission of Pearson Education, Inc., Upper Saddle River, NJ.

●40. The maximum inside diameter of a brake drum is 11.340 in. The brake drum is originally 11.305 in. How much metal can be machined while staying within the maximum diameter?

●41. What is the valve stem clearance on a valve whose stem measures 0.472 in. while the valve guide measures 0.488 in.?

●●42. The labor charges for a customer's bill was $310.50. The parts charges were $279.49. Tax came to $13.97. If the customer writes a check for $450.00, how much does he still owe on the repair?

●●43. A customer buys 5 qt of oil for $1.79, a filter for $4.25, and a set of wiper blades for $8.39. Tax is $1.30, and the customer is given a discount of $2.16 (10% of the merchandise). What is the total amount due?

DECIMAL OPERATIONS WITH APPLICATIONS TO MEASUREMENT AND FINANCE

●●44. If the customer in problem 43 pays with two $20 bills, how much change does she get?

●●45. A customer pays for the following items with a $100 dollar bill: grease gun, $12.29; 1 box of grease zirks, $3.50; serpentine belt, $21.09; air filter, $8.69; two gallons of antifreeze, each of which cost $6.49. The tax on this purchase was $3.22. How much change will he receive?

Reading a Micrometer

Determine the measurement on each of the 0–1 in. English micrometers pictured below.

●1.

●2.

●3.

●4.

●5.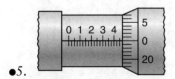

●6.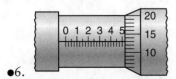

● 7.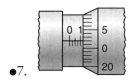

● 8.

● 9.

● 10.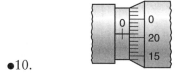

Determine the measurement on each of the 0–25 mm metric micrometers below.

● 11.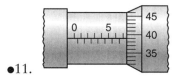

● 12.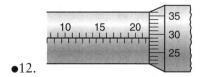

● 13.

DECIMAL OPERATIONS WITH APPLICATIONS TO MEASUREMENT AND FINANCE   **37**

●14.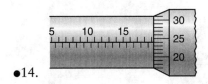

●15.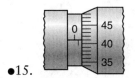

●16.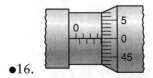

●17.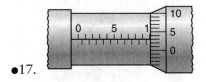

●18.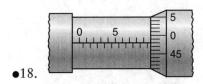

●19.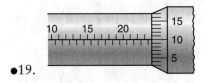

●20.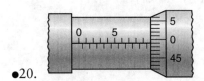

Multiplication of Decimal Numbers

Put the decimal point in the correct place in the products below.

●1. 13.8
     × 7
     ―――
     966

●2. 43.71
     × 8.2
     ―――
     8742
     349680
     ―――――
     358422

●3. 0.375
     × 1.8
     ―――
     3000
     3750
     ―――
     6750

●4. 2.892
     × 0.74
     ―――
     11568
     202440
     ―――――
     214008

Find the product for each of the problems below.

- •5. 13.8 × 6
- •6. 24.6 × 13
- •7. 2.4 × 0.6
- •8. 51.4 × 3.7

- •9. 4.57 × 9
- •10. 2.98 × 6.3
- •11. 0.83 × 0.62
- ••12. 62.7 × 0.93

- ••13. 1.88 × 4.25
- ••14. 9.64 × 3.09
- ••15. 3.862 × 7.4
- ••16. 7.413 × 1.86

- ••17. 5.307 × 2.97
- ••18. 7.635 × 6.55
- •••19. 4.5986 × 6.37
- •••20. 71.4382 × 4.602

Round each of the following to the nearest thousandth.

- ••21. 8.561 × 2.8
- ••22. 6.83 × 0.55
- •23. 8.108 × 2
- ••24. 4.857 × 2.960

•25. Over a two week period, a technician works 78.6 hours. Based on an hourly rate of $18.40, what is this tech's gross pay?

•26. During a slow week, a tech works 31.2 hours at $16.55 per hour. What is his gross pay for that week?

••27. A tech gets paid time-and-a-half for any time worked beyond 40 hours in a single week. Suppose he works 44.2 hours at an hourly rate of $17.80.
  a. The tech's base pay is $17.80 for the first 40 hours. How much is this?
  b. How many hours of overtime does he work?
  c. What will the rate of pay be for those overtime hours?
  d. The overtime pay will be his overtime rate times the number of overtime hours. How much is this?
  e. What is the total gross pay for this technician?

••28. Another tech also gets paid time-and-a-half for hours worked beyond 40 in one week. Suppose she works 46.1 hours at an hourly rate of $18.34.
  a. What is the base pay for this week?
  b. How many hours of overtime does she work?
  c. What will the pay rate be for the overtime hours?
  d. What is the overtime pay for this week?
  e. What is the total gross pay for this week?

•29. The fuel tank on a certain truck holds 16.4 gallons. Assuming the truck gets 21 miles per gallon, how far can it drive on a tank of gasoline?

•30. A motorcycle tank holds 4.9 gallons. If the cycle can get 47.5 miles per gallon, how far could it go on a tank of gas?

DECIMAL OPERATIONS WITH APPLICATIONS TO MEASUREMENT AND FINANCE

●31. A certain vehicle's engine produces 192 ft·lb of torque. The overall drive ratio in first gear is 9.57:1. How much torque is there at the drive wheels in first gear?

●32. A truck's diesel engine produces a maximum torque of 513 ft·lb. The overall drive ratio in first gear is 14.39:1. How much torque is available at the drive wheels of this truck?

●33. The labor rate at a certain shop is $49.50 per hour. If a brake job takes 1.7 hours, what will the labor charge be for this job?

●34. It takes 9.3 hours to replace a transmission at a shop rate of $55.75 per hour. What is the total labor charge for this job?

●35. A case of oil contains 12 qt. The per quart price is $1.89. What is the case price?

●36. What is the 12-qt case price of 10W-40 oil is if the per quart price is $1.55?

●37. Synthetic motor oil is usually sold in 6-qt cases. What is the price for a case of synthetic motor oil if the price per quart is $5.19?

●●38. A parts store buys oil at $0.84 per quart, and sells it for $1.39.
   a. What is the profit, per quart?
   b. What is the profit, per 12-qt case?

●●●39. After losing a bet, you need to buy everyone in the shop doughnuts and juice for breakfast. There are 17 total people in the shop, and doughnuts are $0.55 each. Juice is $3.89 per gallon, and you'll need three gallons. How much did this lost bet cost you?

●●40. A brake job takes 1.3 hours to complete. The labor rate is $52.75 per hour, and the total cost of the brakes was $27.35. What is the total bill, before tax?

●●41. It takes 3.4 hours to replace a timing belt at a rate of $51.50 per hour. The timing belt was $21.49, and no other parts were needed. What is the total bill before tax?

●●●42. It takes 5.7 hours to replace a head gasket, 1.4 hours to repair a radiator, and 1.3 hours to replace a front engine seal. The gasket was $23.50, the engine seal was $12.38, and the radiator needed no parts, except two hose clamps at $1.75 each. If the labor rate is $57.75 per hour, what is the total bill, before tax?

Division of Decimal Numbers

Put the decimal point in the correct place in the quotients below.

●1. $2.8 \overline{)4.48}$ quotient 16

●2. $0.37 \overline{)1.369}$ quotient 37

●3. $3.856 \overline{)0.320048}$ quotient 000083

●4. $17.83 \overline{)12.1244}$ quotient 00068

Find the quotient for each of the problems below.

●5. 6)13.2   ●6. 7)62.3   ●7. 12)32.4

●8. 15)85.5   ●9. 0.8)14.4   ●10. 0.2)7.6

●11. 0.4)15.28   ●12. 1.9)12.73   ●13. 6.4)23.68

●●14. 0.24)7.68   ●●15. 0.35)9.1   ●16. 3.00)22.50

●●17. 0.37)3.6334   ●●18. 2.62)10.2442   ●●●19. 0.627)28.3404

●20. 11.18 ÷ 1.3 =   ●21. 112.52 ÷ 9.7 =   ●●22. 3.51 ÷ 0.65 =

●●23. 88.35 ÷ 0.62 =   ●●24. 78.37 ÷ 8.5 =   ●●●25. 7.995 ÷ 1.23 =

Find the quotient for each of the problems below, rounding to the nearest hundredth.

●26. 8.7 ÷ 4 =   ●27. 22.3 ÷ 7 =   ●28. 13.5 ÷ 4.2 =

●29. 33.7 ÷ 9.3 =   ●●30. 13.952 ÷ 4.5 =   ●●31. 100.55 ÷ 8.7 =

●32. A case contains 12 qt of motor oil, and sells for $28.68. What is the price per quart?

●33. A box of 25 assorted fuses costs $5.69. What is the price per fuse? Round to the nearest penny.

●34. The price of a 50-ft roll of three-quarter in. heater hose is $35.50. What is the price per foot?

DECIMAL OPERATIONS WITH APPLICATIONS TO MEASUREMENT AND FINANCE

••35. The price of a 250-ft spool of heater hose is $77.42. What is the price per foot, rounded to the nearest cent?

•36. A 1250 sq. ft shop space rents for $2,700 per month. What is the price per square foot?

••37. An oil change shop orders oil by the pallet. Each pallet contains 48 cases of oil, and each case contains 12 qt. The cost per pallet is $437.76.
   a. How many quarts are there on a pallet?
   b. What is the price per case?
   c. What is the price per quart?

••38. The same shop in problem 31 can buy oil by the drum. Each drum contains 55 gallons, and sells for $149.60.
   a. How many quarts does a drum of oil contain (*note:* there are 4 qt in 1 gallon)?
   b. When oil is bought by the drum, what is the price per gallon?
   c. What oil is bought by the drum, what is the price per quart?

•39. A truck drives 311.6 miles while using 16.4 gallons. What is the average fuel economy?

•40. A car is able to drive 308.2 miles while using 11.5 gallons of fuel. What is the average fuel economy?

••41. A driver wants to check her fuel economy. She fills the tank and notes the odometer reading of 42,117.4. After driving some distance, the odometer reading is 42,347.7 and it takes 9.8 gallons to fill the tank. What is her average fuel economy?

••42. A few years later, the driver in problem 41 wants to recheck her fuel economy. Again, she fills the tank, and records the odometer reading of 97,962.7. She fills the tank with 8.5 gallons a few days later, and notes the odometer reading is 98,129.3.
   a. What is her fuel economy now?
   b. By how much has her fuel economy decreased since she calculated it in problem 33?

••43. A diesel powered car is able to drive 362.8 miles while using 9.1 gallons of fuel. What is the car's fuel economy, rounded to the nearest tenth?

•44. On a long trip, including necessary stops, a driver covers 578.4 miles in 12 hours. What is her average speed?

##45. On a vacation, a driver drives from St. Paul, Minnesota to Glacier National Park in Montana, a distance of 1154.6 miles. It takes him 18 hours and 30 minutes of total driving time. What is his average speed?

##46. During a certain week, a technician works 39.6 hours. After taxes and other deductions her net pay is $487.93. What is the tech's hourly wage, after taxes?

##47. A technician at a different shop works 38.7 hours in a certain week. His pay after taxes and other deductions is $603.72. What is this tech's hourly wage, after taxes?

Square Roots

Find the value of each square root below without using a calculator.

●1. $\sqrt{9} =$ ●2. $\sqrt{49} =$ ●3. $\sqrt{100} =$

●4. $\sqrt{4} =$ ●5. $\sqrt{81} =$ ●6. $\sqrt{144} =$

Use a calculator to find the value of each square root below. Round to the nearest thousandth.

●7. $\sqrt{19} =$ ●8. $\sqrt{57} =$ ●9. $\sqrt{31} =$

●10. $\sqrt{107} =$ ●11. $\sqrt{1250} =$ ●12. $\sqrt{397} =$

Find the value of each expression below using your calculator. Round to the nearest thousandth.

●13. $4.00 \times \sqrt{43} =$ ●14. $6.50 \times \sqrt{57} =$ ●15. $17.87 + \sqrt{299} =$

●16. $9.683 + \sqrt{468} =$ ●17. $\sqrt{630} + \sqrt{135} =$ ●18. $\sqrt{397} + \sqrt{961} =$

CHAPTER 3

Fractional Operations with Applications to Measurement

EQUIVALENT FRACTIONS

Fractions provide an easy way to express parts of a whole unit. When breaking a whole unit into equal pieces, the first step is to decide how many equal pieces the whole will be broken into. For example, instead of buying a full gallon of oil, we can buy a quart, which is one-fourth of a gallon. Here, the gallon was divided into four equal parts called quarters or fourths. In any fraction, the bottom number, called the **denominator,** shows how many parts a whole unit is broken into. Figure 3.1 shows examples of the denominator.

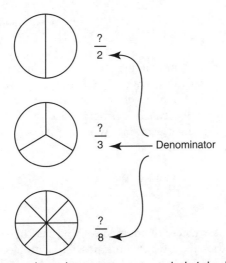

FIGURE 3.1 The denominator shows how many parts a whole is broken into.

The denominator alone, though, is not enough to describe a quantity. When a gallon is divided into fourths called quarts, we know the denominator of the fraction will be 4. Unless we know how many of these quarts or fourths we have, dividing the gallon into parts is meaningless. The **numerator** is the top number in a fraction, and tells how many of the pieces we have. For example, if we have 3 qt, and each quart is a fourth, we know that we have 3/4 of a gallon. Figure 3.2 shows examples of the numerator.

Notice the three diagrams in Figure 3.2. In the fraction 1/2, the whole circle is divided into two parts (the denominator), and one of them (the numerator) is shaded. In the fraction 2/3, there are three equal pieces (the denominator) and two of them (the numerator) are shaded. Finally, in 5/8, there are eight equal parts (the denominator) and five of them (the numerator) are shaded.

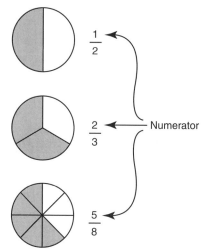

FIGURE 3.2 The numerator tells how many of the pieces we have.

There's a special case that shows up right about now. What if a whole unit is broken into five equal pieces (the denominator is 5) and we have 5 of the pieces (the numerator is also 5)? Then the fraction could be written as 5/5. By looking at the diagram in Figure 3.3, however, we can see that the whole circle is shaded. This means that when the numerator and denominator are the same, the fraction is actually equivalent to a whole, or one.

FIGURE 3.3 An example of the fraction for which the numerator and the denominator are the same.

We can say, then, that any time the numerator (top number) is smaller than the bottom number, the fraction has a value less than one. In fact, these are the fractions we'll be dealing with almost all the time. Any fraction in which the numerator is smaller than the denominator (bottom number) is called a **proper fraction,** and has value less than one. Here are some examples of proper fractions. What numbers could you fill in for the missing values to make proper fractions?

$$\frac{1}{3}, \frac{2}{5}, \frac{4}{9}, \frac{13}{127}, \frac{401}{402}, \frac{?}{6}, \frac{7}{?}$$

Note that there are cases where the numerator is larger than the denominator. That is, we have plenty of pieces to make a whole unit, possibly with pieces left over. For example, suppose you have 7 qt of oil. How many gallons do you have? Well, the numerator must be 7, and since each quart is a fourth, the denominator must be four. Then we have 7/4 of a gallon. An **improper fraction** is a fraction with a numerator that's equal to or larger than the denominator. Improper fractions are awkward, though, because they're hard to look at and quickly understand. Let's look at the improper fraction 7/4 shown in Figure 3.4. Visually, picture each quart as a fourth of a whole gallon.

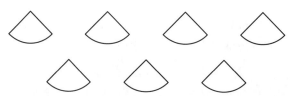

FIGURE 3.4 These seven quarters of a whole represent the fraction 7/4.

FRACTIONAL OPERATIONS WITH APPLICATIONS TO MEASUREMENT

The denominator is four. That means for every four parts, we can make a whole. Here, we can make a whole, and have three left over as shown in Figure 3.5.

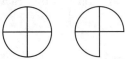

FIGURE 3.5 The seven quarters of a whole from Figure 3.4 now look like 1 3/4.

The easier way to express this number is by writing 1 3/4. A **mixed number** is a number that has a whole part and a fractional part. The number 1 3/4 is an example of a mixed number. Improper fractions can be written as mixed numbers (or as whole numbers if there is no fraction left over).

Example 3-1   Write 5/3 as a mixed number.

Because 3 goes into 5 one time, our whole number will be 1.
That uses up 3 of the 5 pieces, so there are 2 left. The fractional part is 2/3.
Then 5/3 = 1 2/3.

$$\frac{5}{3} \quad \textcircled{1} \text{ Fits into 5 } \textit{one} \text{ time.}$$

$$\textcircled{2} \text{ Then } 5 - 3 = 2, \text{ so the fractional part is } \frac{2}{3}.$$

$$\frac{5}{3} = 1\frac{2}{3}$$

Example 3-2   Write 19/5 as a mixed number.

Because 5 goes into 19 three times, our whole number will be 3.
That uses up 15 of the 19 pieces, so there are 4 left. The fractional part is 4/5.
Then 19/5 = 3 4/5.

$$\frac{19}{5} \quad \textcircled{1} \text{ Fits into 19 } \textit{three} \text{ times.}$$

$$\textcircled{2} \text{ Then } 19 - 15 = 4, \text{ so the fractional part is } \frac{4}{5}.$$

$$\frac{19}{5} = 3\frac{4}{5}$$

Later, to multiply and divide, we'll need to reverse this process and convert mixed numbers to improper fractions.

Example 3-3   Write 4 1/2 as an improper fraction.

Since the denominator is 2, each of the four whole units will be made into two parts. That means there are 2 × 4 = 8 parts, plus the 1 in the fraction, or a total of 9 parts. Then 4 1/2 = 9/2.

$$4\frac{1}{2}$$

$$4 \times 2 = 8 \text{ parts} + 1 \text{ in numerator} = 9 \text{ parts}$$

$$4\frac{1}{2} = \frac{9}{2}$$

Example 3-4 Write 6 5/8 as an improper fraction.

Here, the denominator is 8, and we have 6 whole units, which gives us 48 parts to begin with. Added to the 5 in the fraction, there are a total of 53 parts. Then 6 5/8 = 53/8.

$$6\frac{5}{8}$$

$$6 \times 8 = 48 \text{ parts} + 5 \text{ in numerator} = 53 \text{ parts}$$

$$6\frac{5}{8} = \frac{53}{8}$$

The last basic thing we need to do with fractions is change the denominator. Any fraction can be written hundreds of different ways, all without changing the actual value. For example, if you have half of a dollar, you could have any of the following:

5 dimes: Ten dimes make a whole, we have five of them, or 5/10 of a dollar.

2 quarters: Four quarters make a whole, we have two of them, or 2/4 of a dollar.

10 nickels: Twenty nickels make a whole, we have ten of them, or 10/20 of a dollar.

In each of these representations, we still have 1/2 of a dollar. We can say then, that

$$\frac{1}{2} = \frac{5}{10} = \frac{2}{4} = \frac{10}{20}$$

We can write the same fraction in different ways by multiplying the numerator and denominator by the same number.

Example 3-5 Write 3/4 in three different ways.

Let's multiply both the numerator 3, and the denominator 4, by 2. Then

$$\frac{3}{4} = \frac{3}{4} \times \frac{2}{2} = \frac{6}{8} \quad \text{That's one way.}$$

For a different way, we'll multiply both by 5. Then

$$\frac{3}{4} = \frac{3}{4} \times \frac{5}{5} = \frac{15}{20} \quad \text{That's another way.}$$

Finally, let's multiply both by 10. Then

$$\frac{3}{4} = \frac{3}{4} \times \frac{10}{10} = \frac{30}{40} \quad \text{We could go on and on.}$$

Example 3-6 Write 7/8 using a denominator of 16.

Well, we already know the new denominator must be 16. We need to find the new numerator.

$$\frac{7}{8} = \frac{?}{16}$$

Since the old denominator, 8, is multiplied by 2 to get 16, we need to multiply the old numerator 7 by 2 also. Then our new numerator will be 14.

$$\frac{7 \,\textcircled{\times 2}}{8 \,\textcircled{\times 2}} = \frac{14}{16}$$

FRACTIONAL OPERATIONS WITH APPLICATIONS TO MEASUREMENT

Example 3-7 Write 3 5/9 using a denominator of 27.

The whole number will not change, and the new denominator is 27.

$$3\frac{5}{9} = 3\frac{?}{27}$$

The old denominator gets multiplied by 3 to get the new denominator of 27, so we had better do the same for the old numerator.

$$3\frac{5 \,(\times 3)}{9 \,(\times 3)} = 3\frac{15}{27}$$

If there are so many ways to write any fraction, which way is the best? Usually, it's best to write a fraction in lowest terms. A fraction is in **lowest terms** if there is no whole number (other than 1) that divides both the numerator and denominator.

Example 3-8 Write 6/9 in lowest terms.

You may notice that both 6 and 9 are divisible by 3. That means we're reversing what we did in the last few examples, and dividing both the numerator and denominator by the same number.

$$\frac{6 \,(\div 3)}{9 \,(\div 3)} = \frac{2}{3}$$

After dividing both numerator and denominator by 3, we get the fraction 2/3. Now, can we reduce this number any more? Is there any number besides 1 that divides both 2 and 3? No, there isn't. Then 2/3 is in lowest terms.

Example 3-9 Write 9 12/20 in lowest terms.

Again, the whole number does not change. What is the largest number that will divide both 12 and 20? The number 4 divides both, so divide both 12 and 20 by 4.

$$9\frac{12 \,(\div 4)}{20 \,(\div 4)} = 9\frac{3}{5}$$

COMMON DENOMINATORS

The next logical thing to do when working with fractions is to add and subtract. When adding and subtracting parts of a whole, we need to make sure those parts are all the same size. Therefore, we need to be able to write two different fractions using the same denominator. Since all fractions can be written using different denominators, we try to find a denominator that both fractions can use. The **lowest common denominator**, or LCD, is the lowest number that can be used as the denominator in both fractions. While there are many possible common denominators that will work just fine, we usually try to use the lowest common denominator to avoid additional work in reducing our answer.

Example 3-10 Find the lowest common denominator for the fractions $\frac{3}{4}$ and $\frac{5}{6}$.

First, think about ways we can write $\frac{3}{4}$. From what we've just done,

$$\frac{3}{4} = \frac{6}{8} = \frac{9}{12} = \frac{12}{16}\ldots$$

Do the same with $\frac{5}{6}$. Here, $\frac{5}{6} = \frac{10}{12} = \frac{15}{18} = \frac{20}{24}\ldots$

The lowest number that works as a denominator for both fractions is 12. Then our LCD is 12. Now, we can write $\frac{3}{4}$ and $\frac{5}{6}$ using the LCD as $\frac{9}{12}$ and $\frac{10}{12}$, respectively.

More easily, though, notice that 12 is the smallest number that our original denominators, 4 and 6, divide into evenly.

Example 3-11 Find the lowest common denominator for the fractions $\frac{5}{8}$ and $\frac{7}{12}$.

The denominator 8 could be changed to 16, 24, 32, and so on. The denominator 12 could be changed to 24, 36, 48, and many more. The lowest number that both 8 and 12 divide evenly is 24. Then our LCD is 24. Finally, $\frac{5}{8}$ and $\frac{7}{12}$ can be written using the LCD as $\frac{15}{24}$ and $\frac{14}{24}$ respectively.

Example 3-12 Find the LCD for the fractions $\frac{2}{3}, \frac{4}{5},$ and $\frac{1}{2}$.

The LCD for these fractions will be the smallest number that 3, 5, and 2 divide into evenly. In this case, 30 is the smallest number that works. Now, rewrite each of the fractions using a denominator of 30.

$$\frac{2}{3} = \frac{20}{30}, \frac{4}{5} = \frac{24}{30}, \text{ and } \frac{1}{2} = \frac{15}{30}$$

Another thing common denominators can be used for is to compare the relative size of two fractions that have different denominators.

Example 3-13 Which is larger, $\frac{1}{3}$ or $\frac{5}{16}$?

It's not immediately clear which of these two values is the larger one. To make it more apparent, write the two fractions using a common denominator.

First, find the LCD. It must be 48 since that is the smallest number that both 3 and 16 divide into evenly.

Next, rewrite both of these fractions using the LCD.

$$\frac{1}{3} = \frac{16}{48}$$

$$\frac{5}{16} = \frac{15}{48}$$

Finally, compare the two fractions. Since $\frac{16}{48}$ is more than $\frac{15}{48}$, it is true that $\frac{1}{3}$ is larger than $\frac{5}{16}$.

Example 3-14 Put in order from smallest to largest: $\frac{3}{4}, \frac{11}{16},$ and $\frac{5}{8}$.

First, find the LCD. For these fractions the LCD is 16.

Second, rewrite each of the fractions using the LCD.

$$\frac{3}{4} = \frac{12}{16}$$

$$\frac{11}{16}$$

$$\frac{5}{8} = \frac{10}{16}$$

Now, we can see that $\frac{10}{16}$ or $\frac{5}{8}$ is the smallest, and $\frac{12}{16}$ or $\frac{3}{4}$ is the largest. So putting the fractions in order we have $\frac{5}{8}, \frac{11}{16}, \frac{3}{4}$.

ADDITION OF FRACTIONS

Now that we can write two fractions using a common denominator, we can add and subtract them, because both fractions will be counting parts that are the same size.

Example 3-15 Find the sum of $\frac{7}{12} + \frac{1}{12}$.

The first step is to find the LCD, then rewrite the problem using the LCD. Here, both fractions already have a common denominator of 12, so there is nothing to do.

Next, add the numerators. Since $7 + 1 = 8$, we have

$$\frac{7}{12} + \frac{1}{12} = \frac{8}{12}$$

Remember that the denominators do not get added; they simply say what the size of the fractional parts are.

The last step is to reduce the fraction, and get rid of improper fractions. This is a proper fraction, but it is not reduced. Since both 8 and 12 are divisible by 4, we can conclude

$$\frac{7}{12} + \frac{1}{12} = \frac{2}{3}$$

Example 3-16 Add $\frac{2}{5} + \frac{1}{3}$.

Begin by finding the LCD. Here, the LCD is 15 because it is the smallest number divisible by 3 and 5.

Now, rewrite the problem using the LCD.

$$\frac{2}{5} + \frac{1}{3} \text{ becomes } \frac{6}{15} + \frac{5}{15}$$

Adding the numerators we have

$$\frac{6}{15} + \frac{5}{15} = \frac{11}{15}$$

which will not reduce.

Then we can conclude that

$$\frac{2}{5} + \frac{1}{3} = \frac{11}{15}$$

It is worth noting that any common denominator will work when adding or subtracting fractions. For example, a common denominator of 30 could have been used in the previous example. If the common denominator used is not the lowest one possible, however, your answer will need to be reduced.

Example 3-17 Find the sum of $\frac{4}{9} + \frac{3}{4}$.

Start by finding the LCD. The LCD here must be 36.
Next, rewrite the problem using the LCD.

$$\frac{4}{9} + \frac{3}{4} \text{ becomes } \frac{16}{36} + \frac{27}{36}$$

Adding the numerators, we have

$$\frac{16}{36} + \frac{27}{36} = \frac{43}{36}$$

Notice, though, that this is an improper fraction. The number 36 fits into 43 just once, leaving 7 for the numerator in the mixed number.

$$\frac{43}{36} = 1\frac{7}{36}$$

We can conclude that

$$\frac{4}{9} + \frac{3}{4} = 1\frac{7}{36}$$

Example 3-18 Add $2\frac{1}{6} + 4\frac{5}{8}$.

As always, start by finding the LCD. Here the LCD is 24.
Now rewrite the problem using the LCD.

$$2\frac{1}{6} + 4\frac{5}{8} \text{ becomes } 2\frac{4}{24} + 4\frac{15}{24}$$

Since the whole numbers will stay whole numbers, we can add them at the same time we add the numerators. We find that

$$2\frac{4}{24} + 4\frac{15}{24} = 6\frac{19}{24}$$

This cannot be reduced, and the fractional part is proper. We can then conclude that

$$2\frac{1}{6} + 4\frac{5}{8} = 6\frac{19}{24}$$

Example 3-19 Add $7\frac{8}{9} + 5\frac{7}{12}$.

The LCD here is 36.
Rewriting the problem,

$$7\frac{8}{9} + 5\frac{7}{12} \text{ becomes } 7\frac{32}{36} + 5\frac{21}{36}$$

Add the whole number parts, and add the numerators. We get

$$7\frac{32}{36} + 5\frac{21}{36} = 12\frac{53}{36}$$

Since the fractional part of this mixed number is improper, we must convert $\frac{53}{36}$ to a mixed number. Since $\frac{53}{36}$ can be written $1\frac{17}{36}$, we gain a whole number. We can

FRACTIONAL OPERATIONS WITH APPLICATIONS TO MEASUREMENT

add that to the 12 we already had and see that

$$12\frac{53}{36} = 13\frac{17}{36}$$

Then

$$7\frac{8}{9} + 5\frac{7}{12} = 13\frac{17}{36}$$

Example 3-20 A technician worked on a vehicle for part of the day for three days one week. On Monday he worked $3\frac{3}{10}$ hours, on Tuesday he worked $4\frac{1}{2}$ hours, and on Wednesday he finished the job and worked $1\frac{3}{4}$ hours. What was the total time spent on this job?

Our goal here is to add

$$3\frac{3}{10} + 4\frac{1}{2} + 1\frac{3}{4}$$

The LCD for these fractions is 20, so we'll rewrite the problem.

$$3\frac{3}{10} + 4\frac{1}{2} + 1\frac{3}{4} \text{ becomes } 3\frac{6}{20} + 4\frac{10}{20} + 1\frac{15}{20}$$

Add the whole numbers, and add the numerators to find that

$$3\frac{6}{20} + 4\frac{10}{20} + 1\frac{15}{20} = 8\frac{31}{20}$$

The fractional part is improper, though, so we'll convert it to a mixed number. From our work earlier, note that $\frac{31}{20} = 1\frac{11}{20}$. Adding to the 8 whole hours we already had,

$$8\frac{31}{20} = 9\frac{11}{20}$$

Then the total time spent on that job is

$$3\frac{3}{10} + 4\frac{1}{2} + 1\frac{3}{4} = 9\frac{11}{20} \text{ hours}$$

SUBTRACTION OF FRACTIONS

Subtracting fractions follows the same basic idea used when adding fractions. Instead of adding the numerators, we'll subtract them.

Example 3-21 Subtract $\frac{11}{12} - \frac{7}{12}$.

Normally, the first step would be to make sure the denominators are the same. That's already the case, so we can just subtract the numerators. Since $11 - 7 = 4$, we can say

$$\frac{11}{12} - \frac{7}{12} = \frac{4}{12}$$

Since $\frac{4}{12}$ can be reduced, we can write our solution as

$$\frac{11}{12} - \frac{7}{12} = \frac{1}{3}$$

Example 3-22 Find the difference for $\frac{5}{8} - \frac{1}{10}$.

Like addition, begin by finding the LCD. Here, the LCD is 40.
Now, rewrite the problem using the LCD.

$$\frac{5}{8} - \frac{1}{10} \text{ becomes } \frac{25}{40} - \frac{4}{40}$$

Third, subtract the numerators and reduce if possible.

$$\frac{25}{40} - \frac{4}{40} = \frac{21}{40}$$

Since $\frac{21}{40}$ is not reducible, we can conclude

$$\frac{5}{8} - \frac{1}{10} = \frac{21}{40}$$

Example 3-23 Subtract $2\frac{1}{4} - \frac{3}{5}$.

The LCD here is 20.
Rewriting,

$$2\frac{1}{4} - \frac{3}{5} \text{ becomes } 2\frac{5}{20} - \frac{12}{20}$$

It looks like we have a problem here. We can't simply subtract the numerators, since 12 is larger than 5. We need to take one of the whole units from $2\frac{5}{20}$ and break it up into parts to make the numerator larger.

Since $1 = \frac{20}{20}$, we can rewrite $2\frac{5}{20}$ as $1\frac{20}{20} + \frac{5}{20}$ or simply $1\frac{25}{20}$. Now, rewrite the problem as

$$2\frac{1}{4} - \frac{3}{5} = 2\frac{5}{20} - \frac{12}{20} = 1\frac{25}{20} - \frac{12}{20}$$

Now we can subtract the numerators, then the whole number part, and we see that

$$1\frac{25}{20} - \frac{12}{20} = 1\frac{13}{20} \text{ which we cannot reduce.}$$

Apparently, then,

$$2\frac{1}{4} - \frac{3}{5} = 1\frac{13}{20}$$

Example 3-24 Subtract $7\frac{3}{8} - 2\frac{1}{2}$.

The LCD is 8, so we'll rewrite to see that

$$7\frac{3}{8} - 2\frac{1}{2} \text{ becomes } 7\frac{3}{8} - 2\frac{4}{8}$$

FRACTIONAL OPERATIONS WITH APPLICATIONS TO MEASUREMENT 53

Again, the numerators can't simply be subtracted because the 4 is larger than the 3.

We need to make the 3 larger. Since $7\frac{3}{8} = 6 + 1\frac{3}{8} = 6 + \frac{11}{8} = 6\frac{11}{8}$, we write $7\frac{3}{8}$ as $6\frac{11}{8}$. Then our problem becomes

$$7\frac{3}{8} - 2\frac{1}{2} = 7\frac{3}{8} - 2\frac{4}{8} = 6\frac{11}{8} - 2\frac{4}{8}$$

Subtract the numerators, then the whole numbers, and we find

$$6\frac{11}{8} - 2\frac{4}{8} = 4\frac{7}{8} \text{ which cannot be reduced.}$$

To summarize,

$$7\frac{3}{8} - 2\frac{1}{2} = 4\frac{7}{8}$$

Example 3-25 Find the difference for $9\frac{3}{10} - 5\frac{5}{12}$.

The LCD in this case is 60. This problem then becomes $9\frac{18}{60} - 5\frac{25}{60}$.

Again the first numerator is not large enough, so we need to borrow a whole unit. Doing this, $9\frac{18}{60}$ becomes $8\frac{78}{60}$. So far then, we have

$$9\frac{3}{10} - 5\frac{5}{12} = 9\frac{18}{60} - 5\frac{25}{60} = 8\frac{78}{60} - 5\frac{25}{60}$$

Subtracting the numerators, then the whole numbers, we end up with

$$8\frac{78}{60} - 5\frac{25}{60} = 3\frac{53}{60}$$

Since that does not reduce, our final result is

$$9\frac{3}{10} - 5\frac{5}{12} = 3\frac{53}{60}$$

Example 3-26 A cooling system holds 2 3/4 gallons. A tech drains 1 1/2 gallons from the system. How much coolant remains?

We need to find the solution to $2\frac{3}{4} - 1\frac{1}{2}$.

The LCD is 4, so the problem becomes $2\frac{3}{4} - 1\frac{2}{4}$.

In this case, the first numerator is large enough, so there is no need to borrow. Subtract the numerators, then the whole numbers, and we find that

$$2\frac{3}{4} - 1\frac{2}{4} = 1\frac{1}{4}$$

There are 1 1/4 gallons still in the system.

MULTIPLICATION OF FRACTIONS

Suppose a cooling system holds 13 1/2 qt of coolant; half of this is water, the other half is antifreeze. How much of each is needed? This question could be answered by simply multiplying 13 1/2 qt by 1/2. We'll now explore the process for multiplying fractions.

The product of two fractions is found by multiplying numerator by numerator and denominator by denominator.

Example 3-27 Multiply $\frac{3}{4} \times \frac{5}{8}$.

First, multiply across. The product of the numerators is 15, the product of the denominators is 32.

$$\frac{3}{4} \times \frac{5}{8} = \frac{3 \times 5}{4 \times 8} = \frac{15}{32}$$

The product of these fractions is $\frac{15}{32}$, which does not reduce.

Example 3-28 Find the product of $\frac{2}{3} \times \frac{7}{10}$.

Proceeding as we did in the last problem, we could multiply across to find that

$$\frac{2}{3} \times \frac{7}{10} = \frac{2 \times 7}{3 \times 10} = \frac{14}{30}$$

Since $\frac{14}{30}$ reduces to $\frac{7}{15}$, we can conclude that $\frac{2}{3} \times \frac{7}{10} = \frac{7}{15}$.

There is a shortcut that can often be used in multiplication problems. Before multiplying across, see if there is a common factor in the numerator of one fraction and the denominator of another. Reducing in this way is called **canceling common factors**. Can the previous example use this shortcut?

Example 3-29 Again, multiply $\frac{2}{3} \times \frac{7}{10}$. This time, use the method of canceling common factors.

Notice that 2 and 10 have a common factor of 2. Divide both of these by 2, and rewrite the problem.

$$\frac{\cancel{2}^1}{3} \times \frac{7}{\cancel{10}_5}$$

Then $\frac{2}{3} \times \frac{7}{10}$ becomes $\frac{1}{3} \times \frac{7}{5}$. There is no other canceling to do, so we'll multiply.

$$\frac{2}{3} \times \frac{7}{10} = \frac{1}{3} \times \frac{7}{5} = \frac{7}{15}$$

Example 3-30 Multiply $\frac{5}{14} \times \frac{21}{25}$.

Begin by canceling common factors. The 5 and 25 can be divided by 5.

$$\frac{\cancel{5}^1}{14} \times \frac{21}{\cancel{25}_5}$$

14 and 21 can both be divided by 7.

$$\frac{\cancel{5}^1}{\cancel{14}_2} \times \frac{\cancel{21}^3}{\cancel{25}_5}$$

FRACTIONAL OPERATIONS WITH APPLICATIONS TO MEASUREMENT

Then $\frac{5}{14} \times \frac{21}{25}$ becomes $\frac{1}{2} \times \frac{3}{5}$. Now, we can multiply to find that

$$\frac{5}{14} \times \frac{21}{25} = \frac{1}{2} \times \frac{3}{5} = \frac{3}{10}$$

Now we are ready to work on the problem we used to open this section.

Example 3-31 What is 1/2 of 13 1/2 qt?

To answer this question, we simply need to multiply $\frac{1}{2} \times 13\frac{1}{2}$.

When multiplying we must always convert mixed numbers to improper fractions.

Since $13\frac{1}{2} = \frac{27}{2}$, we can write the problem as

$$\frac{1}{2} \times \frac{27}{2}$$

There are no common factors to cancel, so we can simply multiply across.

$$\frac{1}{2} \times \frac{27}{2} = \frac{27}{4}$$

Now, we simply need to write our solution as a mixed number. 4 fits into 27 six times, with 3 left over for the numerator. Then

$$\frac{1}{2} \times 13\frac{1}{2} = \frac{1}{2} \times \frac{27}{2} = \frac{27}{4} = 6\frac{3}{4} \text{ qt}$$

Example 3-32 Multiply $5\frac{1}{3} \times 2\frac{1}{4}$.

First, convert each mixed number to an improper fraction. Then

$$5\frac{1}{3} \times 2\frac{1}{4} \text{ becomes } \frac{16}{3} \times \frac{9}{4}$$

Next, cancel common factors where possible.

$$\frac{\cancel{16}^4}{\cancel{3}_1} \times \frac{\cancel{9}^3}{\cancel{4}_1}$$

Now, multiply across and reduce.

$$5\frac{1}{3} \times 2\frac{1}{4} = \frac{16}{3} \times \frac{9}{4} = \frac{4}{1} \times \frac{3}{1} = \frac{12}{1} = 12$$

Example 3-33 Multiply $4\frac{3}{8} \times 2\frac{1}{5}$.

Convert mixed numbers to improper fractions.

$$4\frac{3}{8} \times 2\frac{1}{5} \text{ becomes } \frac{35}{8} \times \frac{11}{5}$$

The 35 and 5 have a common factor of 5 and will cancel,

$$\frac{\cancel{35}^7}{8} \times \frac{11}{\cancel{5}_1}$$

and we can now multiply across.

$$4\frac{3}{8} \times 2\frac{1}{5} = \frac{35}{8} \times \frac{11}{5} = \frac{7}{8} \times \frac{11}{1} = \frac{77}{8}$$

Since $\frac{77}{8}$ should be written as $9\frac{5}{8}$, our conclusion is

$$4\frac{3}{8} \times 2\frac{1}{5} = 9\frac{5}{8}$$

Example 3-34 Three steel plates, each 7/16 in. thick, are to be bolted together. What is the total thickness?

We need to multiply $3 \times \frac{7}{16}$. Begin by writing the whole number 3 as the fraction $\frac{3}{1}$. Now, multiply as usual.

$$3 \times \frac{7}{16} = \frac{3}{1} \times \frac{7}{16} = \frac{21}{16} = 1\frac{5}{16} \text{ in.}$$

DIVISION OF FRACTIONS

In the previous section we wanted to find how much of the 13 1/2 qt of coolant was antifreeze, knowing that water and antifreeze each made up 1/2 of the solution. We answered the question by multiplying 13 1/2 by 1/2. We could also have answered the question by dividing 13 1/2 by 2. This means multiplication and division are closely related.

To divide fractions, we need a new term. Every fraction has a related number called its **reciprocal,** which is found by inverting the fraction.

Example 3-35 Find the reciprocal of $\frac{3}{4}$.

Reversing the numerator and denominator, we see that the reciprocal of $\frac{3}{4}$ is $\frac{4}{3}$.

Example 3-36 Find the reciprocal of $3\frac{5}{8}$.

First, write $3\frac{5}{8}$ as an improper fraction, then reverse the numerator and denominator. $3\frac{5}{8}$ becomes $\frac{29}{8}$, and the reciprocal is $\frac{8}{29}$.

To divide fractions, we simply multiply the dividend by the reciprocal of the divisor.

Example 3-37 Divide $\frac{5}{8} \div \frac{2}{3}$.

We need to change this to a multiplication problem. The divisor is always the second fraction, which is $\frac{2}{3}$ in this case. The reciprocal of $\frac{2}{3}$ is $\frac{3}{2}$, so that

$$\frac{5}{8} \div \frac{2}{3} \text{ becomes } \frac{5}{8} \times \frac{3}{2}$$

We already know how to multiply fractions, so we can find the solution.

$$\frac{5}{8} \div \frac{2}{3} = \frac{5}{8} \times \frac{3}{2} = \frac{15}{16}$$

Example 3-38 Divide $2\frac{3}{4} \div 1\frac{1}{6}$.

Convert both numbers to improper fractions.

$$2\frac{3}{4} \div 1\frac{1}{6} \text{ becomes } \frac{11}{4} \div \frac{7}{6}$$

FRACTIONAL OPERATIONS WITH APPLICATIONS TO MEASUREMENT

Now, find the reciprocal of $\frac{7}{6}$ and multiply. Note that we can cancel a common factor of 2.

$$\frac{11}{4} \times \frac{6}{7} = \frac{11}{2} \times \frac{3}{7} = \frac{33}{14}$$

Since $\frac{33}{14}$ should be written as the mixed number $2\frac{5}{14}$, we can see that

$$2\frac{3}{4} \div 1\frac{1}{6} = \frac{11}{4} \times \frac{6}{7} = \frac{33}{14} = 2\frac{5}{14}$$

Example 3-39 Divide $7\frac{3}{5} \div 3$.

Write both numbers as improper fractions.

$$7\frac{3}{5} \div 3 \text{ is written as } \frac{38}{5} \div \frac{3}{1}$$

Using the reciprocal to multiply,

$$\frac{38}{5} \div \frac{3}{1} \text{ becomes } \frac{38}{5} \times \frac{1}{3} = \frac{38}{15}$$

Finally, convert the solution to a mixed number.

$$7\frac{3}{5} \div 3 = \frac{38}{5} \div \frac{3}{1} = \frac{38}{5} \times \frac{1}{3} = \frac{38}{15} = 2\frac{8}{15}$$

Example 3-40 The rear toe on some vehicles is changed by turning an adjusting nut. This rear toe adjusting nut can be seen in Figure 3.6. If each turn of the nut changes the rear toe by 1/16 in., how many turns will be needed to change the toe 5/32 in.?

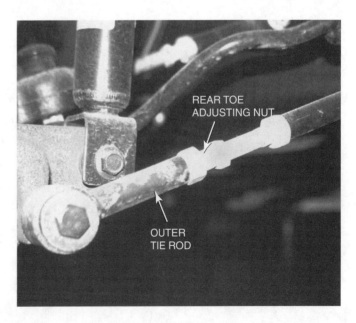

FIGURE 3.6 Turning the adjusting nut will change the rear toe setting on some vehicles.

Source: James D. Halderman and Chase D. Mitchell, Jr., *Automotive Steering, Suspension, and Alignment,* Second Edition © 2000. Reprinted by permission of Pearson Education, Inc., Upper Saddle River, NJ.

We need to divide $\frac{5}{32} \div \frac{1}{16}$ to see how many turns are needed.

Changing to a multiplication problem, we have

$$\frac{5}{32} \div \frac{1}{16} = \frac{5}{32} \times \frac{16}{1}$$

Canceling, we can see that

$$\frac{5}{32} \times \frac{16}{1} = \frac{5}{2} \times \frac{1}{1} = \frac{5}{2}$$

Then

$$\frac{5}{32} \div \frac{1}{16} = \frac{5}{32} \times \frac{16}{1} = \frac{5}{2} = 2\frac{1}{2} \text{ turns}$$

CONVERTING FRACTIONS TO DECIMALS

Suppose the choke clearance on a carburetor could be set using a 5/16-in. drill bit. It would probably be useful to know the decimal equivalent of this measurement because the choke clearance specification shown in a repair manual may be given as a decimal part of an inch.

Example 3-41 Convert the fraction $\frac{5}{8}$ to a decimal.

Since a fraction symbol indicates division, we'll simply divide 5 by 8.
Don't forget to add zeros to the 5 as placeholders.

$$5 \div 8 =$$

$$\begin{array}{r} 0.625 \\ 8 \overline{\smash{)}5.000} \\ -48 \\ \hline 20 \\ -16 \\ \hline 40 \\ -40 \\ \hline 0 \end{array}$$

$$\frac{5}{8} = 0.625$$

The fraction $\frac{5}{8}$ is equal to 0.625.

Example 3-42 Convert the fraction $4\frac{11}{16}$ to a decimal.

The whole number 4 will also be a whole number when we convert this number to a decimal. We only need to perform the conversion on the fractional part.
We'll divide 11 by 16.

$$11 \div 16 =$$

$$\begin{array}{r} 0.6875 \\ 16 \overline{\smash{)}11.0000} \\ -96 \\ \hline 140 \\ -128 \\ \hline 120 \\ -112 \\ \hline 80 \\ -80 \\ \hline 0 \end{array}$$

$$\frac{11}{16} = 0.6875$$

Don't forget that we had a whole number in this value.
Then $4\frac{11}{16}$ is equal to 4.6875. Rounded to the nearest thousandth, we have 4.688.

FRACTIONAL OPERATIONS WITH APPLICATIONS TO MEASUREMENT

Example 3-43 Convert the fraction $6\frac{3}{7}$ to a decimal, rounding to three decimal places.

Again, we only need to convert the fractional part, $\frac{3}{7}$ to a decimal.

Since we want to round to three decimal places, our division must include four decimals to decide if the third digit is rounded up.

Divide 3 by 7.

$$3 \div 7 =$$

$$\begin{array}{r} 0.4285 \\ 7\overline{)3.0000} \\ \underline{28} \\ 20 \\ \underline{14} \\ 60 \\ \underline{56} \\ 40 \\ \vdots \end{array}$$

$$\frac{3}{7} \approx 0.4285$$

Notice that this division could go on because it has not divided evenly.

Again, replace the whole number part of the value, and round to three decimals. $6\frac{3}{7}$ is *approximately* equal to 6.429.

CHAPTER 3 *Practice Problems*

Equivalent Fractions

Identify each of the following as a proper fraction, improper fraction, or mixed number.

- •1. $\frac{3}{7}$
- •2. $\frac{13}{4}$
- •3. $1\frac{4}{9}$
- •4. $\frac{4}{4}$
- •5. $\frac{15}{20}$
- •6. $6\frac{17}{25}$

Write each of the improper fractions below as a mixed number in lowest terms or a whole number where appropriate.

- •7. $\frac{3}{2}$
- •8. $\frac{7}{3}$
- •9. $\frac{9}{5}$
- •10. $\frac{13}{6}$
- •11. $\frac{24}{8}$

- •12. $\frac{31}{10}$
- •13. $\frac{45}{4}$
- ••14. $\frac{27}{6}$
- ••15. $\frac{38}{8}$
- ••16. $\frac{60}{9}$

- ••17. $\frac{24}{16}$
- ••18. $\frac{40}{32}$
- ••19. $\frac{52}{20}$
- ••20. $\frac{84}{21}$
- ••21. $\frac{120}{64}$

Write each of the whole or mixed numbers below as an improper fraction in lowest terms.

- •22. $6\frac{1}{2}$
- •23. $2\frac{3}{7}$
- •24. $4\frac{8}{9}$
- •25. $1\frac{17}{20}$
- •26. $10\frac{3}{4}$

●27. 9   ●28. $5\frac{4}{11}$ ●29. 13   ●30. $33\frac{1}{3}$ ●●31. $12\frac{1}{64}$

●●32. $2\frac{17}{29}$ ●●33. $7\frac{10}{11}$ ●●34. $10\frac{21}{42}$ ●●35. $17\frac{5}{8}$ ●●36. $31\frac{15}{32}$

Write each of the fractions below using the new denominator given.

●37. $\frac{1}{2}$ using 4ths   ●38. $\frac{1}{6}$ using 12ths   ●39. $\frac{5}{8}$ using 16ths

●40. $\frac{3}{10}$ using 50ths   ●41. $\frac{2}{3}$ using 9ths   ●42. $\frac{2}{3}$ using 12ths

●43. $\frac{2}{3}$ using 21sts   ●44. $\frac{3}{4}$ using 16ths   ●45. $7\frac{3}{8}$ using 64ths

●●46. $4\frac{2}{9}$ using 54ths   ●●47. $1\frac{6}{11}$ using 99ths   ●●48. $3\frac{12}{13}$ using 104ths

Write each of the fractions below in lowest terms.

●49. $\frac{6}{8}$ ●50. $\frac{4}{10}$ ●51. $\frac{8}{12}$ ●52. $\frac{15}{18}$ ●53. $\frac{20}{30}$

●54. $\frac{10}{15}$ ●55. $\frac{8}{32}$ ●56. $\frac{14}{35}$ ●57. $2\frac{27}{36}$ ●58. $4\frac{25}{35}$

●●59. $7\frac{26}{36}$ ●●60. $13\frac{13}{52}$ ●●61. $9\frac{11}{15}$ ●●62. $1\frac{34}{51}$ ●●63. $41\frac{60}{64}$

Common Denominators

Find the LCD for each group of fractions given below and rewrite each fraction using the LCD.

●1. $\frac{1}{2}, \frac{3}{4}$ ●2. $\frac{1}{6}, \frac{2}{3}$ ●3. $\frac{1}{2}, \frac{5}{8}$ ●4. $\frac{5}{6}, \frac{11}{12}$

●5. $\frac{1}{6}, \frac{1}{4}$ ●6. $\frac{3}{5}, \frac{4}{7}$ ●●7. $\frac{5}{8}, \frac{7}{10}$ ●●8. $\frac{3}{10}, \frac{5}{12}$

●●9. $\frac{11}{12}, \frac{9}{16}$ ●●10. $\frac{1}{2}, \frac{3}{4}, \frac{2}{5}$ ●●11. $\frac{5}{6}, \frac{1}{8}, \frac{9}{10}$ ●●12. $\frac{1}{32}, \frac{7}{16}, \frac{37}{48}$

Write each group of fractions below from least to greatest.

●13. $\frac{7}{8}, \frac{3}{4}$ ●14. $\frac{5}{6}, \frac{2}{3}$ ●●15. $\frac{2}{3}, \frac{5}{8}$ ●●16. $\frac{4}{7}, \frac{3}{4}$

●●17. $\frac{1}{3}, \frac{11}{32}$ ●●18. $\frac{5}{8}, \frac{3}{4}, \frac{4}{5}$ ●●19. $\frac{7}{12}, \frac{1}{2}, \frac{3}{5}$ ●●20. $\frac{9}{32}, \frac{3}{8}, \frac{5}{16}$

FRACTIONAL OPERATIONS WITH APPLICATIONS TO MEASUREMENT

Addition of Fractions

Find the sum for each problem below. Write your answer in lowest terms.

●1. $\dfrac{4}{7} + \dfrac{2}{7} =$ ●2. $\dfrac{7}{16} + \dfrac{3}{16} =$ ●3. $\dfrac{11}{15} + \dfrac{8}{15} =$

●4. $\dfrac{3}{8} + \dfrac{1}{2} =$ ●5. $\dfrac{1}{12} + \dfrac{3}{4} =$ ●6. $\dfrac{13}{18} + \dfrac{2}{3} =$

●7. $\dfrac{2}{7} + \dfrac{2}{3} =$ ●●8. $5\dfrac{1}{6} + 2\dfrac{3}{4} =$ ●●9. $11\dfrac{7}{10} + 3\dfrac{2}{15} =$

●●10. $5\dfrac{5}{12} + 4\dfrac{7}{8} =$ ●●11. $2\dfrac{7}{8} + 2\dfrac{3}{16} =$ ●●12. $6\dfrac{5}{14} + \dfrac{16}{21} =$

●●13. $8 + 11\dfrac{3}{13} =$ ●●●14. $23\dfrac{13}{18} + 14\dfrac{5}{12} =$ ●●●15. $8\dfrac{13}{24} + 3\dfrac{17}{18} =$

●●●16. $2\dfrac{1}{6} + 7\dfrac{1}{4} + 6\dfrac{1}{3} =$ ●●●17. $4\dfrac{3}{5} + 8\dfrac{1}{2} + 9\dfrac{2}{3} =$ ●●●18. $56\dfrac{7}{8} + 19\dfrac{5}{6} + 21\dfrac{4}{5} =$

●19. A technician works on a truck for 5 1/2 hours. After ordering parts, she works on it for an additional 3 1/4 hours. What is the total time she spends on the truck?

●20. To replace the transmission on a car, a tech works 4 7/10 hours on Monday, and 3 1/4 hours on Tuesday. What is the total time spent on the car?

●●21. To replace a hard-to-reach turn-signal lamp, a technician spends 4 1/4 hours on Wednesday, 6 3/10 hours on Thursday, and 5 1/4 hours on Friday. How much time did he spend trying to change that bulb?

●●22. During a four-day workweek, a tech works 10 hours on Monday, 9 1/2 hours on Tuesday, 10 3/10 hours on Wednesday, and 8 7/10 hours on Thursday. How many hours does he work that week?

●23. Three pieces of steel plate are to be bolted together. Each is 3/8 in. thick. What is the total thickness of the three when bolted together?

●24. A 1/2 in. bracket is to be welded to a 1/8 in. steel frame. What is the total thickness once they are welded?

●●25. A 3/4 in. spacer is on top of an intake manifold. The gasket below it is 1/16 in. thick, and the gasket above is 1/8 in. thick. What is the total thickness of these three when used together?

●●26. A water pump mounting hole is 1 7/16 in. thick, and the timing cover it bolts to is 2 1/4 in. thick. The bolt must be at least 3/4 in. longer than those two combined. How long must the bolt be?

Subtraction of Fractions

Find the difference for each problem below.

●1. $\dfrac{5}{9} - \dfrac{1}{9} =$

●2. $\dfrac{7}{10} - \dfrac{3}{10} =$

●3. $\dfrac{7}{8} - \dfrac{1}{4} =$

●4. $2\dfrac{5}{6} - \dfrac{2}{3} =$

●5. $6\dfrac{4}{5} - \dfrac{3}{4} =$

●6. $1\dfrac{4}{7} - \dfrac{6}{7} =$

●7. $5\dfrac{1}{8} - \dfrac{5}{8} =$

●●8. $9\dfrac{1}{2} - \dfrac{5}{6} =$

●●9. $4\dfrac{1}{4} - \dfrac{7}{12} =$

●●10. $7\dfrac{3}{5} - 2\dfrac{11}{15} =$

●●11. $5\dfrac{4}{9} - 2\dfrac{21}{36} =$

●●12. $8\dfrac{2}{7} - 5\dfrac{3}{4} =$

●●●13. $7\dfrac{1}{6} - 6\dfrac{5}{8} =$

●●●14. $10\dfrac{7}{10} - 10\dfrac{5}{12} =$

●●●15. $14\dfrac{7}{9} - 10\dfrac{11}{12} =$

●16. To check an engine's oil consumption, the new oil added to an engine is carefully measured. Later, at the end of the oil's life, it is measured again when it is drained. In January, exactly 5 qt are added after an oil change. 3000 miles later, that oil is drained, but only 4 1/4 qt remain. How much oil was lost over this interval?

●17. As in problem 16, the oil consumption for a truck needs to be measured. At one oil change, 5 1/2 qt of fresh oil are added. When this oil is later drained, only 4 1/4 qt remain. How much oil was consumed?

●●18. The owner's manual for a certain vehicle says that the engine requires 4 3/4 qt of oil, but this can be reduced by 1/2 qt if the filter is not being changed. If the filter is not being replaced, how many quarts of oil are necessary?

●●19. Tubing sizes are usually measured by the inside diameter (I.D.), outside diameter (O.D.), and wall thickness. If any two of these three are known, the third dimension can be found. If a certain piece of tubing has a 3/4 in. O.D. and 1/8 in. wall thickness, what is the I.D.?

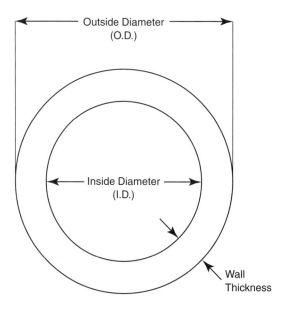

FRACTIONAL OPERATIONS WITH APPLICATIONS TO MEASUREMENT 63

●●20. A piece of tubing has wall thickness 1/16 in. and O.D. 13/16 in. What is the I.D.?

●●21. The O.D. of a piece of tubing is 1 in. and the I.D. is 3/4 in. What is the wall thickness?

●●22. Find the missing dimension A in the diagram below.

●●●23. Find the missing dimension B in the diagram below.

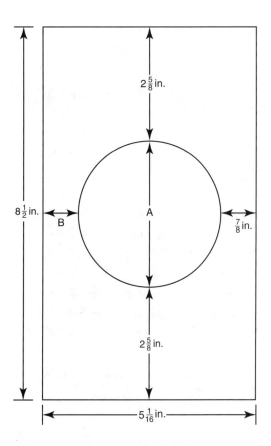

Multiplication of Fractions

Find the product for each problem below.

●1. $\dfrac{1}{2} \times \dfrac{3}{7} =$

●2. $\dfrac{4}{5} \times \dfrac{2}{3} =$

●3. $\dfrac{5}{6} \times \dfrac{7}{10} =$

●4. $\dfrac{7}{8} \times \dfrac{6}{11} =$

●5. $\dfrac{8}{15} \times \dfrac{5}{16} =$

●6. $\dfrac{3}{8} \times \dfrac{2}{21} =$

●7. $4 \times \dfrac{3}{5} =$

●8. $\dfrac{5}{9} \times 8 =$

●9. $8 \times \dfrac{3}{4} =$

●10. $\dfrac{7}{12} \times 4 =$

●●11. $4\dfrac{1}{2} \times \dfrac{3}{10} =$

●●12. $6\dfrac{2}{3} \times \dfrac{2}{5} =$

••13. $3\frac{1}{7} \times 5\frac{3}{5} =$ ••14. $1\frac{7}{16} \times 2\frac{2}{3} =$ ••15. $9\frac{5}{8} \times 2\frac{4}{11} =$

••16. $1\frac{1}{2} \times 12\frac{1}{8} =$ ••17. $2\frac{1}{4} \times 1\frac{3}{4} =$ •••18. $9\frac{11}{15} \times 4\frac{1}{9} =$

•19. Four washers, each 3/16 in. thick, are stacked together. What is the total thickness of the four washers when stacked?

•20. To install a wiring harness, four 16-guage wires are used, and each wire is 6 1/2 ft long. How much total wire is needed?

•21. The weight of fuel varies, but suppose a gallon weighs 5 7/8 lb. How much will 15 gallons of fuel weigh?

•22. Water weighs very close to 8 1/3 lb per gallon. How much will 7 gallons of water weigh?

••23. A certain tire for a certain driver wears at a rate of about 3/32 in. per year.
   a. How much tread will wear off in 3 years?
   b. How much tread will wear off in 4 1/2 years?

••24. A transmission leaks oil at a rate of about 1/2 qt per month.
   a. How much fluid will be lost in 5 months?
   b. How much fluid will be lost in 2 1/2 months?

•••25. A certain adjustment bolt has 16 threads per inch. That means that for every full turn, the bolt advances 1/16 in.

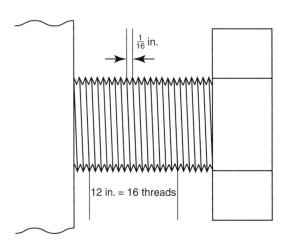

   a. How far will the bolt advance if it is turned 1/2 turn?
   b. How far will the bolt advance if it is turned 2/3 turn?
   c. How far will the bolt advance if it is turned 1 3/4 turns?

FRACTIONAL OPERATIONS WITH APPLICATIONS TO MEASUREMENT 65

••26. On a certain truck, every full turn of a tie rod sleeve changes the toe-in on a vehicle by 1/8 in.

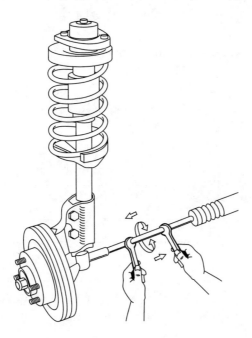

   a. How much will the toe-in change if the sleeve is rotated 1/2 turn?
   b. How much will the toe-in change if the sleeve is rotated 3/4 turn?
   c. How much will the toe-in change if the sleeve is rotated 1 1/3 turns?

••27. On a small pickup, one turn of the tie rod sleeve changes the toe-in setting by 3/16 in.
   a. If the tie rod sleeve is turned 1/3 turn, how much will the toe-in change?
   b. If the tie rod sleeve is turned 3/4 turn, how much will the toe-in change?
   c. If the tie rod sleeve is turned 1 1/4 turns, how much will the toe-in change?

Division of Fractions

Find the quotient for each problem below.

•1. $\dfrac{1}{3} \div \dfrac{5}{8} =$ •2. $\dfrac{2}{5} \div \dfrac{7}{12} =$ •3. $\dfrac{3}{4} \div \dfrac{1}{6} =$

•4. $\dfrac{11}{15} \div \dfrac{4}{5} =$ •5. $\dfrac{8}{21} \div \dfrac{4}{7} =$ •6. $\dfrac{6}{25} \div \dfrac{3}{5} =$

••7. $4\dfrac{1}{2} \div \dfrac{1}{2} =$ ••8. $2\dfrac{3}{7} \div 4 =$ ••9. $5\dfrac{1}{6} \div 3 =$

••10. $5\dfrac{1}{4} \div 2\dfrac{5}{8} =$ ••11. $5\dfrac{1}{7} \div 3\dfrac{1}{7} =$ ••12. $2\dfrac{1}{12} \div 1\dfrac{3}{10} =$

••13. $8\dfrac{2}{3} \div 2\dfrac{1}{6} =$ •••14. $9\dfrac{7}{8} \div 4\dfrac{9}{10} =$ •••15. $6\dfrac{7}{12} \div 4\dfrac{11}{16} =$

●●16. How many 7 1/2 ft pieces of heater hose can be taken from a 50-ft roll?

●●17. A 25-ft piece of fuel line is unrolled, and several 3 1/4 ft pieces are cut off. How many pieces can be cut from the original 25 ft?

●●18. A piece of 2 1/2 in. exhaust tubing is 67 3/4 in. long. It needs to be cut in half. How long will each piece be?

●●19. An aluminum tube 46 1/2 in. long is to be cut into 3 equal pieces. What should be the length of each piece?

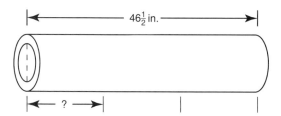

●●20. The rear toe on a vehicle is changed 1/16 in. for every full turn of the adjusting nut. How many turns are needed to change the toe 3/32 in. See Example 3-40.

●●21. Each time the adjusting nut is turned, the toe changes 3/32 in. How many turns are needed to change the toe 1/8 in.?

●22. The *throw* of a crankshaft is the distance from the centerline of the main journal to the centerline of the connecting rod journal. The *stroke* of an engine is the distance the piston travels from top-dead-center (TDC) to bottom-dead-center (BDC). The throw is always 1/2 of the stroke.

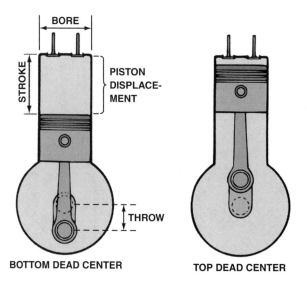

  a. If the throw of a crankshaft is 2 in., what is the stroke?
  b. If the stroke of a crankshaft is 10 cm, what is the throw?

●23. A crankshaft has a stroke of 3 1/2 in. What is the throw? (See problem 22.)

FRACTIONAL OPERATIONS WITH APPLICATIONS TO MEASUREMENT

●24. A crankshaft has a throw of 1 7/8 in. What is the stroke? (See problem 22.)

●25. A crankshaft has a stroke of 3 13/16 in. What is the throw? (See problem 22.)

●●●26. $2\frac{1}{2} + 3\frac{3}{4} \times \frac{4}{5} \div 2 =$

●●●27. $\frac{2}{3} \times \frac{3}{4} + \left(4\frac{3}{8} \div \frac{7}{16}\right) =$

●●●28. $5\frac{1}{2} - 2\frac{3}{16} + \left(10 \div 1\frac{1}{4}\right) =$

●●●29. $12\frac{1}{2} \div \left(1\frac{1}{4} + 2\frac{1}{4} \div 3\right) =$

Converting Fractions to Decimals

Convert each of the fractions below to a decimal. Round to three decimal places if necessary.

●1. $\frac{4}{5}$

●2. $\frac{3}{8}$

●3. $\frac{7}{10}$

●4. $\frac{13}{20}$

●5. $\frac{2}{3}$

●6. $\frac{6}{7}$

●7. $6\frac{3}{15}$

●8. $1\frac{1}{6}$

●9. $4\frac{2}{11}$

●10. $8\frac{14}{25}$

●11. $7\frac{11}{13}$

●12. $5\frac{6}{17}$

●13. A $\frac{7}{16}$ in. drill bit is used to set a choke clearance. What is the decimal equivalent of this fraction?

●14. A connecting rod was designed to be $5\frac{3}{32}$ in. long. How long is this, as a decimal?

Convert each of the fractions below to the nearest ten-thousandth, and perform the operations. Round your final answer to the nearest thousandth.

●●15. $\frac{3}{8} + 0.128 + \frac{7}{16} + 2.70 =$

●●16. $12.630 - \frac{3}{16} + 2\frac{7}{8} - 3.007 =$

●●17. $8\frac{9}{20} \times 2.7 - 6\frac{7}{8} =$

●●18. $2\frac{1}{3} \div \left(4\frac{2}{5} - 4.081\right) + 6.5 =$

●●●19. $2.785 \times 5\frac{3}{4} - \left[2.367 + \left(4\frac{9}{13} \div 2\frac{1}{11}\right)\right] =$

●●●20. $\left[4\frac{11}{15} - 3\frac{31}{35} \times \left(8\frac{5}{9} - 8.204\right)\right] \times 2\frac{15}{17} =$

CHAPTER 4

Ratio and Proportion

EQUIVALENT RATIOS

If you've ever mixed the fuel for a two-cycle engine, you know that the oil and gasoline must be mixed properly to provide proper lubrication for the engine. The manufacturer of the engine has determined how much oil should be mixed with the gasoline to provide the proper mixture. The proper mixture is described using a **ratio,** which is a comparison of two quantities. Often the two quantities have the same units.

Example 4-1 A certain lawnmower is designed to operate with an oil and gasoline mixture as its fuel. For every quart (32 ounces) of gasoline, one ounce of oil should be added. What is the gasoline to oil ratio for this engine?

Here, we want the ratio of gasoline to oil. Since there will be 32 ounces of gasoline for each ounce of oil, we can say the ratio is

32 ounces *to* 1 ounce

or simply

32 *to* 1.

Usually, though, the word "to" is not used to describe ratios. A colon is more common. Then we could write

32:1 is the gasoline to oil ratio.

Notice that this is NOT the same as a 1:32 ratio. The order of the two quantities being compared must be made clear. A typical two-cycle engine oil container is shown in Figure 4.1.

FIGURE 4.1 Oil and gasoline must be mixed in the proper ratio for proper lubrication.

RATIO AND PROPORTION

It is often useful to write ratios as fractions. For example, the gasoline:oil ratio in Example 4-1 could be written any of the following ways given the order shown.

gasoline *to* oil	gasoline:oil	$\dfrac{gasoline}{oil}$
32 *to* 1	32:1	$\dfrac{32}{1}$

All ratios can be written several ways without changing the actual value of the ratio.

Example 4-2 A driver averages 45 miles per hour on a trip. This ratio can be written as

45 miles *to* 1 hour	45 miles:1 hour	$\dfrac{45 \text{ miles}}{1 \text{ hour}}$

Write the ratio in several other ways.

The driver would travel 90 miles in 2 hours, 135 miles in 3 hours, and so on. Then the driver's speed is described by any of the following:

90 miles to 2 hours	135 miles:3 hours	$\dfrac{450 \text{ miles}}{10 \text{ hours}}$

Example 4-3 Most types of antifreeze have their maximum freeze protection when the antifreeze and water are mixed in a 3 to 2 ratio. A typical bottle of antifreeze is shown in Figure 4.2. This means three parts of antifreeze should be mixed with two parts water. Write this ratio as a fraction in several ways.

First, clarify the order of the two quantities. We'll put antifreeze on top.

$$\dfrac{antifreeze}{water} \qquad \dfrac{3 \text{ gal}}{2 \text{ gal}} \quad \text{or} \quad \dfrac{3 \text{ qt}}{2 \text{ qt}}$$

Multiplying the numerator and denominator by the same number will give an **equivalent ratio**, as it does with fractions.

$$\dfrac{3 \text{ gal} \times 4}{2 \text{ gal} \times 4} = \dfrac{12 \text{ gal}}{8 \text{ gal}}$$

FIGURE 4.2 Mixing antifreeze and water in a 1:1 ratio does not maximize freeze protection. The best freeze protection is achieved with an antifreeze to water ratio of 3 to 2.

Notice that the ratio $\frac{3 \text{ gal}}{2 \text{ gal}}$ cannot be reduced. It is said to be in **lowest terms**.

The ratio $\frac{12 \text{ gal}}{8 \text{ gal}}$ can be reduced in the same way a fraction can. It is *not* in lowest terms.

Since the units in both quantities are the same, they cancel. This leaves us with a ratio that contains no units.

$$\frac{12 \cancel{\text{ gal}}}{8 \cancel{\text{ gal}}} = \frac{12}{8} = \frac{3}{2}$$

We'll use that idea more in Chapter 7 when we convert units.

The last type of ratio we need to consider is called a unit rate. A **unit rate** is a ratio in which the second quantity is always 1.

Example 4-4 A case of 12 quarts of motor oil sells for $13.80. Write this as a ratio and as a unit rate.

As a ratio, it will make more sense to find the number of dollars per quart than it will to find the number of quarts per dollar. Then

$$\frac{\text{dollars}}{\text{quantity of oil}} = \frac{\$13.80}{1 \text{ case}} = \frac{\$13.80}{12 \text{ qt}}$$

As a unit rate, we can divide both numerator and denominator by 12 to make the bottom quantity 1.

$$\frac{\$13.80 \div 12}{12 \text{ qt} \div 12} = \frac{\$1.15}{1 \text{ qt}}$$

We write the unit rate as $1.15/qt or $1.15 per quart.

Example 4-5 Write the ratio 3/4 in.:6 in. as a fraction in lowest terms.

First, write the ratio as a fraction. The first quantity should go on top.

$$3/4 \text{ in.}:6 \text{ in.} \quad \text{becomes} \quad \frac{3/4 \text{ in.}}{6 \text{ in.}}$$

Now, cancel the units and simplify, treating the ratio as a fraction.

$$\frac{3/4 \cancel{\text{ in.}}}{6 \cancel{\text{ in.}}} = \frac{3/4}{6} = \frac{3}{4} \div 6 = \frac{3}{4} \times \frac{1}{6} = \frac{3}{24} = \frac{1}{8}$$

This ratio could also be written 1 *to* 8 or 1:8.

Example 4-6 Write the ratio 2.7 to 4.5 as a fractional ratio in lowest terms using whole numbers.

First, write the ratio as a fraction.

$$2.7 \text{ to } 4.5 = \frac{2.7}{4.5}$$

Eliminate the decimals by multiplying numerator and denominator by 10.

$$\frac{2.7 \times 10}{4.5 \times 10} = \frac{27}{45}$$

Now, reduce.

$$2.7 \text{ to } 4.5 = \frac{27}{45} = \frac{3}{5}$$

RATIO AND PROPORTION

WRITING AND SOLVING PROPORTIONS

When two or more ratios are equal they can be used to write a **proportion**. For example, if the ratio 100 miles:2 hours is written in lowest terms, an equivalent ratio is formed.

$$\frac{100 \text{ mi}}{2 \text{ hours}} = \frac{50 \text{ mi}}{1 \text{ hour}}$$

This is an example of a proportion. We say that the two ratios are proportional. Notice that each ratio uses the same units in the numerator and in the denominator.

We can multiply the numerator of the one ratio by the denominator of the other; when this is done in both directions, it is called **cross-multiplying**. If the two ratios are equal, the products we find when we cross-multiply will be equal. If the ratios are not equal, the proportion cannot be true, so the products we find when cross-multiplying will be different. Cross-multiplying is demonstrated in Figure 4.3.

$$\frac{100 \text{ mi}}{2 \text{ hours}} = \frac{50 \text{ mi}}{1 \text{ hour}}$$

Units are the same

Then

$$\frac{100}{2} \times \frac{50}{1}$$

$$50 \times 2 = 100 \qquad 100 \times 1 = 100$$

Cross products are equal

FIGURE 4.3 If the products of cross-multiplying are equal, then the proportion is true.

This fact can be used to check if two ratios are equal.

Example 4-7 Determine if the ratios 40:15 and 8:5 are equal.

Begin by writing the ratios as a possible proportion. Then, cross-multiply.

$$\frac{40}{15} \stackrel{?}{=} \frac{8}{5}$$

$$\frac{40}{15} \stackrel{?}{\times} \frac{8}{5}$$

$$8 \times 15 = 120 \qquad 40 \times 5 = 200$$

Since the products are not the same, the ratios are not equal.

Example 4-8 A two-cycle engine is designed to operate on a gasoline:oil ratio of 40:1. The operator pours an 8-ounce can of oil into 1 gallon of gas. Has he mixed the correct ratio? A gallon contains 128 ounces.

The operator mixes a gasoline:oil ratio of 128 ounces:8 ounces. Set up a possible proportion to check if this is the same as 40:1.

$$\frac{\text{gasoline}}{\text{oil}} = \frac{1 \text{ gal}}{8 \text{ ounces}} = \frac{128 \text{ ounces}}{8 \text{ ounces}}$$

$$\frac{128}{8} \stackrel{?}{=} \frac{40}{1}$$

$$\frac{128}{8} \stackrel{?}{\times} \frac{40}{1}$$

$$40 \times 8 = 320 \qquad 128 \times 1 = 128$$

Since the cross products are not equal, the fuel has not been mixed in the correct ratio.

Once we have determined what is not the correct ratio, the question arises: How much gasoline should an operator mix with an 8-ounce can of oil? Proportions can be used to find missing values in a ratio. In the following example, we solve this problem using a variable. A **variable** is a letter or symbol that represents some unknown quantity.

Example 4-9 How much gasoline should be mixed with 8 ounces of oil to create a gasoline:oil ratio of 40:1?

Making sure the units in our two ratios match, we'll write our proportion with one unknown value.

$$\frac{40 \text{ ounces of gasoline}}{1 \text{ ounce of oil}} = \frac{? \text{ ounces of gasoline}}{8 \text{ ounces of oil}}$$

All units match, so we'll rewrite the proportion without them. We'll use the variable n in the place of our unknown value.

$$\frac{40}{1} = \frac{n}{8}$$

Now, cross-multiply.

$$1 \times n = 40 \times 8$$
$$n = 320$$

The unknown value is 320, so 320 ounces (or 2.5 gallons) of gasoline should be used.

Example 4-10 Antifreeze should be mixed with water in a 3:2 ratio for maximum cooling. How many gallons of water should be mixed with 5 gallons of antifreeze if this ratio is used?

Write the two ratios with an unknown value, making sure the units match.

$$\frac{3 \text{ gal antifreeze}}{2 \text{ gal water}} = \frac{5 \text{ gal antifreeze}}{? \text{ gal water}}$$

Now, rewrite and cross-multiply.

$$\frac{3}{2} = \frac{5}{n}$$

$$10 = 3 \times n$$

Note that since we want to find the value of n we must divide both sides by 3, which reverses our recent multiplication by 3.

$$\frac{10}{3} = \frac{3 \times n}{3}$$

$$\frac{10}{3} = n$$

$$3\frac{1}{3} = n$$

We should use 3 1/3 gallons of water to create the correct ratio.

Example 4-11 Find the missing value in the proportion

$$\frac{2.7}{8.1} = \frac{?}{16.5}$$

RATIO AND PROPORTION

Write the proportion using a variable, and cross-multiply.

$$\frac{2.7}{8.1} = \frac{n}{16.5}$$

$$\frac{2.7}{8.1} \times \frac{n}{16.5}$$

$$8.1 \times n = 44.55$$

$$\frac{8.1 \times n}{8.1} = \frac{44.55}{8.1}$$

$$n = 5.5$$

Again, we needed to divide both sides by 8.1 to isolate n.
The missing value is 5.5.

Example 4-12 Find the missing value in the proportion

$$\frac{2\frac{1}{2}}{8\frac{3}{4}} = \frac{n}{4\frac{3}{8}}$$

First, cross-multiply:

$$2\frac{1}{2} \times 4\frac{3}{8} = 8\frac{3}{4} \times n$$

$$\frac{5}{2} \times \frac{35}{8} = \frac{35}{4} \times n$$

$$\frac{175}{16} = \frac{35}{4} \times n$$

Divide both sides by $\frac{35}{4}$ (which is the same as multiplying by $\frac{4}{35}$).

$$\frac{4}{35} \times \left(\frac{175}{16}\right) = \frac{4}{35} \times \left(\frac{35}{4} \times n\right)$$

$$\frac{5}{4} = n$$

$$n = 1\frac{1}{4}$$

Example 4-13 Solve the same proportion as in Example 4-12 using decimals.

We can rewrite the proportion

$$\frac{2\frac{1}{2}}{8\frac{3}{4}} = \frac{n}{4\frac{3}{8}}$$

using decimals as

$$\frac{2.5}{8.75} = \frac{n}{4.375}$$

Cross-multiply.

$$2.5 \times 4.375 = 8.75 \times n$$

$$10.9375 = 8.75 \times n$$

Divide both sides by 8.75.

$$\frac{10.9375}{8.75} = \frac{\cancel{8.75} \times n}{\cancel{8.75}}$$

$$1.25 = n$$

We'll solve many more proportions when we work with transmission and differential ratios in Chapter 11.

FINDING NEAREST FRACTIONAL PART

In the previous chapter we practiced converting fractions to decimals. Suppose, however, that we have a decimal, and we're looking for a fraction that has about the same value. We need a different process to find such a fraction.

For example, say we wanted to round 0.550 to the nearest 1/16 of an inch. Trial and error shows that

$$\frac{8}{16} = \frac{1}{2} = 0.500 \text{ and } \frac{9}{16} = 0.5625$$

Although not exact, $\frac{9}{16}$ is closest to 0.550.

Then 0.550 rounded to the nearest 1/16 is $\frac{9}{16}$.

This is called rounding to the **nearest fractional part** of a decimal.

We'll use proportions to find the nearest fractional part of a decimal. Consider the simple decimal 0.25. We know that $0.25 = \frac{1}{4}$. Since a decimal can always be written as a fraction with a denominator of 1, we can write the proportion

$$\frac{0.25}{1} = \frac{1}{4}$$

Cross-multiplying will verify that these two are equal. This idea can be applied to more complicated fractions.

Example 4-14 Round the decimal 0.380 in. to the nearest 1/8 of an inch.

Set up a proportion, with the numerator of our fraction unknown.

$$\frac{0.380}{1} = \frac{n}{8}$$

Cross-multiply to solve for n.

$$0.380 \times 8 = 1 \times n$$
$$3.04 = n$$

This rounds to the whole number 3, so we know the closest numerator is 3.

$$0.380 \text{ in.} \approx \frac{3}{8} \text{ in.}$$

Example 4-15 Round 4.863 in. to the nearest 1/16 of an inch.

As before, we know our measurement will have a whole number value of 4, so we need to focus on the decimal part only.

RATIO AND PROPORTION

Then we need to round 0.863 to the nearest 1/16 of an inch. Set up a proportion. Notice that the denominator in our fraction must be 16, and we must find the best numerator.

$$\frac{0.863}{1} = \frac{n}{16}$$
$$0.863 \times 16 = 1 \times n$$
$$13.808 = n$$

Rounding, we see that the best numerator is 14. Then we can conclude that

$$0.863 \approx \frac{14}{16}, \text{ and that}$$
$$4.863 \text{ in.} \approx 4\frac{14}{16} \text{ in.}$$

We can reduce this to see that

$$4.863 \text{ in.} \approx 4\frac{7}{8} \text{ in.}$$

CHAPTER 4 *Practice Problems*

Equivalent Ratios

Write a ratio describing the comparisons below. Express the ratio using words, a colon, and as a fraction.

1. Fuel is mixed in a ratio of 40 parts gasoline to 1 part oil. Write the gasoline *to* oil ratio.

2. The drive shaft of a vehicle turns 3 times for every 1 revolution of the rear wheels. Write the drive shaft rotations *to* wheel rotations ratio.

3. A tech works 4 days and has 3-day weekends. Write the days on *to* days off ratio.

4. The input shaft of a transmission turns 2.79 times for every revolution of the output shaft. Write the input *to* output ratio.

5. The volume of the combustion chamber at BDC is 528 cm^3 while the volume at TDC is 63 cm^3. Write the volume at BDC *to* volume at TDC ratio.

6. An amount of 14.7 lb of air are burned with every pound of fuel. Write the air *to* fuel ratio.

Write each of the ratios below in lowest terms (with whole numbers) using a colon *and* as a fraction.

●7. 12 *to* 6   ●8. 30 *to* 5   ●9. 18 *to* 20   ●10. 34 *to* 40

●11. 6 *to* 1.5   ●12. 15 *to* 2.5   ●13. 1.4 *to* 3.5   ●14. 2.8 *to* 5.6

●●15. 1/2 *to* 4   ●●16. 4/5 *to* 15   ●●17. 6.4 *to* 1 5/8   ●●18. 8.5 *to* 2 1/4

Write each ratio below as a unit rate.

●19. 55 mi:1 hour

●20. 160 km:2 hours

●21. 9000 revolutions:3 minutes

●22. 35 lb:7 gal

●23. $7.00 to 4 gal

●24. $9.25 to 37 ft

●25. $316.64 to 4 tires

●26. 330 oil changes to 30 days

Writing and Solving Proportions

Determine if each of the ratios below are equal. Answer with either "equal" or "not equal."

●1. $\dfrac{3}{4} = \dfrac{9}{12}$ ●2. $\dfrac{5}{8} = \dfrac{15}{24}$ ●3. $\dfrac{4}{5} = \dfrac{10}{15}$ ●4. $\dfrac{8}{3} = \dfrac{12}{5}$

●5. $\dfrac{3}{2} = \dfrac{24}{16}$ ●6. $\dfrac{6}{4} = \dfrac{12}{6}$ ●7. $\dfrac{7.5}{2.5} = \dfrac{6}{3}$ ●8. $\dfrac{4.2}{5.6} = \dfrac{3.6}{4.8}$

●●9. $\dfrac{1\tfrac{2}{3}}{2\tfrac{3}{4}} = \dfrac{2}{3}$ ●●10. $\dfrac{3\tfrac{1}{2}}{5} = \dfrac{2\tfrac{1}{10}}{3}$

Use your calculator to find the missing value in each of the proportions below. Round decimal answers to the nearest thousandth.

●11. $\dfrac{3}{4} = \dfrac{9}{n}$ ●12. $\dfrac{5}{4} = \dfrac{20}{n}$ ●13. $\dfrac{13}{6} = \dfrac{n}{8}$

RATIO AND PROPORTION

●14. 3:5 = 9:*n*

●15. 3:1 = 5:*n*

●16. 10 to 3 = *n* to 6

●17. 16 to 5 = *n* to 8

●18. $\dfrac{11}{4} = \dfrac{n}{10}$

●19. $\dfrac{16}{1} = \dfrac{2.5}{n}$

●●20. $\dfrac{14.4}{4.8} = \dfrac{n}{6.0}$

●●21. $\dfrac{0.39}{1.30} = \dfrac{9.9}{n}$

●●22. $\dfrac{6.25}{10.625} = \dfrac{9.3}{n}$

●●23. $\dfrac{2.470}{5.681} = \dfrac{n}{8.220}$

●●24. $\dfrac{2\frac{1}{2}}{8} = \dfrac{n}{10}$

●●25. $\dfrac{3}{4\frac{5}{8}} = \dfrac{n}{2\frac{5}{16}}$

●●●26. $\dfrac{4\frac{1}{5}}{12\frac{3}{5}} = \dfrac{3\frac{2}{7}}{n}$

●●●27. $\dfrac{2\frac{5}{6}}{4\frac{1}{4}} = \dfrac{5\frac{1}{2}}{n}$

●●28. For an outboard boat motor, the gasoline:oil ratio should be 50:1.
   a. How many ounces of gasoline should be mixed with a 4-ounce container of oil?
   b. How many gallons is this? (There are 128 ounces in a gallon.)

●●29. A lawnmower requires that the gasoline:oil ratio be 32:1.
   a. How many ounces of gasoline should be mixed with an 8-ounce container of oil?
   b. How many gallons is this? (There are 128 ounces in a gallon.)

●●30. Air and fuel are mixed in the induction system in a 14.7:1 ratio. How many pounds of air are needed to completely burn 6 lb of fuel?

●●31. An engine using lean-burn technology mixes air and fuel in an 18.0:1 ratio. How many pounds of fuel are used while the engine consumes 50 lb of air?

●●32. A driver covers a distance of 210 miles in 4.2 hours. How far will she drive in 6.8 hours if she continues this pace? Use a proportion to answer this question.

●●33. A truck driver on the Interstate covers a distance of 418 miles in 6.3 hours. How far will he drive in another 1.2 hours at this pace? Again, use a proportion to answer this question.

Finding Nearest Fractional Part

Find the value of each decimal below to the nearest 1/8 of an inch.

●●1. 0.625 in.   ●●2. 0.891 in.   ●●3. 4.443 in.

●●4. 6.200 in.   ●●5. 1.310 in.   ●●6. 22.573 in.

Find the value of each decimal below to the nearest 1/16 of an inch.

●●7. 0.3125 in.   ●●8. 0.671 in.   ●●9. 8.402 in.

●●10. 12.400 in.   ●●11. 2.796 in.   ●●12. 6.714 in.

●●13. You want to use a drill bit to set the choke pulloff on a carburetor. The setting should be 0.190 in. What size drill bit, to the nearest 1/16, should you use?

●●14. A hole should be drilled to have a diameter of about 0.450 in. What size drill bit, to the nearest 1/32 of an inch, should you use?

CHAPTER 5

Percent and Percent Applications

CONVERTING FRACTIONS, DECIMALS, AND PERCENTS

Percent values are used in almost every aspect of the automotive field. As a technician, you're likely to see percents used to express an acceptable range of measurements. For example, it is widely accepted that when a compression test is done on all the cylinders of an engine, the lowest compression reading should be within 15% of the highest compression reading. Also, percents are used to deal with every financial aspect of the automotive trades from calculating the amount of sales tax on a part to determining how expenses for a shop should be budgeted.

Percent literally means per hundred. If we knew that 25% of all tires were not properly inflated, then we could conclude that 25 of every 100 tires are not inflated to the correct pressure. See Figure 5.1.

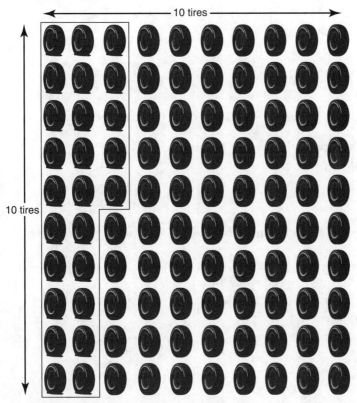

FIGURE 5.1 Twenty-five percent of all tires are not properly inflated means that if we had 100 tires, 25 of them would not be properly inflated.

This ratio of 25 per 100 could also be written as a fraction, and reduced as we did in Chapter 3.

$$25\% = \frac{25}{100} = \frac{5}{20} = \frac{1}{4}$$

Example 5-1 Suppose using the air conditioner on a vehicle reduces gas mileage by 8%. What fraction of the gas mileage is lost when the air conditioner is used?

Convert 8% to a fraction and reduce.

$$8\% = \frac{8}{100} = \frac{2}{25}$$

This means a vehicle that usually gets 25 MPG will lose 2 MPG of fuel economy.

Example 5-2 Convert 56.4% to a fraction, and reduce.

Begin by writing the percent value as a fraction.

$$56.4\% = \frac{56.4}{100}$$

Then, remove decimals

$$\frac{56.4}{100} = \frac{564}{1000}$$

and reduce.

$$\frac{564}{1000} = \frac{141}{250} \qquad \text{Then } 56.4\% = \frac{141}{250}$$

Example 5-3 Convert 135% to a fraction in lowest terms.

Notice that 135% will result in a fraction greater than 1. Percentages over 100% represent numbers larger than 1.

$$135\% = \frac{135}{100} = 1\frac{35}{100} = 1\frac{7}{20}$$

Example 5-4 Convert $16\frac{2}{3}\%$ to a fraction in lowest terms.

Begin by writing the percentage as a fraction.

$$16\frac{2}{3}\% = \frac{16\frac{2}{3}}{100}$$

Next, simplify this complex fraction by removing mixed numbers.

$$\frac{16\frac{2}{3}}{100} = 16\frac{2}{3} \div 100 = \frac{50}{3} \div 100$$

Finally, invert and multiply.

$$\frac{50}{3} \div 100 = \frac{50}{3} \times \frac{1}{100} = \frac{1}{6}$$

Then

$$16\frac{2}{3}\% = \frac{1}{6}$$

PERCENT AND PERCENT APPLICATIONS

Since dividing a number by 100 always makes the number smaller and moves the decimal point two places, we can convert percents to decimals by moving the decimal two places to the left.

Example 5-5 Convert 87.3% to a decimal.

$$87.3\% = \frac{87.3}{100} = 0.873$$

Or more simply,

$$87.3\% = .873 = 0.873$$

Example 5-6 Convert 7% to a decimal.

$$7\% = \frac{7}{100} = 0.07 \text{ or,}$$

$$7\% = .7 = 0.07$$

Example 5-7 Convert $12\frac{1}{2}\%$ to a decimal.

$$12\frac{1}{2}\% = 12.5\% = 0.125$$

Since a decimal can usually be thought of as a fraction with a denominator of 1, it makes sense that we must multiply the decimal by 100 to convert it to a fraction.

Example 5-8 Convert 0.872 to a percent.

$$0.872 = \frac{0.872}{1} = \frac{87.2}{100} = 87.2\%$$

Because multiplying a number by 100 always makes the number larger and moves the decimal point two places, we can convert decimals to percents by moving the decimal two places to the right.

Example 5-9 Convert 0.00355 to a percent.

$$0.00355 = \frac{0.00355}{1} = \frac{0.355}{100} = 0.355\% \text{ or more simply,}$$

$$0.00355 = 0.00355 = 0.355\%$$

Notice that this decimal was so small to begin with that it results in a percent value that is less than 1%.

Finally, we need to be able to convert fractions to percents. We already know how to do this though, since we can convert fractions to decimals, and now decimals to percents.

Example 5-10 Convert $\frac{5}{8}$ to a percent.

First, convert $\frac{5}{8}$ to a decimal.

$$\frac{5}{8} = 5 \div 8 = 0.625$$

Now, move the decimal place to convert this value to a percentage.

$$\frac{5}{8} = 0.625 = 62.5\%$$

Example 5-11 Convert $4\frac{11}{16}$ to a percent.

First, convert the fractional part of this number to a decimal.

$$\frac{11}{16} = 0.6875$$

Therefore

$$4\frac{11}{16} = 4.6875 = 468.75\%$$

Example 5-12 Convert $\frac{3}{500}$ to a percent.

Begin by converting this fraction to a decimal. Notice that the fraction is very small, so the percent value could be less than 1%.

$$\frac{3}{500} = 0.006$$

Now, convert the decimal to a percent.

$$\frac{3}{500} = 0.006 = 0.6\%$$

BASE, RATE, AND AMOUNT

Many percent problems can be reduced to simple statements and solved using a proportion. Every basic percent statement has three parts. Let's consider such a statement, then examine the parts.

10% of 80 is 8

- The **base** is usually the beginning value or the complete value. You can usually identify the base because it is often after the word "of." The base in this example is 80.
- The **rate** is the percent value that is given. The rate is identified by the percent sign (%). The rate in this example is 10%.
- The **amount** is calculated by taking some percentage of the base. The amount is typically next to the word "is." The amount in this example is 8.

Example 5-13 Identify the base, rate, and amount in the following statement.

15 is 75% of 20

The base is 20. It's the whole amount, and it is after the word "of."
The rate is 75%. It's the percent value given.
The amount is 15. It's a percentage of the base, and it's next to the word "is."

When the base, rate, and amount are found, the following proportion is always true:

$$\frac{\text{Amount}}{\text{Base}} = \frac{\text{Rate}}{100\%} \quad \text{or simply} \quad \frac{A}{B} = \frac{R}{100\%}$$

We can use this proportion to find one of the three values if the other two are known.

Example 5-14 Use a proportion to solve the following. What is 16% of 200?

Identify the base, rate, and amount.

The base is 200 (it is after the word "of"). $B = 200$

PERCENT AND PERCENT APPLICATIONS

The rate is 16% (it is a percent value). $R = 16\%$
The amount is not known. $A = ?$

Substitute into the proportion given earlier and solve for A.

$$\frac{A}{B} = \frac{R}{100\%} \qquad \frac{A}{200} = \frac{16\%}{100\%} \qquad \frac{A}{200} = \frac{16}{100}$$

$$\frac{A}{200} \times \frac{16}{100}$$

$$3200 = 100A$$

$$\frac{3200}{100} = \frac{100A}{100}$$

$$32 = A$$

The amount is 32. Then 32 is 16% of 200.

Example 5-15   Use a proportion to solve the following. 105 is 35% of what number?

Identify the base, rate, and amount.

The rate is 35%. $R = 35\%$
The amount is 105 (it's next to the word "is"). $A = 105$
The base is not known. $B = ?$

Substitute into the proportion given earlier and solve for A.

$$\frac{A}{B} = \frac{R}{100\%} \qquad \frac{105}{B} = \frac{35\%}{100\%} \qquad \frac{105}{B} = \frac{35}{100}$$

$$\frac{105}{B} \times \frac{35}{100}$$

$$35B = 10{,}500$$

$$\frac{35\,B}{35} = \frac{10{,}500}{35}$$

$$B = 300$$

The base is 300. We can conclude that 105 is 35% of 300.

Example 5-16   Identify the base, rate, and amount, then solve the following using a proportion. A major car manufacturer noticed that the price of a tire was going to increase by 9% from the current price of $54.30. How much is the increase?

The problem can be written as: 9% of $54.30 is how much?
The base is the original value of $54.30.
The rate is 9%.
The amount is unknown.

$$\frac{A}{B} = \frac{R}{100\%} \qquad \frac{A}{\$54.30} = \frac{9\%}{100\%} \qquad \frac{A}{\$54.30} = \frac{9}{100}$$

$$\frac{A}{\$54.30} = \frac{9}{100}$$

$$\$488.70 = 100\,A$$

$$\frac{\$488.70}{100} = \frac{100\,A}{100}$$

$$\$4.887 = A$$

$$\$4.89 \approx A$$

The amount of the increase is $4.89.

Example 5-17 Identify the base, rate, and amount and solve the following. Suppose you deposit some money into a savings account drawing 4.5% annual interest, and you hope to have $3000 in that account one year from now. How much must you deposit?

The base is unknown, since we don't know how much to deposit.
The rate is 104.5%, because you'll have 100% of your savings and 4.5% interest.
The amount is $3000 since it depends on the original investment (the base).

$$\frac{A}{B} = \frac{R}{100\%} \qquad \frac{\$3000}{B} = \frac{104.5\%}{100\%}$$

$$\frac{\$3000}{B} = \frac{104.5}{100}$$

$$104.5B = \$300{,}000$$

$$\frac{104.5\,B}{104.5} = \frac{\$300{,}000}{104.5}$$

$$B = \$2{,}870.81$$

The amount of the deposit must be $2,870.81

DISCOUNT, MARKUP, AND SALES TAX

In this section we'll consider some common examples in which percents are used. There are two techniques for solving percent problems. If the base and rate are known, the amount can be found by converting the percent to a decimal, then multiplying by the base. If the base or rate are missing, it is often easier to use the proportion we just used. It is important to note that any problem can be solved using either method, however. In the following examples, we'll decide which method is best and solve the problem.

When an item is discounted or put on sale, the amount of the discount is often given as a percentage of the regular price. Then the original price is the base, and the discount percentage is the rate. We'll solve this first example using each method.

Example 5-18 Battery current output is rated in cold cranking amps (CCA). A 650CCA battery is normally $73.99. A customer pays cash, and receives a 5% discount. How much does the customer save?

Method 1: Use a proportion as we did in the last section. The base is $73.99. The rate is 5%.

$$\frac{A}{B} = \frac{R}{100\%} \qquad \frac{A}{\$73.99} = \frac{5\%}{100\%}$$

Here, A represents the amount discounted from the price.
Solve for A by cross-multiplying.

$$100 \times A = 5 \times \$73.99$$

$$A = \frac{5 \times \$73.99}{100}$$

$$A = \$3.6995$$

Then the discount is about $3.70.

Method 2: We can convert 5% (the rate) to a decimal and multiply by the original price (the base).

$$5\% = 0.05$$

$$0.05 \times \$73.99 = \$3.6995$$

Again, we see that the discount is about $3.70.

Example 5-19 A P215/65R15 tire is usually $84.50. When a complete set of four is bought, they're 15% off. Find the sale price for the set of four.

The full price (the base) is known, and the discount percent (rate) is given. It will be easiest to find the amount saved by converting the percent to a decimal and multiplying by the base.

First, find the cost of four tires.

$$4 \times \$84.50 = \$338.00$$

Next find the discount by multiplying the rate (as a decimal) by the base.

$$15\% = 0.15$$
$$0.15 \times \$338.00 = \$50.70$$

To find the sale price, subtract this amount from the original price.

$$\$338.00 - \$50.70 = \$287.30$$

The **retail** price of a part is the amount the customer pays. This amount is usually more than the amount the shop or parts supplier pays for the part, called the **cost**. The amount of profit is called the **markup**. Then it must be true that

$$\text{Cost} + \text{Markup} = \text{Retail}$$

The amount of markup is usually a percentage of the cost of the part. By knowing the percent markup on a part, the retail price can be calculated.

Example 5-20 A parts dealer buys a water pump at a cost of $57.60. The dealer typically uses a markup of 40% of cost. What is the markup, and what is the retail price of the pump?

First, we need to calculate the markup. The cost ($57.60) is the base, and the markup (40%) is the rate. Convert 40% to a decimal, and multiply by the cost.

$$40\% = 0.40$$
Markup: $0.40 \times \$57.60 = \$23.04$
Retail = Cost + Markup
$\phantom{\text{Retail}} = \$57.60 + \$23.04 = \$80.64$

The retail price will be $80.64.

Example 5-21 The cost of a radiator hose is $8.15, while the retail price is $13.50. What is the percent markup, based on cost?

The rate here is not known. Let's solve this problem using a proportion.
The base is $8.15.
The amount is the markup, which is $13.50 - \$8.15 = \$5.35$.

$$\frac{A}{B} = \frac{R}{100\%} \qquad \frac{\$5.35}{\$8.15} = \frac{R}{100\%}$$

Solve this proportion and you'll find that $R = 65.6\%$.

Example 5-22 The cost to a dealership of a headlight assembly is $96.17. The dealership uses a markup rate of 65% for this part. If a customer is given a 10% discount off the retail price, what is the customer's price?

We'll use the shorter method of treating the percents as decimals because the base price of $96.17 and the rate of markup and discount are known.

Begin by calculating the markup.

$$65\% = 0.65$$
Markup: $0.65 \times \$96.17 = \$62.51$

Then, find the usual retail price.

$$\text{Retail} = \text{Cost} + \text{Markup}$$
$$= \$96.17 + \$62.51 = \$158.68$$

The retail price will be $158.68.
Finally, take off the 10% discount.

$$10\% = 0.10$$
$$0.10 \times \$158.68 = \$15.87$$
$$\$158.68 - \$15.87 = \$142.81$$

The customer will pay $142.81.

Example 5-23 The retail price of a windshield is $522.80. This includes a markup based on cost of 40%. What was the cost of the windshield?

We'll use a proportion to find the solution because we do not know the base.

The markup is 40% based on cost, but the cost is not known.

We know that the retail price of $522.80 includes the cost *and* the markup, which is an additional 40% of the cost. Then the retail price is actually 140% of the cost. The rate is 140%, while the amount is $522.80. Find the base.

$$\frac{A}{B} = \frac{R}{100\%} \qquad \frac{\$522.80}{B} = \frac{140\%}{100\%}$$

Cross-multiply to solve.

$$B = \frac{100 \times \$522.80}{140} = \$373.43$$

The cost of the windshield is $373.43. It should make sense that this amount is less than the retail price. Note that it is a very common error to find 40% of the retail price, $522.80, then subtract this amount from the total. This is wrong because 40% of the cost is not equal to 40% of the total price. Be careful to avoid this type of error.

Sales tax is calculated based on a percentage of the final price of an item. If there is no discount, tax is based on the retail price. If there is a discount, as in the previous problem, the discount is taken before the sales tax is calculated. It is usually easiest to find sales tax by converting the tax rate to a decimal. We'll do that for Examples 5-24 through 5-27.

Example 5-24 The sales tax in a certain state is 5.5%. If the retail price of an air cleaner is $11.69, what is the price with tax included?

Begin by calculating the amount of tax on the air cleaner. The base is $11.69.

$$5.5\% = 0.055$$
$$\text{Sales tax} = 0.055 \times \$11.69 = \$0.64295 \text{ or about } \$0.64$$

The total, with tax = $11.69 + $0.64 = $12.33

Example 5-25 The retail price of an ignition switch is $38.75. With tax, the total is $41.27. What is the sales tax rate?

The base is the price before tax, or $38.75.
The amount of tax added to the item was $41.27 − $38.75 = $2.52.
As a percentage of the retail price, this is

$$\frac{\$2.52}{\$38.75} = \frac{R}{100\%}$$

Solve for *R*, and we'll see that the sales tax rate is 6.5%.

PERCENT AND PERCENT APPLICATIONS

Example 5-26 The retail price per wheel of 16-in. chrome rims is $287.20. Assuming these rims are discounted 18%, and sales tax is 6.5%, find the customer's price for a complete set of four rims.

Begin by finding the total price for four rims.

$$4 \times \$287.20 = \$1{,}148.80.$$

Now, take 18% off. The base is $1,148.80, and the rate is 18%.

$18\% = 0.18$
Discount $= 0.18 \times 1{,}148.80 = \$206.78$
Price, without tax $= \$1{,}148.80 - \$206.78 = \$942.02$

Finally, calculate and add on the sales tax.

$6.5\% = 0.065$
Sales tax $= 0.065 \times \$942.02 = \$61.23$
Total price, with tax $= \$942.02 + \$61.23 = \$1{,}003.25$

Example 5-27 A dealership's cost for an intake plenum is $472.50. The markup on this item is 35% of cost. If the sales tax rate is 4%, what is the total amount due from the customer for this part?

Begin by calculating the markup on the plenum.

$35\% = 0.35$
$0.35 \times \$472.50 = \$165.38$

Add the markup to the cost to calculate the retail price.

Retail $= \$472.50 + \$165.38 = \$637.88$

Finally, calculate and add on the sales tax.

$4\% = 0.04$
Sales tax $= 0.04 \times \$637.88 = \$25.52$
Total price, with tax $= \$637.88 + \$25.52 = \$663.40$

Example 5-28 The price of a torque wrench, including 6% tax, is $200.87. What was the price before tax?

We'll use a proportion for this problem because the base, which is the price before tax, is not known.

The base would be the original price of the wrench, which we don't know.
The amount is $200.87.
The rate includes the full price of the wrench (100%) and tax (another 6%). That means the amount is 100% + 6% = 106% of the base. The rate is 106%.

$$\frac{A}{B} = \frac{R}{100\%} \qquad \frac{\$200.87}{B} = \frac{106\%}{100\%}$$

Solve this proportion for B. The price before tax was $189.50.

TECHNICAL APPLICATIONS

Tolerances for many specifications are given as percent values. A few examples of this include compression test specifications and engine performance specifications.

Example 5-29 When a compression test is performed on an engine, the compression on the lowest cylinder should not be more than 15% lower than the compression on the highest cylinder. Compression readings are taken on a V-8 engine and are shown in Figure 5.2. Are the cylinders all within 15% of the highest value?

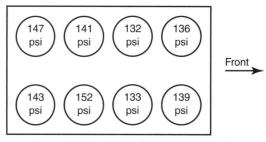

FIGURE 5.2 Compression readings on the eight cylinders of the V-8 engine, from Example 5-29.

Begin by identifying the highest reading. Here, that reading is 152 psi.
Next, take 15% of 152 to determine the range of lower values.

$$15\% = 0.15$$
$$0.15 \times 152 \text{ psi} = 22.8 \text{ psi}$$

Now, find the lowest allowable value by subtracting.

$$152 \text{ psi} - 22.8 \text{ psi} = 129.2 \text{ psi}$$

The lowest cylinder has a value of 132 psi, so the range is acceptable.

Example 5-30 A turbocharger provides increased intake pressure to increase engine output. Suppose an engine under load typically operates at a manifold pressure of 14.0 psi. If a turbocharger provides a 20% pressure boost, what would the manifold pressure then be?

Begin by finding 20% of 14.0 psi.

$$20\% = 0.20$$
$$0.20 \times 14.0 \text{ psi} = 2.80 \text{ psi}$$

Now add this to the normal intake manifold pressure.

$$14.0 \text{ psi} + 2.80 \text{ psi} = 16.8 \text{ psi}$$

Example 5-31 A tire manufacturer claims that driving a vehicle with underinflated tires can reduce fuel economy by up to 8%. If a vehicle usually gets 32 miles per gallon, what would be the fuel economy when the tires are underinflated?

Find 8% of 32 miles per gallon, and reduce by this amount.

$$8\% = 0.08$$
$$0.08 \times 32 \text{ MPG} = 2.56 \text{ MPG}$$
$$32 \text{ MPG} - 2.56 \text{ MPG} = 29.44$$

The reduced fuel economy would be about 29.4 MPG.

Example 5-32 A 12.6-volt electrical system is powered by an alternator that should keep the system voltage about 10% above the battery voltage with the vehicle operating. What is the system voltage while the vehicle is running?

Add an additional 10% to the battery's voltage.

$$0.10 \times 12.6 \text{ volts} = 1.26 \text{ volts}$$
$$12.6 \text{ volts} + 1.26 \text{ volts} = 13.86 \text{ volts}$$

PERCENT AND PERCENT APPLICATIONS

Example 5-33 A vehicle's weight distribution tells how much of the vehicle's weight is present on the front wheels compared to the rear wheels. For example, a 52/48 weight distribution means that the front wheels carry 52% of the vehicle's weight, while the rear wheels carry 48% of the vehicle's weight. How much weight is on the front wheels if a 3,140-lb vehicle has a 56/44 weight distribution? How much weight is on the rear wheels?

The front wheels carry 56% of the 3,140 lb vehicle weight

$$56\% = 0.56$$
$$0.56 \times 3{,}140 \text{ lb} = 1{,}758.4 \text{ lb}$$

The rear wheels carry 44% of the 3,140 lb vehicle weight.

$$44\% = 0.44$$
$$0.44 \times 3{,}140 \text{ lb} = 1{,}381.6 \text{ lb}$$

Example 5-34 An engine produces 170 peak horsepower (hp) when the air conditioner is running, and 178 hp when the air conditioner is off. What is the percent decrease when the air conditioner is running?

The decrease in power here is 178 hp − 170 hp = 8 hp.
When compared to the original 178 hp, this becomes

$$\frac{8 \text{ hp}}{178 \text{ hp}} = 0.0449 \quad \text{or about } 4.5\%$$

Example 5-35 On a trip, a vehicle drives 651 miles and uses 26.7 gallons of fuel. On the return trip, the vehicle travels 668 miles and uses 31.3 gallons of fuel. What is the percent decrease in fuel economy?

We'll start by calculating the fuel economy in each direction, as done in Chapter 1.

$$\text{To destination:} \quad \frac{651 \text{ mi}}{26.7 \text{ gal}} = 24.4 \text{ MPG}$$

$$\text{Return trip:} \quad \frac{668 \text{ mi}}{31.3 \text{ gal}} = 21.3 \text{ MPG}$$

This means the fuel economy decreased by 24.4 MPG − 21.3 MPG = 3.1 MPG Compared to the original fuel economy, this becomes

$$\frac{3.1 \text{ MPG}}{24.4 \text{ MPG}} = 0.127 \quad \text{or about a } 12.7\% \text{ decrease}$$

CHAPTER 5 *Practice Problems*

Converting Fractions, Decimals, and Percents

Convert each of the following percentage values to fractions in lowest terms and decimals. If the percent values are over 100%, write the equivalent fractions using mixed numbers.

- 1. 50%
- 2. 25%
- 3. 40%

- 4. 80%
- 5. 72%
- 6. 28%

- 7. 15%
- 8. 65%
- 9. 278%

- 10. 314%
- 11. 355%
- 12. 210%
- 13. 46.8%
- 14. 32.8%
- 15. 94.35%
- 16. 9.608%
- 17. 0.84%
- 18. 0.0066%
- 19. 0.0112%
- 20. 0.505%
- ●●21. $37\frac{1}{2}$%
- ●●22. $43\frac{3}{4}$%
- ●●23. $83\frac{1}{3}$%
- ●●24. $27\frac{3}{8}$%

Convert each of the following decimals to percents.

- 25. 0.64
- 26. 0.18
- 27. 0.7
- 28. 0.1
- 29. 0.05
- 30. 4
- 31. 1.837
- 32. 2.654
- 33. 0.098
- 34. 0.00635
- 35. 0.000 051
- 36. 0.011 32

Convert each of the following fractions to percents.

- 37. $\frac{47}{100}$
- 38. $\frac{57}{100}$
- 39. $\frac{49}{50}$
- 40. $\frac{14}{25}$
- 41. $\frac{4}{5}$
- 42. $\frac{3}{10}$
- 43. $\frac{213}{300}$
- 44. $\frac{375}{625}$
- 45. $\frac{195}{400}$
- 46. $\frac{99}{880}$
- 47. $3\frac{155}{200}$
- 48. $4\frac{7}{250}$
- 49. $2\frac{327}{600}$
- 50. $1\frac{12}{80}$
- 51. $\frac{9}{4,500}$
- 52. $\frac{1}{200}$
- 53. $\frac{37}{2,000}$
- 54. $\frac{6}{960}$

PERCENT AND PERCENT APPLICATIONS

Base, Rate, and Amount

Identify the base, rate, and amount in each of the statements given below.

	Statement	Base	Rate	Amount
●1.	50% of 40 is 20.			
●2.	30 is 15% of 200.			
●3.	19 is 4.75% of 400.			
●4.	350% of 0.05 is 0.175.			
●5.	0.06% of 6,000 is 3.6.			
●6.	0.25% of 500 is 1.25.			

Use the proportion $\dfrac{\text{Amount}}{\text{Base}} = \dfrac{\text{Rate}}{100\%}$ to solve each of the percent statements given below.

	Statement	Base	Rate	Amount
●7.	What is 54% of 1,500?			
●8.	65% of 8 is how much?			
●9.	22 is what percent of 25?			
●10.	What percent of 1.5 is 0.6?			
●11.	40 is 8% of what number?			
●12.	12% of what number is 14.4?			
●13.	10.92 is 130% of what number?			
●14.	162 is 270% of what number?			
●15.	3 is what percent of 5,000?			
●16.	What percent of 6,500 is 13?			
●17.	What percent of 25 is 45?			
●18.	1,600 is what percent of 40?			

Identify the base, rate, and amount for each of the following problems, then solve.

●●19. A factory has 140 employees, and increases its workforce by 15%. How many jobs are added?

●●20. A gas station sells 4,700 gallons of fuel one Thursday. On Friday, the same station sold 8% more fuel. How many more gallons were sold on Friday?

●●21. A battery's voltage output is normally 12.6 volts. When it is several years old it loses 7% of that output. How many volts can it produce after the decrease? Round your answer to the nearest tenth of a volt.

●●22. A driver is able to obtain an average fuel economy of 23.50 MPG. By replacing the air filter and properly inflating his tires, he can increase this by 5%. What will be the new fuel economy?

●●23. One week a technician performs 44 hours of warranty work. The next week he performs 36 hours. What was the decrease as a percent? Round to the nearest tenth of a percent.

●●24. Oil changes are usually $28.00. Before 10:00, however, they are discounted to $24.00. What is the discount rate? Round to the nearest tenth of a percent.

●●25. On your paycheck stub, you notice that $450 is deducted for health care. The next month you notice that $495 is deducted. What was the percent increase?

●●26. A vehicle weighs 3,400 pounds, of which 590 pounds is the engine. What percent of the overall weight is the engine? Round to the nearest tenth of a percent.

●●27. How much money must be deposited now, at an annual interest rate of 3.0%, so that in one year the account will contain $5,500? Round to the nearest cent.

●●28. In hopes of having $2,500 saved in the next year, a young woman opens a savings account at a fixed 2.5% annual interest rate. How much must she deposit so that she'll reach her goal in a year? No additional deposits will be made.

Discount, Markup, and Sales Tax

For each of the problems below, calculate the reduced price of the item given.

●1. Item: Heater core   Retail price: $94.59   Discount: 5% for cash

●2. Item: Serpentine belt   Retail price: $19.49   Discount: 10% off, on sale

●3. Item: 850 CCA battery   Retail price: $107.99   Discount: 15% off, on sale

●4. Item: CD changer   Retail price: $479.00   Discount: 5%

●5. Item: 12 qt oil   Retail price: $2.19/qt   Discount: 20% off, on sale

●6. Item: 6 Gallons of coolant Retail price: $7.45/gallon   Discount: 3% for cash

Find the retail price for each part listed below.

●7. Part: Alternator   Cost: $62.30   Markup on cost: 60%

●8. Part: Master cylinder   Cost: $57.28   Markup on cost: 48%

●9. Part: Brake shoes   Cost: $19.84   Markup on cost: 70%

●10. Part: Headlight   Cost: $8.50   Markup on cost: 30%

PERCENT AND PERCENT APPLICATIONS

- 11. Part: Brake line   Cost: $14.98   Markup on cost: 100%

- 12. Part: Thermostat   Cost: $2.90   Markup on cost: 110%

Find the total price with tax included of each item below.

- 13. Item: Starter   Retail: $84.70   Sales tax rate: 6%

- 14. Item: Gasket   Retail: $12.15   Sales tax rate: 6%

- 15. Item: O_2 Sensor   Retail: $43.81   Sales tax rate: 4.5%

- 16. Item: Bracket   Retail: $14.50   Sales tax rate: 4.5%

- 17. Item: Muffler   Retail: $37.29   Sales tax rate: 6.5%

- 18. Item: Brake rotor   Retail: $59.28   Sales tax rate: 6.5%

Calculate the percent markup on cost for each of the items below. Round to the nearest tenth of a percent.

- 19. Item: Fuel pump   Cost: $89.77   Retail: $138.99

- 20. Item: Fuel filter   Cost: $19.50   Retail: $26.99

- 21. Item: Exhaust pipe   Cost: $21.70   Retail: $33.15

- 22. Item: EGR valve   Cost: $134.15   Retail: $182.50

Calculate the sales tax rate on the items below. Round to the nearest tenth of a percent.

- 23. Item: Throttle body   Retail price: $138.65   Price including tax: $146.97

- 24. Item: Airflow sensor   Retail price: $58.90   Price including tax: $61.85

- 25. Item: Brake cable   Retail price: $64.59   Price including tax: $68.14

- 26. Item: Alternator   Retail price: $207.90   Price including tax: $220.37

••27. The retail price of a manifold is $125.64. This includes a markup of 32% based on cost. What is the dealership's cost for this manifold?

••28. A gallon of coolant sells for $7.59. This includes a markup of 15% based on cost. What was the parts supplier's cost for the coolant?

••29. The price of a set of plug wires is $27.13 with 6% tax included. What is the price of the wires before tax was added on?

••30. The total price with tax included of a power window motor is $71.73 with 5.5% tax. What was the price before tax?

••31. A receiver hitch is normally $245.50. It is on sale for 10% off. What is the total price when 5% sales tax is added on?

••32. A quart of synthetic motor oil sells for $5.29. If the oil is bought in a full case of 12 qt, a 5% discount is given. What is the total price with tax of one case if the sales tax rate is 6.5%?

••33. The cost to a dealership of an oil pan is $96.17. If the dealership uses a markup based on cost of 24% for this oil pan, and 4.5% tax is added to the retail price, what is the total amount due from the customer for this part?

••34. A parts supplier buys jumper cables at a cost of $12.05 per set. If a markup of 35% based on cost is used, and the sales tax rate is 6%, what is a customer's total price for a set of these cables?

•••35. The cost of an all-terrain tire to a small shop is $68.90. That shop uses a markup of 25% based on cost for all tires. If the tires are on sale for 10% off when a complete set of four are purchased, and the sales tax rate is 5.5%, what is the final price, with tax, for a customer who buys a complete set of four tires? Ignore the price of mounting and balancing.

•••36. A dealership's cost for a radiator is $310.40. The dealership sells the radiator with a 30% markup on cost, but gives the customer a 5% discount on the retail price. If sales tax in that area is 7%, what is the total price, with tax, for the radiator?

Technical Applications

For problems 1 through 4, determine if the compression of all cylinders are within 15% of the cylinder with the highest compression.

•1.

Cylinder Number	1	2	3	4	5	6
Compression Reading	130 psi	129 psi	114 psi	138 psi	126 psi	119 psi

What is the lowest allowable pressure?_____

Are all cylinders within 15% of the highest value?_____

PERCENT AND PERCENT APPLICATIONS

●2.

Cylinder Number	1	2	3	4	5	6
Compression Reading	148 psi	161 psi	155 psi	156 psi	147 psi	149 psi

What is the lowest allowable pressure?_____

Are all cylinders within 15% of the highest value?_____

●3.

Cylinder Number	1	2	3	4	5	6
Compression Reading	150 psi	137 psi	142 psi	139 psi	154 psi	148 psi

What is the lowest allowable pressure?_____

Are all cylinders within 15% of the highest value?_____

●4.

Cylinder Number	1	2	3	4	5	6
Compression Reading	167 psi	136 psi	144 psi	165 psi	151 psi	134 psi

What is the lowest allowable pressure?_____

Are all cylinders within 15% of the highest value?_____

For problems 5 through 8, determine how many pounds are on the front wheels of the vehicle, and how many pounds are on the rear wheels.

●5. Vehicle weight: 2,690 lb   Weight distribution: 51/49

   Front load: _____ lb   Rear load: _____ lb

●6. Vehicle weight: 3,560 lb   Weight distribution: 59/41

   Front load: _____ lb   Rear load: _____ lb

●7. Vehicle weight: 3,246 lb   Weight distribution: 55/45

   Front load: _____ lb   Rear load: _____ lb

●8. Vehicle weight: 3,905 lb   Weight distribution: 57/43

   Front load: _____ lb   Rear load: _____ lb

●9. A battery manufacturer claims that a cold-weather battery has "20% more cold cranking amps than the competitor's 650 CCA battery." How many CCA must this improved battery have?

- 10. Suppose it is claimed that a certain synthetic motor oil lasts 25% longer than regular engine oil. If regular oil must be changed every 3000 miles, how often must synthetic oil be changed?

- 11. A turbocharged engine might produce 20% more horsepower than a similar naturally-aspirated engine. If the non-turbocharged engine normally produces 132 peak hp, what is the peak horsepower output of the turbocharged version?

- 12. By changing camshafts in an engine, you could increase power output by 15% (although the drivability of the engine could decrease). If the engine usually produces 165 hp, what could be the increased output if the camshaft is changed?

- 13. In 1972, a car company produced 225,093 trucks. In 1973, the company produced 256,395 trucks. What was the percent increase in production?

- 14. Suppose that in 1994, a new model of vehicle was introduced, and 198,404 vehicles were sold. The following year, only 172,318 were sold. What was the percent decrease in production?

- 15. In the summer, the owner of a vehicle could drive about 315 miles on 20.0 gallons of fuel. In the winter, though, the vehicle could only travel about 275 miles on 19.0 gallons. What is the percent decrease in fuel economy from summer to winter?

- 16. While against the wind, a vehicle travels 396 miles on 14.2 gallons of fuel. On the return trip, with the wind, the vehicle travels 396 miles but consumes 11.8 gallons of fuel. What is the percent increase in fuel economy when driving with the wind?

The torque of an engine with stock exhaust is measured and charted. Then, the process is repeated with a modified exhaust. Use the chart to answer problems 17–20.

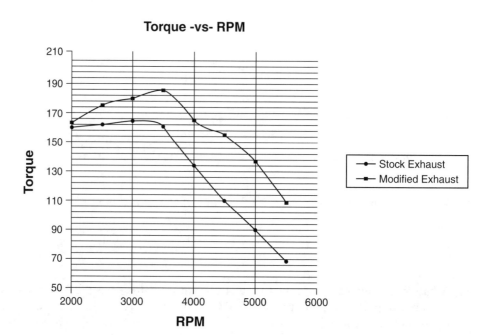

PERCENT AND PERCENT APPLICATIONS

●●17. What is the percent increase in torque with the modified exhaust compared to the stock exhaust at 3,000 RPM?

●●18. What is the percent increase in torque with the modified exhaust compared to the stock exhaust at 4,000 RPM?

●●19. By what percentage does the torque output of the engine with stock exhaust decrease from 4,000 to 5,000 RPM?

●●20. By what percentage does the torque output of the engine with modified exhaust increase from 2,000 to 3,000 RPM?

CHAPTER 6

The Metric System and Unit Conversion

METRIC UNITS OF MEASURE

Almost all vehicles being built today are designed using metric dimensions and fasteners. There was a time when a technician rarely used metric wrenches, but it is likely that the standard wrenches in fractional inch increments will be used less every year. More specifications are being given in metric than ever before, and every technician must be familiar with the metric system.

Table 6.1 describes what metric units are used to make a few common measurements, along with the units used in the customary system.

See Figures 6.1, 6.2, and 6.3 to get a feeling for the amounts of the meter, the kilogram, and the liter.

Example 6-1 Which of these parts is most likely to weigh (or have a mass of) about 1 kg?

a. Cylinder head
b. Lug nut
c. Serpentine belt
d. Piston and connecting rod

One kilogram is about the weight of a book. A cylinder head is much heavier, and a lug nut or belts are probably lighter. A piston and connecting rod would probably be about one kilogram.

TABLE 6.1 Metric and Customary Units with Examples

Quantity	Customary Unit	Metric Unit	Example of Unit Value
Length	Inch (in.) or foot (ft)	Meter (m) or centimeter (cm)	A meter is about 39.4 in., slightly longer than a yardstick. A centimeter is about the thickness of a pencil.
Weight (Mass)*	Ounce (oz) or pound (lb)	Gram (g) or kilogram (kg)	A paperclip has a mass of about one gram. A small textbook has a mass of about one kilogram.
Volume	Quart (qt) or gallon (gal)	Liter (L)	One liter is a common size for soda bottles. A liter is slightly larger than a quart.
Force	Pound (lb)	Newton (N)	A Newton of force is much less than a pound of force.
Pressure	Pounds/inch^2 (lb/in.^{2} or psi)	Pascal (Pa) = 1 Newton/meter^2 (N/m^2)	One Pascal of pressure is much less than 1 psi.
Energy	British thermal unit (Btu)	Joule (J)	A Joule is a very small unit of energy.
Power	Horsepower (hp)	Watt (W)	A Watt is much smaller than a horsepower. One horsepower equals about 746 W.
Torque	Pound · foot (lb · ft)	Newton · meter (N · m)	One Newton · meter is slightly less than a pound · foot.
Temperature	Degrees Fahrenheit (°F)	Degrees Celsius (°C)	Water freezes at 0°C and boils at 100°C.

*Weight and mass do not measure the same quantity. While objects have constant mass, they do not have a constant weight. It is practical, however, to compare an object's mass to its weight. This text will assume all weight values are based on gravity at the earth's surface.

THE METRIC SYSTEM AND UNIT CONVERSION

FIGURE 6.1 A typical motorcycle is about 1 m tall.

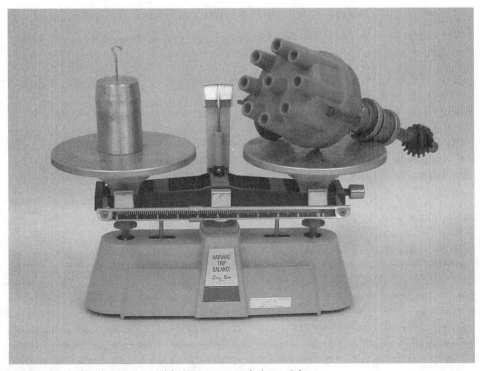

FIGURE 6.2 A distributor assembly has a mass of about 1 kg.

FIGURE 6.3 One liter (1 L) is a common bottle size for soft drinks.

Example 6-2 Which of these would probably hold about 1 L of fluid?

a. Cooling system   c. Oil pan
b. Brake system   d. Automatic transmission

A liter of fluid is about the same as a bottle of soda. Cooling systems, oil pans and automatic transmissions all require much more fluid than this. A braking system might hold about a liter of brake fluid.

Example 6-3 Which dimension is most likely to be about 4 m?

a. Height of a motorcycle   c. Length of a pickup truck
b. Diameter of a tire   d. Bore of a cylinder

Four meters is slightly more than four yards. If four yardsticks were placed end-to-end, the total length would be approximately the length of a pickup truck.

METRIC PREFIXES AND CONVERSIONS

The base units in the metric system can easily be used to measure and express very large or small quantities. Adding a prefix to the base unit can make the units of measure much larger or much smaller.

The length of a head bolt is 0.08 m. Rather than use such a large unit of measure (the meter) to measure the length of a small object, we'll use centimeters. The prefix centi means one-hundredth, or 1/100. Then we could say the length of the bolt is 8/100 of a meter, or 8 cm. Figure 6.4 compares a measurement in meters and centimeters. (The scales shown in Figure 6.4 and other figures in this chapter are not actual size.)

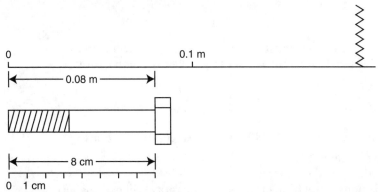

FIGURE 6.4 Measuring 0.04 m with a meter stick would not be very useful. But the same measurement using a centimeter rule gives 8 cm.

Commonly used metric prefixes and the value they carry are given in Table 6.2. Referring to Table 6.2, the prefix centi means 0.01 or 1/100. That means 1 cm = 0.01 m (the base unit), and it takes 100 cm to equal 1 m. Similarly, we see that the prefix kilo means 1000. Then we know that 1 kg is equal to 1000 of the base unit, which is grams. 1 kg = 1000 g.

Table 6.2 can be used to convert easily within the metric system.

Example 6-4 Convert 42 mm to meters.

From Table 6.2, 1 milli = 0.001 base units.
Then 1 mm = 0.001 m.
We have 42 mm, or 42 × 0.001 m = 0.042 m.
There is another way to do this conversion. This conversion can be done more simply by seeing the following from Table 6.2:

1. Milli means 0.001. That's a three decimal-place move from the base unit of 1.0.

$$1.0 \longrightarrow 0.001$$
3 places

2. Since meters are a larger unit, our new measurement will be smaller.
3. Then move the decimal point in 42 three places, making the value smaller.

$$42. \longrightarrow 0.042$$
3 places

42 millimeters = 0.042 meters

TABLE 6.2 Values and Names of Commonly Used Metric Prefixes

Metric Prefix	Value	Exponent Notation
mega (M)	1,000,000	10^6
kilo (k)	1,000	10^3
hecto (h)	100	10^2
deka (da)	10	10^1
BASE UNIT	1.0	
deci (d)	0.1	10^{-1}
centi (c)	0.01	10^{-2}
milli (m)	0.001	10^{-3}
micro (μ)	0.000 001	10^{-6}

Example 6-5 Convert 5.5 kilovolts to volts.

1. Kilo means 1000, which means we need a three decimal-place move to the base unit of volts.

$$1.0 \longrightarrow 1000$$
$$\text{3 places}$$

2. We're going to a smaller unit, so the quantity will increase.
3. Move the decimal three places to make the number larger.

$$5.5 \longrightarrow 5,500$$
$$5.5 \text{ kilovolts} = 5,500 \text{ volts}$$

Example 6-6 Convert 6,580 centigrams (cg) to kilograms.

1. Centi means 0.01, while kilo means 1000. That's a five decimal-place move from centigrams to kilograms.
2. The units are getting larger, so the quantity will get smaller.
3. Move the decimal point five places, making the quantity smaller.

$$6580 \longrightarrow 0.06580$$
$$\text{5 places}$$
$$6,580 \text{ cg} = 0.06580 \text{ kg}$$

Example 6-7 Convert 1.3 megaliters (ML) to deciliters (dL).

1. Notice that even though these units are only five positions apart on Table 6.2, they represent a *seven* decimal-place change. There are no metric prefixes between kilo and mega.
2. The units become smaller, so the quantity will become larger.
3. Move the decimal seven places, making the quantity larger.

$$1.3 \longrightarrow 13,000,000$$
$$\text{7 places}$$
$$1.3 \text{ ML} = 13,000,000 \text{ L}$$

Metric units can also be converted using the metric staircase shown in Figure 6.5.

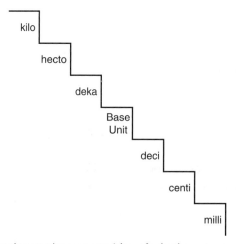

FIGURE 6.5 The metric staircase gives you an idea of whether you are going "up" to bigger units or "down" to smaller units.

Example 6-8 Use the metric staircase to convert 0.470 amps (A) to milliamps (mA).

We're going to convert amps (a base unit) to milliamps.

From the step labeled "base unit" to the step labeled "milli" we move down and to the right three places. This means we're going to a smaller unit, and must move the decimal point three places to the right.

Then 0.470 A become 470 mA.

THE METRIC SYSTEM AND UNIT CONVERSION

Example 6-9 Use the metric staircase to convert 9600 milliliters (mL) to kiloliters.

From the step labeled "milli" to the step labeled "kilo" we must move up and to the left six places. We're then going to a larger unit, and must move the decimal six places to the left.

9600 mL becomes 0.0096 kL.

CONVERSION OF UNITS USING RATIOS

It is often necessary to convert measurements between the customary system and the metric system. One way to do this is by using the unit-fraction conversion method.

Example 6-10 Given that 1 in. = 2.54 cm, convert 8.50 cm to inches.

If we write the conversion factor 1 in. = 2.54 cm as a fractional ratio, we can write either

$$\frac{1 \text{ in.}}{2.54 \text{ cm}} \quad \text{or} \quad \frac{2.54 \text{ cm}}{1 \text{ in.}}$$

Both of these fractional ratios equal 1, since the numerator and denominator are equal in each. These fractional ratios are called *unit-fractions*.

$$\frac{1 \text{ in.}}{2.54 \text{ cm}} = \frac{2.54 \text{ cm}}{1 \text{ in.}} = 1$$

1. Write the unit to be converted as a fraction, using a 1 in the denominator.

$$\frac{8.50 \text{ cm}}{1}$$

2. Multiply this value by a unit-fraction so that cm (the unit you are changing from) cancels.

$$\frac{8.50 \text{ cm}}{1} \times \frac{1 \text{ in.}}{2.54 \text{ cm}}$$

3. Cancel the units of cm, and multiply across.

$$\frac{8.50 \text{ cm}}{1} \times \frac{1 \text{ in.}}{2.54 \text{ cm}} = \frac{8.5 \text{ in.}}{2.54} = 3.35 \text{ in.}$$

As you can see, a conversion factor must be known before any unit conversions can be done. Some common conversions needed in the automotive trades are shown in Table 6.3. Note that in Table 6.3 the symbol ≈ means two values are approximately equal.

Example 6-11 Convert 176.8 miles to kilometers.

1. Write 176.8 miles as a fraction, using a 1 as the denominator.

$$\frac{176.8 \text{ mi}}{1}$$

2. Multiply this value by a unit-fraction with mi (miles) in the denominator, so that it cancels. (The conversion 1 mile = 1.609 km is taken from Table 6.3.)

$$\frac{176.8 \text{ mi}}{1} \times \frac{1.609 \text{ km}}{1 \text{ mi}}$$

3. Cancel the units of mile (mi), and multiply.

$$\frac{176.8 \text{ mi}}{1} \times \frac{1.609 \text{ km}}{1 \text{ mi}} = \frac{284.471 \text{ km}}{1} = 284.471 \text{ km or } 284.5 \text{ km}$$

TABLE 6.3 Exact Conversions and Approximate Conversions Commonly Used in the Automotive Trades

Quantity	Exact Conversions	Approximate Conversions*
Length	1 in. = 2.54 cm 12 in. = 1 foot 3 ft = 1 yard 5280 ft = 1 mi	1 ft ≈ 0.305 m 1 mi ≈ 1.609 km
Area	1 ft^2 = 144 in^2	1 in.^{2} ≈ 6.45 cm^2
Volume	8 fluid oz. = 1 cup 2 cups = 1 pint 2 pints = 1 quart 4 qt = 1 gal 1728 in.^{3} = 1 ft^3 27 ft^3 = 1 yd^3 1 cm^3 = 1 mL 1000 cm^3 = 1 L	1 in.^{3} ≈ 16.39 cm^3 1 gal ≈ 231 in.^{3} 1 L ≈ 1.056 qt
Weight (Mass)	16 oz. = 1 lb 2000 lb = 1 ton	1 kg ≈ 2.2 lb
Power		1 hp ≈ 746 W
Energy		1 Btu ≈ 1055 J
Torque		1 lb·ft ≈ 1.356 N·m
Pressure		1 psi ≈ 6.895 kPa

Although the approximate conversion factors are not exact, a standard equal sign (=) will be used in the conversions in this chapter.

Example 6-12 Convert 110.0 N·m to lb·ft.

1. Write the quantity as a fraction.

$$\frac{110.0 \text{ N·m}}{1}$$

2. Multiply this fraction by the unit conversion so that N·m is in the denominator, and will cancel.

$$\frac{110.0 \text{ N·m}}{1} \times \frac{1 \text{ lb·ft}}{1.356 \text{ N·m}}$$

3. Cancel and multiply.

$$\frac{110.0 \cancel{\text{ N·m}}}{1} \times \frac{1 \text{ lb·ft}}{1.356 \cancel{\text{ N·m}}} = \frac{110 \text{ lb·ft}}{1.356} = 81.12 \text{ lb·ft}$$

Example 6-13 Convert 38.42 cm to feet.

Here, there is not a unit-fraction conversion factor from centimeters to feet. We'll use the fact that 2.54 cm = 1 in. and 12 in. = 1 ft.

1. $\dfrac{38.42 \text{ cm}}{1}$

2. $\dfrac{38.42 \text{ cm}}{1} \times \dfrac{1 \text{ in.}}{2.54 \text{ cm}} \times \dfrac{1 \text{ ft}}{12 \text{ in.}}$

3. $\dfrac{38.42 \cancel{\text{ cm}}}{1} \times \dfrac{1 \cancel{\text{ in.}}}{2.54 \cancel{\text{ cm}}} \times \dfrac{1 \text{ ft}}{12 \cancel{\text{ in.}}} = \dfrac{38.42 \text{ ft}}{30.48} = 1.26 \text{ ft}$

Example 6-14 Convert 794 mL to gallons.

Again, we need to use several conversion factors. We'll use these facts:

$$1000 \text{ mL} = 1 \text{ L} \qquad 1 \text{ L} \approx 1.056 \text{ qt} \qquad 4 \text{ qt} = 1 \text{ gal}$$

1. $\dfrac{794 \text{ mL}}{1}$

2. $\dfrac{794 \text{ mL}}{1} \times \dfrac{1 \text{ L}}{1000 \text{ mL}} \times \dfrac{1.056 \text{ qt}}{1 \text{ L}} \times \dfrac{1 \text{ gal}}{4 \text{ qt}}$

3. $\dfrac{794 \text{ mL}}{1} \times \dfrac{1 \text{ L}}{1000 \text{ mL}} \times \dfrac{1.056 \text{ qt}}{1 \text{ L}} \times \dfrac{1 \text{ gal}}{4 \text{ qt}} = \dfrac{838.464 \text{ gal}}{4000} = 0.210 \text{ gal}$

Example 6-15 Convert 65 miles per hour (MPH) to ft per minute.

Note that the word "per" means division, so we could write 65 MPH (miles per hour) as 65 mi/hr. Then the units of "hour" (hr) must go in the denominator of our original fraction.

1. $\dfrac{65 \text{ mi}}{1 \text{ hr}}$

Now, there are two conversions that must be done:
 a. Miles must be converted to feet, and
 b. Hours must be converted to minutes (min).

2. $\dfrac{65 \text{ mi}}{1 \text{ hr}} \times \dfrac{5280 \text{ ft}}{1 \text{ mi}}$

$\dfrac{65 \text{ mi}}{1 \text{ hr}} \times \dfrac{5280 \text{ ft}}{1 \text{ mi}} \times \dfrac{1 \text{ hr}}{60 \text{ min}}$

Notice that since "hour" was originally in the denominator, it must go in the numerator of the conversion fraction to cancel.

3. $\dfrac{65 \text{ mi}}{1 \text{ hr}} \times \dfrac{5280 \text{ ft}}{1 \text{ mi}} \times \dfrac{1 \text{ hr}}{60 \text{ min}} = \dfrac{343{,}200 \text{ ft}}{60 \text{ min}} = 5720 \text{ ft/min}$

Example 6-16 Convert 25.6 miles per gallon to kilometers per liter.

This conversion will be done in the same way as the last example.

1. $\dfrac{25.6 \text{ mi}}{1 \text{ gal}}$

Now, we'll convert miles to kilometers, and gallons to liters

2. $\dfrac{25.6 \text{ mi}}{1 \text{ gal}} \times \dfrac{1.609 \text{ km}}{1 \text{ mi}}$

$\dfrac{25.6 \text{ mi}}{1 \text{ gal}} \times \dfrac{1.609 \text{ km}}{1 \text{ mi}} \times \dfrac{1 \text{ gal}}{4 \text{ qt}}$

$\dfrac{25.6 \text{ mi}}{1 \text{ gal}} \times \dfrac{1.609 \text{ km}}{1 \text{ mi}} \times \dfrac{1 \text{ gal}}{4 \text{ qt}} \times \dfrac{1.056 \text{ qt}}{1 \text{ L}}$

$\dfrac{25.6 \text{ mi}}{1 \text{ gal}} \times \dfrac{1.609 \text{ km}}{1 \text{ mi}} \times \dfrac{1 \text{ gal}}{4 \text{ qt}} \times \dfrac{1.056 \text{ qt}}{1 \text{ L}} = \dfrac{43.497}{4} = 10.9 \text{ km/L}$

CONVERSION OF UNITS USING TABLES

Another method of performing unit conversions, although more limited, is through the use of conversion tables. An example of a conversion table is given in Table 6.4. You may note the use of exponents in units in Table 6.4. The use of such units as in.^{3} and cm^3 to mean cubic inches and cubic centimeters will be used from this point forward in *Automotive Mathematics*.

TABLE 6.4 Conversion Table

Quantity	Customary	Multiply by	Metric
Length	Inches	→ 25.4 →	Millimeters
	(in.)	← 0.03937 ←	(mm)
	Inches	→ 2.54 →	Centimeters
	(in.)	← 0.3937 ←	(cm)
	Feet	→ 0.3048 →	Meters
	(ft)	← 3.2808 ←	(m)
	Yards	→ 0.9144 →	Meters
	(yd)	← 1.0936 ←	(m)
	Miles	→ 1.6093 →	Kilometers
	(mi)	← 0.6214 ←	(km)
Area	Square inches	→ 6.4516 →	Square centimeters
	(in.^{2})	← 0.1550 ←	(cm^2)
	Square feet	→ 0.0929 →	Square meters
	(ft^2)	← 10.764 ←	(m^2)
Volume	Cubic inches	→ 16.387 →	Cubic centimeters
	(in.^{3})	← 0.06102 ←	(cm^3 or cc)
	Cubic inches	→ 0.01639 →	Liters
	(in.^{3})	← 61.024 ←	(L)
	Quarts	→ 0.9464 →	Liters
	(qt)	← 1.0566 ←	(L)
	Gallons	→ 3.7857 →	Liters
	(gal)	← 0.2642 ←	(L)
	Cubic yards	→ 0.7646 →	Cubic meters
	(cu yd)	← 1.308 ←	(m^3)
Weight (Mass)	Ounces	→ 28.353 →	Grams
	(oz)	← 0.03527 ←	(g)
	Pounds	→ 0.4536 →	Kilograms
	(lb)	← 2.205 ←	(kg)
	Tons	→ 0.9072 →	Metric tons
		← 1.1023 ←	(t)
Power	Horsepower	→ 0.7457 →	Kilowatts
	(hp)	← 1.341 ←	(kW)
Energy	British thermal units	→ 1054.53 →	Joules
	(Btu)	← 0.000948 ←	(J)
Torque	Pound · feet	→ 1.3558 →	Newton · meters
	(lb · ft)	← 0.7376 ←	(N · m)
Pressure	Pounds per square inch	→ 6.895 →	KiloPascals
	(lb/in.^{2} or psi)	← 0.145 ←	(kPa)
Velocity	Miles per hour	→ 1.6093 →	Kilometers per hour
	(mi/hr or MPH)	← 0.6214 ←	(km/hr)
Fuel Economy	Miles per gallon	→ 0.4251 →	Kilometers per liter
	(mi/gal or MPG)	← 2.352 ←	(km/L)

Example 6-17 Convert 4.72 m to feet.

From Table 6.4, we see that we're converting meters (in the right column) to feet (in the left column), and should multiply by 3.2808.

$$4.72 \times 3.2808 = 15.485 \text{ or about } 15.5 \text{ ft}$$

Example 6-18 Convert 9.400 oz to grams.

Here, we're converting ounces (in the left column) to grams (in the right column). We should multiply by 28.353.

$$9.400 \times 28.353 = 266.518 \text{ or about } 266.5 \text{ g}$$

THE METRIC SYSTEM AND UNIT CONVERSION

Example 6-19 Convert 27.83 in. to meters.

There is no direct conversion in Table 6-4. We can, however, convert inches to centimeters, then use our knowledge of the metric system to convert centimeters to meters. To convert from inches to centimeters, we should multiply by 2.54.

$$27.83 \times 2.54 = 70.7 \text{ cm}$$

Now, convert from centimeters to meters by moving the decimal two places. Since meters are larger than centimeters, our numeric value must get smaller.

$$70.7 \text{ cm} = 0.707 \text{ m}$$

TEMPERATURE CONVERSIONS

Converting temperatures between the Fahrenheit and Celsius scales is not as simple as multiplying by a conversion factor. There are two formulas used to convert between temperature scales.

When converting from a Fahrenheit temperature to a Celsius temperature, we use the formula below:

$$T_C = \frac{(T_F - 32°)}{1.8}$$

where T_C = Celsius temperature and T_F = Fahrenheit temperature.

Example 6-20 Convert 194°F (Fahrenheit degrees) to °C (Celsius degrees).

Here, $T_F = 194°$. Use the conversion formula above.

$$T_C = \frac{(194° - 32°)}{1.8} = \frac{162°}{1.8} = 90°C$$

Example 6-21 A thermostat will open at 180°F. What is this temperature on the Celsius scale?

Let $T_F = 180°$, and use the conversion formula.

$$T_C = \frac{(180° - 32°)}{1.8} = \frac{148°}{1.8} = 82.2°C$$

To convert from a Centigrade (another term sometimes used for Celsius) temperature to the Fahrenheit scale, we use a similar formula.

$$T_F = 1.8 \times T_C + 32°$$

Here again, T_F = Fahrenheit temperature and T_C = Celsius temperature.

Example 6-22 Convert 4°C to °F.

We'll use a value of 4° in place of T_C in the conversion formula above.

$$T_F = 1.8 \times 4° + 32° = 7.2° + 32° = 39.2°F$$

Example 6-23 A vehicle requires a 192°F opening temperature for a thermostat. A certain thermostat is labeled with an opening temperature of 73°C. Can it be used?

Convert 73°C to Fahrenheit.

$$T_F = 1.8 \times 73° + 32° = 131.4° + 32° = 163.4°F$$

This thermostat should not be used. It would open at a lower temperature than specified.

CHAPTER 6 *Practice Problems*

Metric Units of Measure

1. Which dimension is most likely to be about 1 m?
   a. Wheelbase of a car
   b. Diameter of a tire
   c. Width of a rim
   d. Height of a fender

2. Which of these would probably hold about 10 L of fluid?
   a. Power steering pump
   b. Cooling system
   c. Rear differential
   d. Small oil can

3. Which of the following probably has a weight (mass) of about 60 kg?
   a. Engine block
   b. Tire
   c. A small car
   d. Water pump

4. Which part is about half of a meter long?
   a. Spark plug
   b. Spark plug wire
   c. Battery
   d. Exhaust system

5. Which of these is about equal to 4 L?
   a. Gallon of antifreeze
   b. Quart of oil
   c. Tube of grease
   d. Bottle of power steering fluid

6. Of the parts below, which would be most likely to have a weight (mass) of 30 g?
   a. Piston
   b. Intake manifold
   c. Radiator
   d. Spark plug

7. Which metric units could be used to measure tire pressure?
   a. Newtons
   b. Pascals
   c. Liters
   d. Joules

8. Which metric units could be used to measure the power output of an engine?
   a. Watts
   b. Pascals
   c. Newton·meters
   d. Joules

9. Which metric units could you find on a torque wrench?
   a. Meters
   b. Newtons
   c. Newton·meters
   d. Joules

10. Which of the following metric units could be used to measure the safe pulling strength of a tow cable?
    a. Pascals
    b. Newtons
    c. Newton·meters
    d. Joules

THE METRIC SYSTEM AND UNIT CONVERSION

Metric Prefixes and Conversions

Convert each of the following metric units using the method of your choice. If needed, refer to Table 6.2 and Figure 6.5.

- 1. Convert 145 mg to grams.

- 2. Convert 8.64 hectoliters (hL) to liters.

- 3. Convert 23 dekameters (dam) to meters.

- 4. Convert 0.073 cm to meters.

- 5. Convert 0.841 decivolts to volts.

- 6. Convert 450 km to meters.

- 7. Convert 9.7 megagrams (Mg) to grams.

- 8. Convert 16.93 A to microamps.

- 9. Convert 520 L to deciliters.

- 10. Convert 4.89 microseconds to seconds.

- 11. Convert 0.005 kL to liters.

- 12. Convert 0.00974 volts to millivolts.

- 13. Convert 0.0064 kilopascals (kPa) to pascals.

- 14. Convert 5.89 hectograms (hg) to grams.

- 15. Convert 8.94 km to centimeters.

- 16. Convert 2,150 micrograms to grams.

- 17. Convert 8.9 decivolts to dekavolts.

- 18. Convert 0.000 000 985 megavolts to millivolts.

- 19. Convert 0.87 megaliters (ML) to kiloliters.

- 20. Convert 157.2 micrometers to millimeters.

- 21. Convert 0.0047 centigrams (cg) to kilograms.

- 22. Convert 3700 milliamps to A.

Conversion of Units using Ratios

Perform each of the conversions below using the unit-fraction conversion method. Use the conversion factors given in Table 6.3 to perform all conversions. Round answers to the nearest hundredth.

- ●1. Convert 1980 ft to miles.
- ●2. Convert 35 psi to kilopascals (kPa).
- ●3. Convert 57,000 W to horsepower.
- ●4. Convert 325 kg to pounds.
- ●5. Convert 430 km to miles.
- ●6. Convert 20 qt to liters.
- ●7. Convert 65.0 cm^3 to in.^{3}.
- ●8. Convert 58.0 cm^3 to in.^{3}.
- ●9. Convert 3.95 in.^{3} to cm^3.
- ●10. Convert 4.31 in.^{3} to cm^3.
- ●●11. Convert 78.4 mL to quarts.
- ●●12. Convert 8.5 gal to pints.
- ●●13. Convert 6,250 m to miles.
- ●●14. Convert 154 in.^{3} to quarts.
- ●●15. Convert 350 in.^{3} to liters.
- ●●16. Convert 302 in.^{3} to liters.
- ●●17. Convert 4.6 L to cubic inches.
- ●●18. Convert 3.0 L to cubic inches.
- ●●19. Convert 3,400 feet per minute to miles per hour.
- ●●20. Convert 68 miles per hour (MPH) to feet per minute.
- ●●21. Convert 35.8 miles per hour (MPH) to feet per second.
- ●●22. Convert 78.2 miles per hour (MPH) to feet per second.
- ●●23. Convert 48.5 feet per second to miles per hour (MPH).
- ●●24. Convert 15.1 feet per second to miles per hour (MPH).
- ●●25. Convert 17.3 miles per gallon (MPG) to kilometers per liter.
- ●●26. Convert 38.6 miles per gallon to (MPG) kilometers per liter.
- ●●27. Convert 12.4 km per L to miles per gallon (MPG).
- ●●28. Convert 8.6 km per L to miles per gallon (MPG).

THE METRIC SYSTEM AND UNIT CONVERSION

Conversion of Units using Tables

Perform each of the conversions below using the conversion factors from Table 6.4. Some conversions will require more than one step. Round to two decimal places when necessary.

- •1. Convert 23.7 mi to kilometers.
- •2. Convert 9.6 lb to kilograms.
- •3. Convert 500 Btu to Joules.
- •4. Convert 15.0 gallons to liters.
- •5. Convert 6.85 m to feet.
- •6. Convert 3.9 L to cubic inches.
- •7. Convert 722 g to ounces.
- •8. Convert 63 kW to horsepower.
- •9. Convert 10.5 in. to millimeters.
- •10. Convert 420 ft^2 to square meters.
- •11. Convert 4.9 yd^3 to cubic meters.
- •12. Convert 125 lb · ft to Newton · meters.
- •13. Convert 245 kPa to pounds per square inch.
- •14. Convert 73 cc (cu cm, or cm^3) to cubic inches.
- •15. Convert 35.0 m to yards.
- •16. Convert 13.4 km/L to miles per gallon.
- ••17. Convert 4.3 yd to centimeters.
- ••18. Convert 3.0 mi to meters.
- ••19. Convert 5.7 L to pints.
- ••20. Convert 9.7 kg to ounces.
- ••21. Convert 0.300 in. to decimeters.
- ••22. Convert 302 in.^{3} to deciliters.
- ••23. Convert 700 mL to quarts.
- ••24. Convert 35,500 W to horsepower.

Temperature Conversions

Convert each temperature below from the Fahrenheit scale to the Celsius scale. Round to two decimal places where necessary.

- 1. 104°F = _____°C
- 2. 149°F = _____°C

- 3. 212°F = _____°C
- 4. 68°F = _____°C

- 5. 183°F = _____°C
- 6. 77°F = _____°C

- 7. −37°F = _____°C
- 8. −15°F = _____°C

Convert these temperatures from the Celsius scale to the Fahrenheit scale. Round to one decimal place where necessary.

- 9. 20°C = _____°F
- 10. 95°C = _____°F

- 11. 0°C = _____°F
- 12. 70°C = _____°F

- 13. 87°C = _____°F
- 14. 108°C = _____°F

- 15. −6°C = _____°F
- 16. −35°C = _____°F

CHAPTER 7

Geometry

MEASUREMENT

In Chapter 2 we used decimals to read a micrometer, the most common measuring tool in the automotive trades. For larger measurements requiring less accuracy, various rulers or tape measures may be used. In this section we'll practice reading common ruler scales.

Most customary rulers are divided into inches, with clear subdivisions such as the one shown in Figure 7.1. (Scales shown in figures in this chapter are not actual size.)

FIGURE 7.1 An inch ruler is divided into half, quarter, eighth, and sixteenth inches.

Before using an inch ruler, it's important to notice how many subdivisions of each inch there are. Rulers commonly divide whole inches into eighths, sixteenths, thirty-seconds, and sixty-fourths or some rulers may divide whole inches into tenths and twentieths. The ruler in Figure 7.1 shows that each inch is divided into sixteenths.

Example 7-1 What is the measurement at each arrow below?

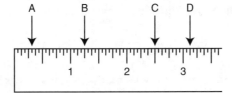

First, note that each inch is divided into 16 parts. Each division then is 1/16 of an inch.

a. The arrow here is located at the fifth mark to the right of zero. Then the measurement at A is 5/16 of an inch.
b. This arrow aligns with the fourth mark past 1 in. This measurement is then 1 4/16 in. This fraction should be reduced to 1 1/4 in.
c. This arrow is located at the eighth mark past 2 in. Since 8/16 also represents 1/2, a larger mark is used to easily identify this spot. The measurement here is then 2 1/2 in.
d. The last arrow is at the second mark past 3. The measurement is then 3 2/16, which reduces to 3 1/8 in.

Example 7-2 What is the measurement at each arrow below?

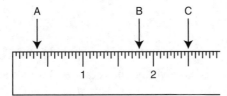

113

Begin by noting that each inch is divided into 20 parts. Each mark represents 1/20 of an inch.

a. The arrow points to the seventh mark past zero. The measurement at this point is then 7/20 of an inch. As a decimal this is written as 0.35 in.
b. Rather than counting all the way from 1, begin at the larger mark that must represent 10/20. Moving to the right, we see that it is the sixteenth mark past the 1, so our measurement is 1 16/20, or 1 8/10 of an inch. A measurement such as this might be referred to as 1.8 in., but probably not 1 4/5 of an inch.
c. This mark is at the large 1/2 indicator past the 2. The measurement at this point is 2 1/2 in.

Metric rulers are typically based on centimeters, which are further divided into tenths. Then each 1/10 of a centimeter is actually a millimeter. A centimeter ruler is shown in Figure 7.2.

FIGURE 7.2 A centimeter ruler is divided into millimeters.

Example 7-3 Find the measurement at each of the arrows below. Report answers using centimeters and millimeters.

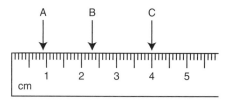

a. This arrow is at the ninth mark past zero, indicating a measurement of 9/10 of a centimeter. This is usually written as 0.9 cm or 9 mm.
b. The arrow here is on the third mark past 2. Then our measurement is 2.3 cm or 23 mm.
c. The last arrow points directly to the mark at 4. The measurement is 4.0 cm or 40 mm.

PERIMETER

As we begin working with two-dimensional (or flat) geometric shapes, we need to be familiar with the concepts of area and perimeter. **Perimeter** is the distance along the outside edge of a two-dimensional figure.

Example 7-4 Find the perimeter of the rectangle below.

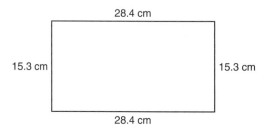

The perimeter in this case will be the total of all four side lengths.
Perimeter = 15.3 cm + 28.4 cm + 15.3 cm + 28.4 cm = 87.4 cm

GEOMETRY

Example 7-5 Find the perimeter of the pentagon below.

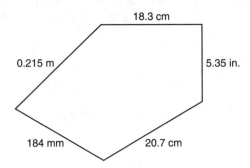

To find the perimeter, we need to add the lengths of the five sides.
Before they can be added, though, all the sides must be converted to the same unit of length.
We'll convert all the lengths to centimeters. We'll start with 5.35 in.

$$\frac{5.35 \text{ in.}}{1} \times \frac{2.54 \text{ cm}}{1 \text{ in.}} = \frac{13.589 \text{ cm}}{1} = 13.6 \text{ cm}$$

$$0.215 \text{ m} = 21.5 \text{ cm}$$
$$184 \text{ mm} = 18.4 \text{ cm}$$

Now, add these lengths together.

Perimeter = 13.9 cm + 21.5 cm + 18.4 cm + 18.3 cm + 20.7 cm = 92.8 cm

The perimeter of a circle is called the **circumference**. The circumference of a circle depends on the **diameter**, which is the distance across a circle when passing through the center. Every circle has a circumference that is about 3.14 times the length of the diameter. The circumference and diameter of a circle are shown in Figure 7.3. The number 3.14 is actually an estimate for the exact ratio of a circle's circumference to its diameter. For any circle, the following ratio is true:

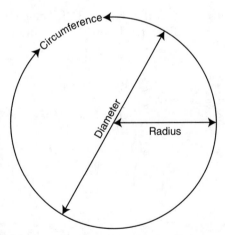

FIGURE 7.3 The diameter, radius, and circumference of a circle.

$$\frac{\text{Circumference}}{\text{Diameter}} = 3.1415926\ldots = \pi \quad \text{We call this ratio } \mathbf{pi}.$$

It follows that

$$\text{Circumference} = \pi \times \text{Diameter} \quad \pi \approx 3.1416 \text{ or } 3.14$$
$$C = \pi \times d$$

Example 7-6 Find the circumference of a tire having a diameter of 29.0 in.

$$\text{Circumference} = \pi \times \text{Diameter}$$
$$C = 3.1416 \times 29.0 \text{ in.} = 91.1 \text{ in.}$$

The **radius** of a circle is the distance from the center of the circle to the edge. The diameter is twice as long as the radius.

$$\text{Diameter} = 2 \times \text{Radius}$$

Example 7-7 Find the perimeter of the figure below.

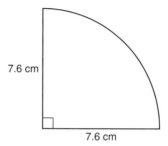

We need to add the two straight sides and the curved portion of the figure to find the perimeter.

The figure appears to be one quarter of a circle, so the curved portion is one quarter of the circumference of a complete circle with a radius of 7.6 cm.

Since the radius of this partial circle is 7.6 cm, the diameter of the complete circle would be 2 × 7.6 cm or 15.2 cm.

$$\text{Curved portion} = \frac{1}{4} \times \text{Circumference} = \frac{1}{4} \times \pi \times 15.2 \text{ cm} = 11.938 \text{ or } 11.94 \text{ cm}$$

$$\text{Perimeter} = 11.94 \text{ cm} + 7.6 \text{ cm} + 7.6 \text{ cm} = 27.14 \text{ cm}$$

Example 7-8 A truck has tires that have a 31 in. diameter. If the wheels turn 650 times per minute, how far will the truck travel each minute?

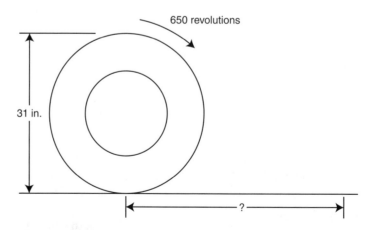

Every time the wheel turns, it will move forward the length of the tire's circumference. Find the circumference of each tire.

$$C = 3.1416 \times 31.0 \text{ in.} = 97.39 \text{ in.}$$

The tire will turn 650 times per minute.

$$650 \times 97.39 \text{ in.} = 63{,}303.5 \text{ in.}$$

Converting this to feet, we have

$$\frac{63{,}303.5 \text{ in.}}{1} \times \frac{1 \text{ ft}}{12 \text{ in.}} = \frac{63{,}303.5 \text{ ft}}{12} = 5275.3 \text{ ft, which is just under a mile.}$$

PYTHAGOREAN THEOREM

Triangles that have one right (90°) angle are called **right triangles.** Triangles of this type are very common and have very useful properties. One property of right triangles that has been known for several thousand years is the Pythagorean theorem.

Consider any right triangle. The two sides that form the right angle are called **legs.** The third side is the longest, and is not adjacent to (does not contact) the right angle. This side is called the **hypotenuse.** This is shown in Figure 7.4.

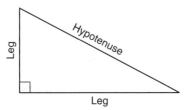

FIGURE 7.4 A right triangle has a hypotenuse, which is the longest side, and two shorter legs.

The three sides of a right triangle are always related by the relationship:

$$(\text{leg})^2 + (\text{leg})^2 = (\text{hypotenuse})^2$$

If we label the legs of the triangle as a and b, and refer to the hypotenuse as c, as in Figure 7.5, the relationship takes on the following form:

$$a^2 + b^2 = c^2$$

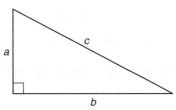

FIGURE 7.5 The right triangle with the legs labeled a and b, and the hypotenuse labeled c.

Note that while it is not important which leg is labeled a and which is labeled b, it is critical that the hypotenuse be labeled c. Remember that c is always the longest side. It's also important to know that this relationship is only true for right triangles.

Consider the triangle in Figure 7.6. The legs have lengths 3 in. and 4 in. The length of the hypotenuse is 5 in.

Example 7-12 Find the area of a rectangle that is 3 ft by 5 ft.

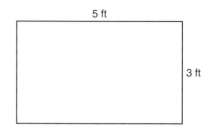

Divide this area up into 1-ft by 1-ft squares.

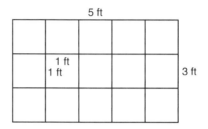

We can see that there are 3 rows of 5, or 15 of these 1-ft by 1-ft squares. We can then say that the area of this rectangle is 15 square feet or 15 ft².
In general, the area of a rectangle is given by the product of the width and length.

Area of a rectangle = Length × Width   or $A_R = l \times w$

Example 7-13 When air filters and oil filters are designed, special care is taken to make sure they have enough area to allow proper flow. A rectangular air cleaner element is 7.5 in. wide by 13.0 in. long. What is the area of the filter?

Area of a rectangle = Length × Width = 13.0 in. × 7.5 in. = 97.5 in.²

Example 7-14 Find the area of a square shop space that is 38 ft by 38 ft.

A square is a special case of a rectangle, so we'll use the same formula.

Area of a rectangle = 38 ft × 38 ft = 1,444 ft²

To find the area of a triangle, we always need a width, which we call the **base** and a length from bottom to top, which we call an altitude or **height**. Any side can be the base, as long as the height is perpendicular (at a right angle) to the base.

Notice that this triangle fits inside a rectangle with the same dimensions. The area of the triangle is half the area of the rectangle, so we can conclude the following about triangles:

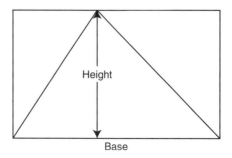

Area of a triangle = $\frac{1}{2}$ Base × Height   or $A_T = \frac{1}{2} b \times h$

Example 7-15 Find the area of the triangle shown below.

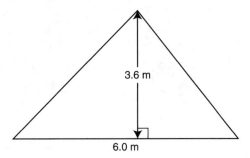

The triangle is 6.0 m wide and 3.6 m tall.
In this case, we have

$$\text{Area of triangle} = \frac{1}{2} \text{ Base} \times \text{Height} = \frac{1}{2} \times 6.0 \text{ m} \times 3.6 \text{ m} = 10.8 \text{ m}^2$$

Example 7-16 Find the area of the following triangle.

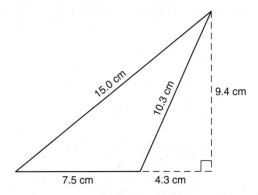

We need to identify the base and height. Notice that the length 4.3 cm is *not* part of the base of this triangle. It is a useless measurement. The base is 7.5 cm.

The height is shown by the 9.4 cm dimension because it forms a right angle with the dashed line extended from the base. Note that the 15.0 cm and 10.3 cm sides do *not* give the height of the triangle. The height then is 9.4 cm.

$$\text{Area of triangle} = \frac{1}{2} \times 7.5 \text{ cm} \times 9.4 \text{ cm} = 35.25 \text{ cm}^2$$

Example 7-17 Find the area of the triangle below.

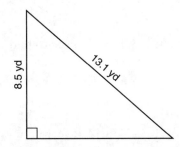

The height of this triangle is 8.5 yards. Unfortunately, the base will be given by the missing side, which is not given. Use the Pythagorean theorem to find the missing side.

Apply the Pythagorean theorem by letting $a = 8.5$, $c = 13.1$. Solve for b.

$$a^2 + b^2 = c^2$$
$$8.5^2 + b^2 = 13.1^2$$
$$72.25 + b^2 = 171.61$$
$$\begin{array}{r} 72.25 + b^2 = 171.61 \\ -72.25 -72.25 \\ \hline b^2 = 99.36 \\ \sqrt{b^2} = \sqrt{99.36} \end{array}$$
$$b \approx 9.97, \text{ we'll round to } 10.0 \text{ yd}$$

Since that missing side is also the base of the triangle, we can now compute the area.

$$\text{Area of triangle} = \frac{1}{2} \times 10.0 \text{ yd} \times 8.5 \text{ yd} = 42.5 \text{ yd}^2$$

The constant π (pi) is not only used to find the circumference of a circle, its also used to find the area. The radius of a circle must be known to find the area.

$$\text{Area of a circle} = \pi \times \text{radius}^2 \quad \text{or} \quad A_C = \pi \times r^2$$

Example 7-18 Find the area of the circle shown below.

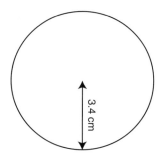

The radius for this circle is shown to be 3.4 cm, so we can find the area.

$$\text{Area of circle} = \pi \times \text{radius}^2 = 3.14 \times (3.4 \text{ cm})^2 \approx 36.3 \text{ cm}^2$$

Example 7-19 Find the area of a circle with diameter 3.4 in.

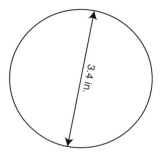

The diameter of the circle is 3.4 in. We need to find the radius.

$$\text{radius} = \frac{1}{2} \times \text{diameter} = \frac{1}{2} \times 3.4 \text{ in.} = 1.7 \text{ in.}$$

Then we can find the area.

$$\text{Area of circle} = 3.14 \times (1.7 \text{ in.})^2 \approx 9.1 \text{ in.}^2$$

It's worth noting that there is a formula to find the area of every common geometric shape. Most other common shapes, then, can be broken into triangles, rectangles and circles, and the total area found.

Example 7-20 Find the area of the figure below.

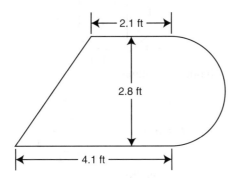

This area can be found by dividing the region into a triangle, a rectangle, and a semicircle as shown.

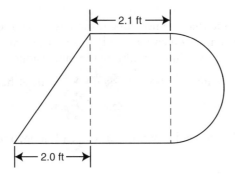

The base of the triangle is 4.1 ft − 2.1 ft = 2.0 ft. The height is 2.8 ft.

$$\text{Area of triangle} = \frac{1}{2} \times 2.0 \text{ ft} \times 2.8 \text{ ft} = 2.8 \text{ ft}^2$$

The rectangle is 2.1 ft wide and 2.8 ft tall.

$$\text{Area of a rectangle} = 2.8 \text{ ft} \times 2.1 \text{ ft} = 5.88 \text{ ft}^2$$

The rounded portion is half of a circle. The diameter of the circle is 2.8 ft, so the radius must be 1.4 ft. Find the area of a complete circle this size.

$$\text{Area of circle} = 3.14 \times (1.4 \text{ ft})^2 \approx 6.15 \text{ ft}^2$$

We only have half of a complete circle, so

$$\text{Area of semi-circle} = \frac{1}{2} \times 6.15 \text{ ft}^2 \approx 3.08 \text{ ft}^2$$

Then the total area of the figure can be found by adding these three regions.

$$\text{Area} = 2.8 \text{ ft}^2 + 5.88 \text{ ft}^2 + 3.08 \text{ ft}^2 = 11.76 \text{ ft}^2$$

VOLUME

You're probably familiar with such terms as "cubic feet per minute" when speaking of a carburetor, or "cubic inches" when referring to engine size, but what exactly is a cubic foot? What is a cubic inch? To understand this, we need some way to measure space. **Volume** is a measure of how much space is occupied by some object. For example, one cubic inch is a cube that is exactly 1 inch × 1 inch × 1 inch as shown in Figure 7.7.

We know the height is 35 in. and the radius is 11.5 in. Then

$$\text{Volume} = 3.14 \times (11.5 \text{ in.})^2 \times 35 \text{ in.} = 14{,}534.28 \text{ in.}^3$$

We'll convert this to gallons and liters using conversion facts from Table 6.3.

$$1 \text{ gal} \approx 231 \text{ in.}^3 \qquad 1 \text{ in.}^3 \approx 16.39 \text{ cm}^3 \qquad 1 \text{ liter} = 1000 \text{ cm}^3$$

Convert 14,534.28 in.³ to gallons:

$$\frac{14{,}534.28 \text{ in.}^3}{1} \times \frac{1 \text{ gal}}{231 \text{ in.}^3} = \frac{14{,}534.28 \text{ gal}}{231} = 62.9 \text{ gal}$$

Convert 14,534.28 in.³ to liters.

$$\frac{14{,}534.28 \text{ in.}^3}{1} \times \frac{16.39 \text{ cm}^3}{1 \text{ in.}^3} \times \frac{1 \text{ L}}{1000 \text{ cm}^3} = \frac{238{,}216.85 \text{ L}}{1000} = 238.2 \text{ L}$$

A **sphere** refers to the shape of a ball. Some vacuum canisters are spherical in shape. The volume of a sphere can be seen in Figure 7.10.

$$V = \frac{4\pi r^3}{3}$$

FIGURE 7.10 The volume of a sphere is given by $4/3 \times \pi \times \text{radius}^3$, or $V_S = 4/3\, \pi r^3$.

$$\text{Volume of a sphere} = \frac{4}{3} \times \pi \times (\text{radius})^3$$

Example 7-25 A spherical vacuum canister is 12 cm in diameter. Find the volume in cm³.

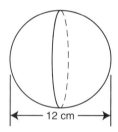

Since the diameter is 12 cm, the radius must be 6 cm.

$$\text{Volume} = \frac{4}{3} \times \pi \times (\text{radius})^3 = \frac{4}{3} \times 3.14 \times (6 \text{ cm})^3 = 904.32 \text{ cm}^3$$

ANGLE MEASURE AND APPLICATIONS

When adjusting the alignment on a vehicle, the technician must make sure that the wheels all form the proper angles with the vehicle and with each other. The standard unit of angle measure is the degree. One **degree** (1°) is 1/360 of a complete revolution.

GEOMETRY   **127**

A protractor is a tool used to measure angles as small as a degree.

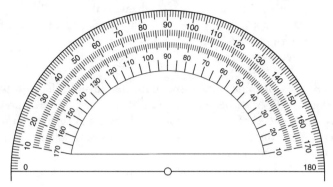

Example 7-26   Measure the angle below.

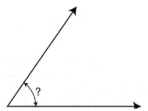

The "corner" of the angle is called the vertex. Place the protractor on the angle as shown. Be sure to read the correct scale, beginning at zero.

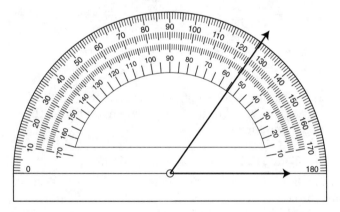

Reading the scale that increases from the right, we see that this is a 54° angle.

Example 7-27   The camber angle for both front wheels on a vehicle is shown below. Read the camber angle for both wheels. A partial protractor is shown above each wheel.

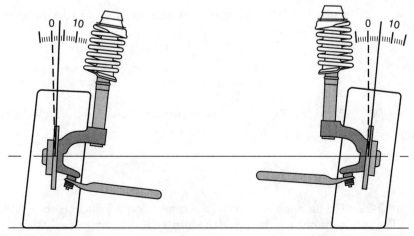

Negative camber angle   Positive camber angle

The left wheel forms an angle of 3° with vertical. Since it leans inward, we say the camber has a negative value. Then the left wheel camber is −3°.

The right wheel forms an angle of 5° with vertical. It leans outward, so it has a positive value. The right wheel camber is +5°.

We'll work more with signed numbers in the next chapter.

Example 7-28 The opening and closing of the intake and exhaust valves is controlled by the camshaft. Both of those events, and some others, are measured in relation to how far past top dead center (TDC) the crankshaft has turned. Suppose the intake valve is open for the duration of 208 crankshaft degrees. What fraction and percent of a complete revolution is this?

Since there are 360° in a complete revolution,

$$\frac{208°}{360°} = \frac{26}{45} = 0.578 = 57.8\% \text{ of a revolution}$$

The intake valve is open for slightly more than half of a complete crankshaft revolution.

Even though a degree seems small, many angles (such as those in wheel alignment) must be measured even more precisely. An angular **minute** (1′) is 1/60 of a degree. A minute is far too small of an angle to be measured using a protractor.

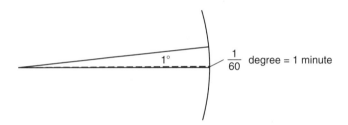

Using these two units, very precise angle measurements can be made. Often, these very precise angles are given as decimal degrees or degrees and minutes. It is useful to be able to convert between the two forms.

Example 7-29 Convert 12.5° to degrees and minutes.

The decimal part, 0.5°, is 1/2 of a degree. Since 1° = 60′ then 1/2° = 30′. Or, using the conversion process,

$$0.5° \times \frac{60'}{1°} = 30'$$

In either case, 12.5° = 12°30′.

Example 7-30 Convert 3.709° to degrees and minutes. Round to the nearest minute.

$$0.709° \times \frac{60'}{1°} = 42.54'$$

Then 3.709° = 3° 42.54′. Rounded to the nearest minute, we have 3°43′.

Example 7-31 Convert 24°23′ to decimal degrees, rounded to the nearest thousandth.

Since each minute is 1/60 of a degree, we have

$$24°23' = 24\frac{23}{60}° = 23.383°$$

Example 7-32 On an alignment rack, the camber on a vehicle is measured to be 1° 39′. The specifications for the vehicle require that the camber be between 0.850° and 1.750°. Is the camber within the acceptable range?

GEOMETRY

We'll convert our measurement to decimal degrees.

$$1°39' = 1\frac{39}{60}° = 1.650°$$

Yes, this is within the acceptable range.
Like any other quantity, angle measurements can be added and subtracted.

Example 7-33 Add $28°38' + 9°40'$.

$$\begin{array}{r} 28°38' \\ +9°40' \\ \hline 37°78' \end{array}$$

Note that our answer, $37°78'$, gives us a number of minutes greater than 60. Since there are only 60 minutes in a degree, we subtract.

$$78' - 60' = 18'$$

and carry the extra degree. So

$$37°78' = 38°18'$$

Example 7-34 Subtract $35°16' - 22°35'$.

$$\begin{array}{r} 35°16' \\ -22°35' \\ \hline \end{array}$$

Since we can't subtract $35'$ from $16'$, we need to borrow one degree in the minuend, the top number. Since one degree equals 60 minutes ($1° = 60'$), when we borrow one degree we add 60 minutes to the original 16 minutes to get 76 minutes. Then

$$\begin{array}{r} 34°76' \\ -22°35' \\ \hline 12°41' \end{array}$$

Example 7-35 The current caster angle on a vehicle is $1.125°$. An adjustment is made that reduces the caster by $15'$. What is the new caster angle, in degrees and minutes?

Convert $1.125°$ to degrees and minutes.

$$0.125° \times \frac{60'}{1°} = 7.5'$$

Then the original caster is $1°7.5'$. Now, subtract the adjustment.

$$\begin{array}{r} 1°7.5' \\ -0°15' \\ \hline \end{array}$$

Again, we need to borrow one degree in the minuend.

$$\begin{array}{r} 0°67.5' \\ -0°15' \\ \hline 0°52.5' \end{array}$$

When very high precision is needed in measuring angles, a smaller unit than the minute is used. An angular **second** ($1''$) is 1/60 of a minute. Angles using this high precision are written in degree-minute-second, or DMS notation.

To convert a DMS angle to decimal degrees, we use a similar process to the one shown in Example 7-31. For example,

$$17°\,26'\,18'' = 17°26\frac{18}{60}' = 17°26.3' = 17\frac{26.3}{60}° = 17.438°$$

To convert a decimal degree angle to DMS, we turn the decimal portion of a degree into minutes as we did in Example 7-30. We repeat the idea to change decimal portion of a minute into seconds. For example,

$$9.372° = 9°(0.372 \times 60)' = 9°22.32' = 9°22'(0.32 \times 60)'' = 9°22'19.2''$$

CHAPTER 7 *Practice Problems*

Measurement

Write the measurement at each location below. Write answers with proper units.

●1. A:_____ B:_____ C:_____ D:_____ E:_____

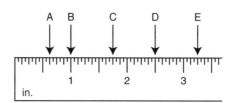

●2. A:_____ B:_____ C:_____ D:_____ E:_____

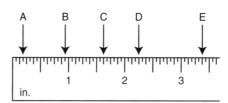

●3. A:_____ B:_____ C:_____ D:_____ E:_____

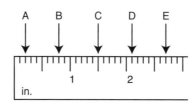

●4. A:_____ B:_____ C:_____ D:_____ E:_____ F:_____

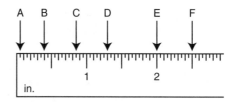

GEOMETRY 131

● 5. A:_____ B:_____ C:_____ D:_____ E:_____ F:_____

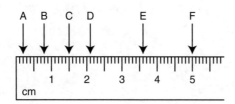

Use a ruler to measure the length of the lines below to the nearest 1/16 in.

● 6. _____

● 7. _____

● 8. _____

● 9. _____

● 10. _____

● 11. _____

● 12. _____

● 13. _____

Use a metric ruler to measure the length of each line below. Report your values using centimeters and millimeters.

● 14. _____

● 15. _____

● 16. _____

● 17. _____

● 18. _____

● 19. _____

● 20. _____

Perimeter

Find the perimeter of each figure below. Round to one decimal place if necessary.

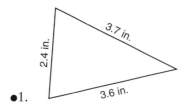

● 1.

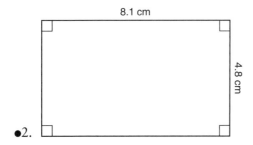

● 2.

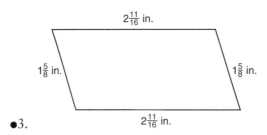

● 3.

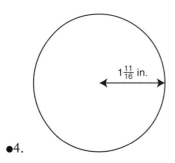

● 4.

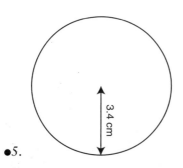

● 5.

GEOMETRY 133

•6.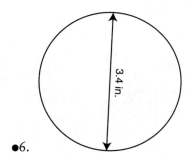

Find the perimeter, in centimeters, of each of the figures below. Round to the nearest tenth of a centimeter. Use 3.14 as the approximation for π.

••7.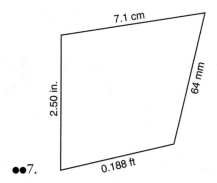

••8.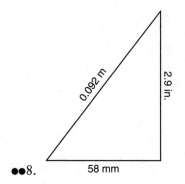

••9.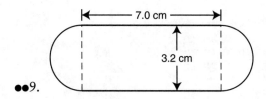

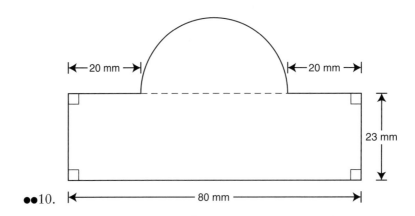

●●10.

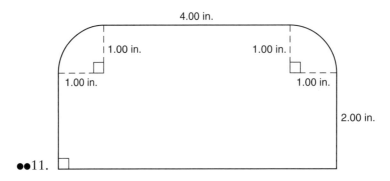

●●11.

After finding the perimeter in inches, convert to centimeters.

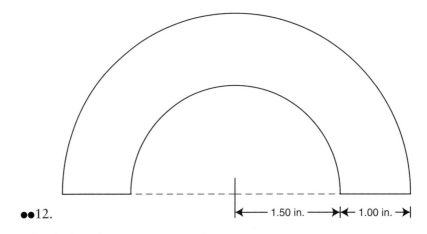

●●12.

After finding the perimeter in inches, convert to centimeters.

GEOMETRY

Use a ruler to find the perimeter, in centimeters and inches, of each of the figures below.

●●13.
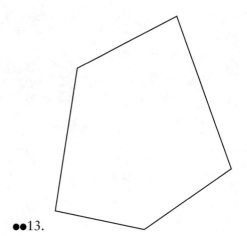

●●14.
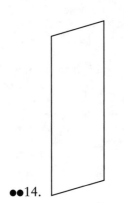

●●15.
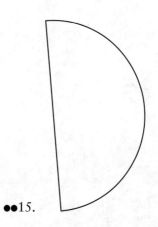

●●●16.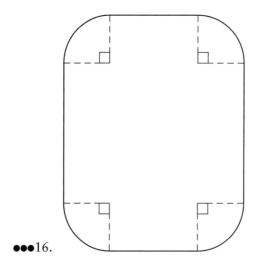

●17. A vehicle has tires that have a 27.0 in. diameter. Approximately how far will this car travel, in feet, if the wheels turn 150 revolutions?

●18. How far will the vehicle in Problem 17 travel if the tires are replaced with a set having a diameter of 28.5 in.?

Pythagorean Theorem

Find the length of the missing side for each triangle shown below. Round to the nearest thousandth of a unit.

●●1.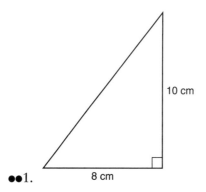

●●2.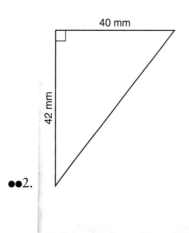

GEOMETRY

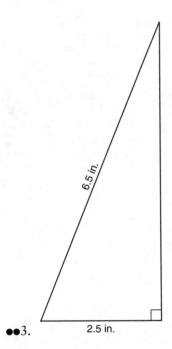
●●3.   2.5 in.

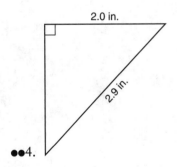
●●4.

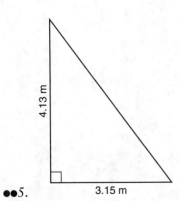
●●5.   3.15 m

●●6.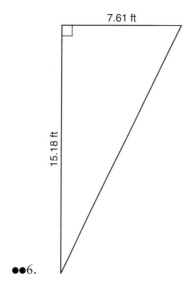

●●7.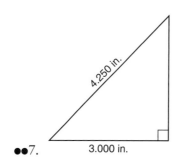

●●8.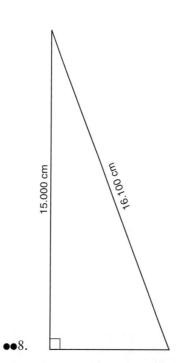

GEOMETRY   **139**

Find the perimeter of each triangle shown below. Round to the nearest hundredth of a unit.

●●9.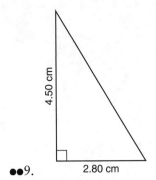

●●10.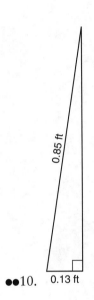

●●11.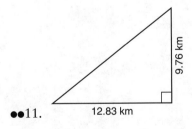

●●12.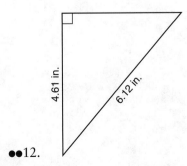

CHAPTER 7

Area

Find the area of each figure below. Round to the nearest hundredth. For the value of π, use 3.14.

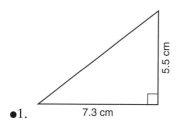

●1.

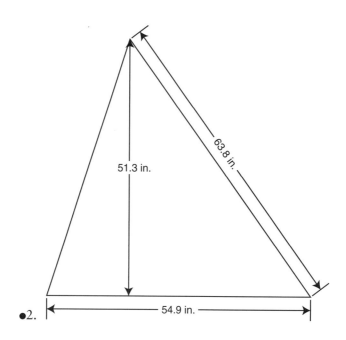

●2.

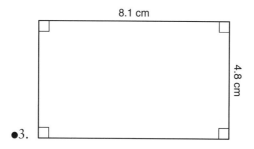

●3.

GEOMETRY

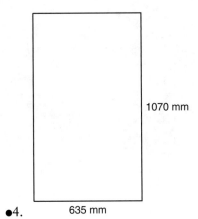
● 4.

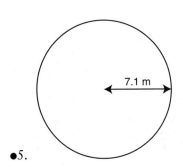
● 5.

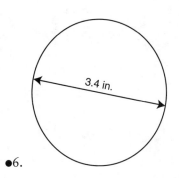
● 6.

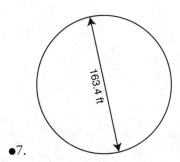
● 7.

●8.
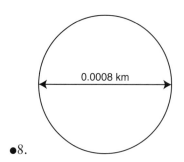

●9.
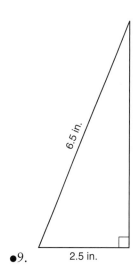

●10.
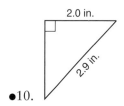

●●11.
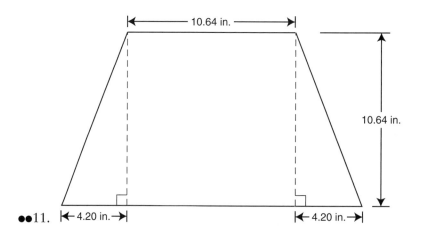

GEOMETRY

••12.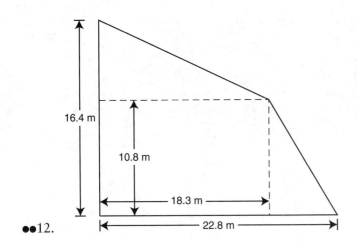

••13.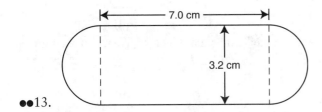

••14.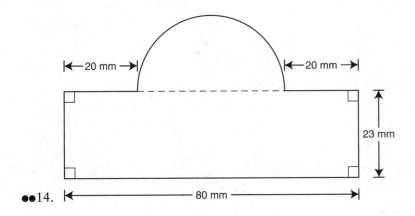

••15.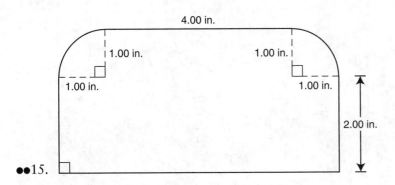

CHAPTER 7

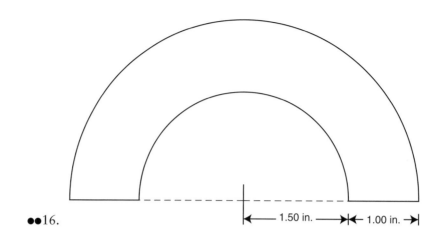

●●16.

Use a ruler to find the area, in square centimeters and square inches, of each of the figures below.

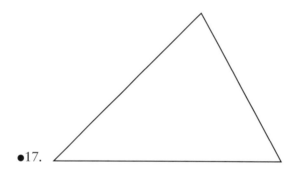

●17.

●18.

GEOMETRY   **145**

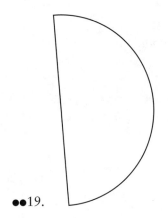

●●19.

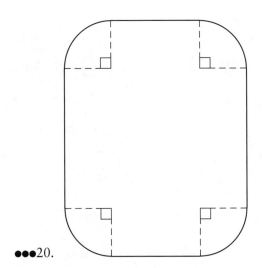

●●●20.

●21. A rectangular air filter is 24 cm by 42 cm. Find the area of this filter.

●22. A square air filter is 4.75 in. along each edge. What is the area of this filter?

●23. Find the area of the top of a piston with a diameter of 4.200 in.

●24. A piston with diameter 92 mm is used in a turbocharged engine. Find the area of the top of that piston.

●●25. Which is larger: An air filter that is rectangular, 14 1/4 in. by 5 1/2 in., or a filter that is round, 11 1/2 in. in diameter?

●●26. Which has the larger cross-sectional area: A square air duct that is 6 3/4 in. on each side, or a round air duct that is 7 in. in diameter?

CHAPTER 7

Volume

Find the volume of each figure below. Round to the nearest hundredth.

1.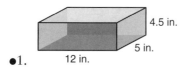

2.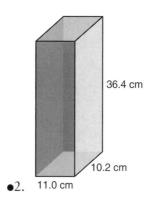

3.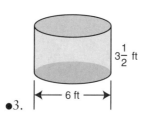

4.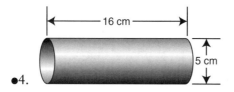

5.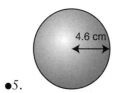

6.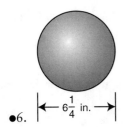

For the following problems, use Table 6.3 for the necessary conversion factors.

7. A propane fuel tank is cylindrical with hemispherical ends, as shown in the figure below. What is the volume of this tank, in cubic inches? What is the volume in gallons?

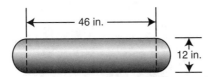

8. A nitrous-oxide tank is cylindrical in shape, often with a hemispherical top, as shown below. Find the volume, in cubic inches and in liters.

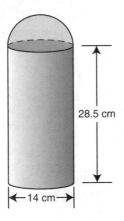

9. An in-bed fuel tank for a pickup truck is 24 in. long, 24 in. tall, and 50 in. long. Find the volume in cubic inches and in gallons.

10. A steel oil pan is shaped like a box, having a depth of 18.5 cm, a width of 28.0 cm, and a length of 58.3 cm. Find the volume of this oil pan, in cubic centimeters and liters.

11. The bed of a pickup truck is approximately 96 in. long, 66 in. wide, and 19 in. deep. Calculate the volume of the bed in gallons. Ignore the wheel-wells.

12. The trunk of a large automobile is 1.7 m wide, 1.2 m long, and 0.45 m deep. Find the volume of the trunk in cubic feet.

●●13. The oil filter on a certain V-8 engine is cylindrical. It is 5.5 in. long and 4 in. in diameter. What is the volume in cubic inches and in quarts?

●●14. A hydraulic cylinder is cylindrical. It is 36 in. long and 5 in. in diameter. Find the volume in cubic inches and in quarts.

●●15. How many quarts of coolant are held in a hose that has an inside diameter of 1 1/2 in., and is 28 in. long? Assume the hose is straight.

●●16. A steel fuel tank is cylindrical. It is 70 cm long and has a diameter of 45 cm. Find the volume of the tank in liters and gallons.

●●17. A spherical vacuum canister is 7 in. in diameter. Find the volume of the canister in quarts.

●●18. Suppose a propane tank is spherical in shape, and has a 1-ft diameter. How many gallons will this tank hold?

●●●19. A "hemi" head engine has combustion chambers that are roughly hemispherical in shape. That is, the combustion chamber resembles a half-sphere. What is the volume of a truly hemispherical combustion chamber if the diameter of the chamber is 4.00 in.? Find the volume in cubic centimeters.

●●●20. What is the volume, in cubic centimeters, of a truly hemispherical combustion chamber if the diameter is 3.5 in.? (See problem 19.)

Angle Measure and Applications

Use a protractor to measure each of the angles below to the nearest degree.

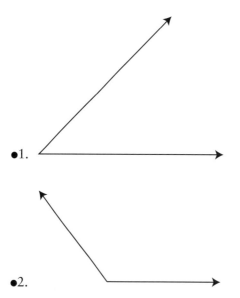

●1.

●2.

GEOMETRY 149

3.

4.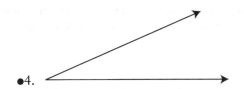

Use a protractor to create angles given below.

5. 137°

6. 48°

7. 90°

8. 180°

9. 21°

10. 162°

11. A certain camshaft has an advertised exhaust duration of 270°. What fraction of a complete crankshaft revolution is this?

CHAPTER 8

Signed Numbers

SIGNED NUMBERS

As we saw in Chapter 7, a technician must ensure that the caster and camber angles are within specifications when aligning the wheels on a vehicle. As you learn more about wheel alignment, however, you'll see that caster and camber angles are always given as being positive or negative. For example, the camber or caster angles could be given as +2°, −1.5°, 0°, or −3.5°. These types of numbers are called **signed numbers.** Signed numbers are often used to represent direction.

The sign on a camber angle determines which direction the wheel tilts, compared to vertical. Vehicles typically have positive (+) camber angles, which means the wheel tilts slightly outward from vertical. Vehicles with excessively worn components may develop negative camber (−) so that the wheel tilts slightly inward from vertical. Positive and negative camber angles for a wheel on the right side of a car are shown in Figure 8.1.

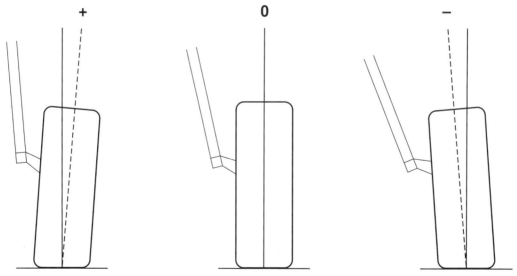

FIGURE 8.1 Signed numbers are used to show the direction of tilt in camber measurements. A camber measurement of +2° would mean the right side wheel tilts outward 2° from vertical. A camber measurement of −2° means the wheel would tilt inward 2° from vertical. A camber angle of 0° means the wheel is truly vertical.

On the number line, signed numbers such as these indicate position. A few signed numbers are shown in Figure 8.2. Numbers that are greater than zero are **positive,** and numbers that are less than zero are **negative.** Note that a number that does not have a sign in front is considered positive.

SIGNED NUMBERS

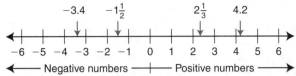

FIGURE 8.2 The number line shows signed numbers.

Numbers of greater value on the number line always appear to the right of numbers with smaller value. The number 4.2 is greater than 0 because it appears to the right, and 0 is greater than −3.4 for the same reason.

Example 8-1 For each pair of numbers below, determine which number is greater.

a. 5 or 3   b. 6 or −2   c. −8 or 0   d. −4 or −1

a. 5 or 3   5 is greater.
b. 6 or −2   6 is greater. It is to the right of −2.
c. −8 or 0   0 is greater than −8.
d. −4 or −1   −1 is greater than −4.

Example 8-2 For each pair of numbers below, determine which number is less.

a. 0 or 2   b. −3 or 5   c. −10 or −15   d. 0 or −2

a. 0 or 2   0 is less than 2.
b. −3 or 5   −3 is less than 5. It is to the left on the number line.
c. −10 or −15   −15 is less than −10.
d. 0 or −2   −2 is less than 0.

Example 8-3 The front left wheel on a vehicle has a camber angle of +3°. The front right wheel has a camber angle of +1.5°. Do these wheels tilt inward or outward? Which wheel tilts more?

Since the camber angles are positive for both wheels, they both tilt outward.

The camber angle of +3° is further from vertical than +1.5°, so the front left wheel tilts more.

Example 8-4 The front left wheel on a truck has a camber angle of −1°. The front right wheel has a camber angle of −2.5°. Which way does each wheel tilt? Which wheel tilts more?

The camber angles are negative, so the wheels tilt inward.

The front left wheel, with its −2.5° camber angle, tilts more because it is farther from 0°.

Example 8-5 On a certain January Monday in Minnesota, the high temperature reaches −8° Fahrenheit. On Tuesday, the high temperature reaches −14°. Which day is warmer?

The temperature of −8 is greater than −14, so it's warmer on Monday.

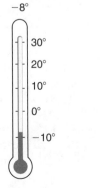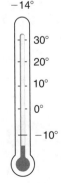

154 CHAPTER 8

ADDITION AND SUBTRACTION OF SIGNED NUMBERS

Suppose the camber angle on a vehicle is +3°. As suspension components wear, the camber angle might decrease by 2°. What is the new camber angle? This is a case where we'll need to be able to work with two signed numbers by adding or subtracting.

When adding two signed numbers, we combine their value. As in football, consider positive numbers to be moves in one direction (to the right), while negative numbers are moves in the opposite direction (to the left). Let's look at some examples.

Example 8-6 Determine the sum for the following.

a. $5 + 3$ b. $-4 + 10$ c. $-2 + (-5)$ d. $6 + (-5)$ e. $-9 + 7$

a. $5 + 3 =$

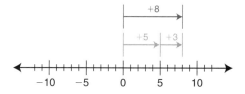

Begin at 5. Adding 3 moves us to the right.

$$5 + 3 = 8$$

b. $-4 + 10 =$

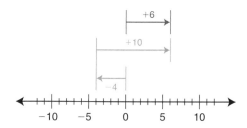

Begin at -4. Adding 10 moves us to the right.

$$-4 + 10 = 6$$

c. $-2 + (-5) =$

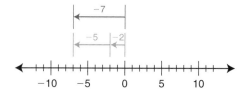

Begin at -2. Move 5 units in the negative direction.

$$-2 + (-5) = -7$$

d. $6 + (-5) =$

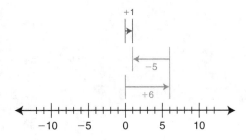

Begin at 6. Move 5 units in the negative direction.

$$6 + (-5) = 1$$

e. $-9 + 7 =$

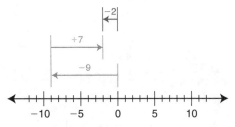

Begin at -9. Move 7 units in the positive direction.

$$-9 + 7 = -2$$

Example 8-7 The camber angle on a vehicle is $+2.5°$. It is adjusted and increased by $1.25°$. What is the new angle?

An increase usually means addition.

$$2.5° + 1.25° = 3.75°$$

Example 8-8 The temperature on a winter morning is $-13°$. The day gradually warms, and the temperature increases by $25°$. What is the new temperature?

We'll start at -13 and add 25.

$$-13 + 25 = 12$$

The new temperature is $12°$ (above zero).

When subtracting two signed numbers, we're looking for a difference. Just as subtraction is the opposite of addition, subtracting a signed number reverses the direction we would move on the number line. Let's consider some examples.

Example 8-9 Determine the difference for the following.

a. $6 - 2$ b. $7 - 9$ c. $-2 - 3$

a. $6 - 2 =$

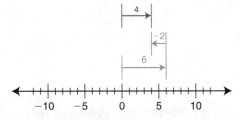

Begin at 6. Move 2 units in the negative direction.

$$6 - 2 = 4$$

b. 7 − 9 =

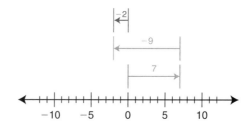

Begin at 7. Move 9 units in the negative direction.

$$7 - 9 = -2$$

c. −2 − 3 =

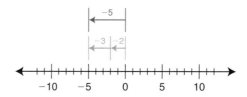

Begin at −2. Move 3 units in the negative direction.

$$-2 - 3 = -5$$

This next example demonstrates a general principle you may have heard before: When dealing with negative numbers, two negative signs can be treated as an addition sign.

Example 8-10 Determine the difference for the following.

a. −7 − (−3)   b. 5 − (−1)

a. −7 − (−3) =

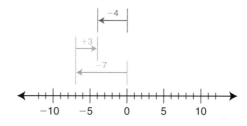

Begin at −7. The negative sign means we'll reverse the sign on the −3. The problem becomes −7 + 3.

$$-7 + 3 = -4$$

b. 5 − (−1) =

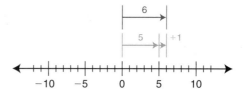

Begin at 5. Again, the negative sign reverses the effect of the −1. This problem becomes 5 + 1.

$$5 + 1 = 6$$

SIGNED NUMBERS

Example 8-11 The camber angle on a vehicle is adjusted from $-2.75°$ to $3.25°$. How much was it changed?

We want to find the change from $-2.75°$ to $3.25°$.

$$3.25° - (-2.75°) = 3.25° + 2.75° = 6°$$

The angle was increased $6°$.

Example 8-12 It's winter again. One afternoon, the temperature is $-4°$. That night, the temperature drops another $13°$. What is the new temperature?

The problem reduces to $-4° - 13°$.

$$-4° - 13° = -17°$$

MULTIPLICATION AND DIVISION OF SIGNED NUMBERS

To complete our work with signed numbers, we'll consider multiplication and division. When multiplying two positive numbers, the product is positive. When multiplying a positive number by a negative number, the product will be negative. An example will help us see why.

Example 8-13 Find the product: $3 \times (-2)$

We're really just moving two places in the negative direction, but we're making that move three times.

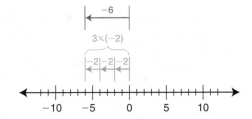

As you can see on the number line, our answer will be negative.

$$3 \times (-2) = -6$$

When multiplying two negative numbers, the negative signs reverse each other, and we're left with a positive product. These guidelines are all true for division as well, and so we have a simple rule for multiplying and dividing signed numbers.

> When multiplying or dividing two signed numbers:
> - If the signs of both numbers are the same (both positive or both negative) the result is positive
> - If the signs of both numbers are different (one negative, one positive) the result is negative.

Example 8-14 Evaluate the following.

a. 5×6 b. -10×4 c. $9 \times (-3)$ d. $-7 \times (-11)$
e. $72 \div 6$ f. $15 \div (-5)$ g. $-56 \div (-8)$

a. $5 \times 6 =$

The factors are both positive, so the product will be positive.

$$5 \times 6 = 30$$

b. -10×4

The factors have opposite signs, so the product will be negative.
$$-10 \times 4 = -40$$

c. $9 \times (-3) =$

The factors have opposite signs, so the product will be negative.
$$9 \times (-3) = -27$$

d. $-7 \times (-11) =$

The factors are both negative. The product must be positive.
$$-7 \times (-11) = 77$$

e. $72 \div 6 =$

The signs are the same. The quotient must be positive.
$$72 \div 6 = 12$$

f. $15 \div (-5) =$

The signs are not the same. The quotient must be negative.
$$15 \div (-5) = -3$$

g. $-56 \div (-8) =$

Since both signs are negative, the quotient must be positive.
$$-56 \div (-8) = 7$$

Example 8-15 Each turn of an adjusting nut changes the camber angle by $-0.5°$. If the nut is turned 3 times, how much will the angle change?

The problem becomes $3 \times (-0.5°)$.
$$3 \times (-0.5°) = -1.5°$$

CHAPTER 8 *Practice Problems*

Signed Numbers

1. Which number is greater: -3 or 6?

2. Which number is greater: 9 or -9?

3. Which number is less: 0 or -5?

4. Which number is less: -2 or 0?

5. Which number is greater: -7.6 or -9.2?

6. Which number is greater: -138 or -183?

7. Which number is less: -21.6 or -60?

SIGNED NUMBERS 159

- 8. Which number is less: 215 or −220?

- 9. Which number is greater: 16 or 65?

- 10. Which number is greater: 0.7 or −0.9?

- 11. Which number is less: $-1/2$ or $-1/4$?

- 12. Which number is less: −5 2/3 or −5 1/8?

- 13. Which temperature is warmer: −6°F or −13°F?

- 14. Which temperature is warmer: 5°F or −2°F?

- 15. Which temperature is colder: −10°F or −8°F?

- 16. Which temperature is colder: 37°F or −38°F?

Addition and Subtraction of Signed Numbers

Evaluate each of the following.

- 1. 5 + 8 =
- 2. 16 + 3 =
- 3. −9 + 15 =
- 4. −20 + 13 =
- 5. 7 − 18 =
- 6. 13 − 60 =
- 7. −9 − 9 =
- 8. −20 − 51 =
- 9. 27 + (−15) =
- 10. 55 + (−30) =
- 11. 17 + (−22) =
- 12. 80 + (−100) =
- 13. −16 − (−15) =
- 14. 0 − (−40) =
- 15. 9 − (−10) =
- 16. 15 − (−23) =
- 17. −28 + (−35) =
- 18. −26 + (−40) =
- 19. −6.7 + 9.3 =
- 20. −16.2 + 31.0 =
- 21. −8.1 + (−29.7) =
- 22. −22.2 + (−3.7) =

●23. $-6.7 - (-1.1) =$

●24. $-13.5 - (-13.5) =$

●25. $-18.6 + (-39.3) =$

●26. $-213 - (-0) =$

●●27. $-3\frac{1}{2} + 6 =$

●●28. $-8 + (-2\frac{1}{4}) =$

●●29. $13\frac{1}{2} - (-38\frac{1}{3}) =$

●●30. $-3\frac{1}{8} - (-4\frac{1}{6}) =$

●●31. $-18 + 31 - (-2.8) - 7.0 =$

●●32. $250 + (-325) - (-50) + 80 =$

●●33. $0 - 8.69 - 13.8 - (-6.8) + (-21) =$

●●34. $3.14 + (-3.14) - (-3.14) + 3.14 =$

●35. The air temperature one morning is −2°C. It later warms to 17°C. How much did the temperature increase?

●36. The coldest temperature one day is recorded to be −15°F. The highest temperature that day was 11°F. How much did the temperature increase between those times?

●37. One morning, the temperature outside is found to be −11°. If it warms up 25° that day, how warm will it get?

●38. The outside temperature one afternoon is −3°F. Inside a vehicle, it was 70° warmer. What was the temperature in the vehicle?

The caster angle on a vehicle is the angle formed from the forward or backward tilt of the steering axis (spindles) to a vertical line. Positive, negative, and zero caster are shown in Figure 8.3.

●39. The initial caster angle on a vehicle is +2.5°. It is decreased 0.75°. What is the new caster angle?

●40. Before alignment, a caster angle is found to be −1.75°. It is moved 0.5° closer to vertical. What is the new caster angle?

●41. The caster angle on a vehicle is moved from −1.25° to +2.5°. How much was the angle changed?

●42. The caster angle on a vehicle was found to be −0.75°. After being corrected, the caster angle is 2.0°. How much was the angle changed?

SIGNED NUMBERS

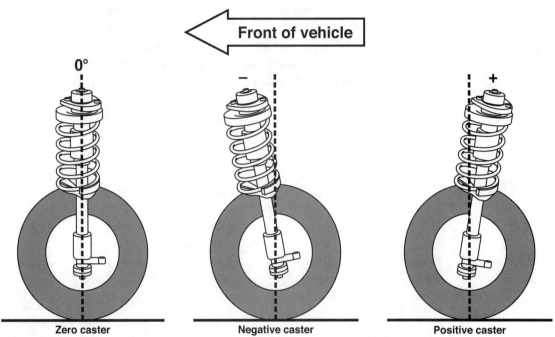

FIGURE 8.3 Signed numbers are also used to show the direction of tilt for caster measurements. A caster measurement of 0° means the spindle and shock are vertical. A positive caster angle means the spindle and strut tilt away from the front of the vehicle, while a negative caster angle means the spindle and strut tilt toward the front of the vehicle.

Multiplication and Division of Signed Numbers

Evaluate each of the following.

●1. $6 \times 7 =$

●2. $3 \times 15 =$

●3. $-9 \times 4 =$

●4. $17 \times (-2) =$

●5. $80 \div 16 =$

●6. $65 \div 5 =$

●7. $-9 \div 9 =$

●8. $-25 \div (-5) =$

●9. $-10 \times (-4) =$

●10. $-8 \times (-30) =$

●11. $-32 \div (-4) =$

●12. $-16 \div (-4) =$

●13. $-1.6 \times (-2.5) =$

●14. $18.7 \times (-4.5) =$

●15. $13.8 \div (-2.4) =$

●16. $-2.89 \div (-17) =$

●●17. $-\dfrac{1}{2} \times 3\dfrac{3}{4} =$

●●18. $-4\dfrac{3}{8} \times \left(-2\dfrac{2}{5}\right) =$

●●19. $9\dfrac{1}{2} \div \left(-1\dfrac{1}{4}\right) =$

●●20. $-20 \div \left(-2\dfrac{3}{10}\right) =$

●21. Each mark on a strut support represents −2° of caster angle. What angle is represented by four marks?

●22. Every 1500 lb of weight changes the camber angle by 1.5°. How much will the angle change if 3000 lb of weight are added to the vehicle?

●23. Each turn of an adjustment nut changes the camber angle by −0.25°. How much will the angle change if the nut is turned five times?

●24. Every 100 lb of weight removed from a race car changes its quarter-mile time by 0.2 s. How much will the time be reduced if 400 lb are removed from the vehicle?

CHAPTER 9

Engine

GENERAL ENGINE MEASUREMENTS

There are dozens of situations where basic math skills are needed in assessing the condition of an engine. One example is determining the taper of a cylinder. Others are finding the camshaft lobe height and the valve lift.

When examining an engine for wear, the cylinder taper is measured. As a cylinder wears, the top of the cylinder typically wears more than the bottom, causing the bore to be slightly larger near the top, just below the ridge than at the bottom. The **taper** is given by finding the difference between measurements A and B shown in Figure 9.1.

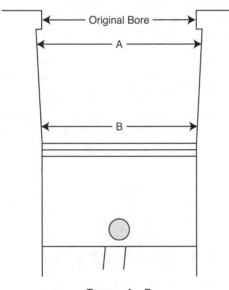

FIGURE 9.1 The bore of a cylinder needs to be measured for taper.

Example 9-1 For a 2.3 liter engine, a T-gauge is used to measure dimensions A and B in the diagram above. Dimension A is 3.154 in., while dimension B is 3.149 in. What is the cylinder taper in this case?

The difference between these two dimensions is $3.154 - 3.149 = 0.005$ in.

Because of the direction of crankshaft rotation, a cylinder bore tends to wear less along the centerline of the crankshaft (dimension D), and more perpendicularly to the crankshaft's centerline (dimension C). **Out-of-round** is the difference between these two measurements. Figure 9.2 shows the crankshaft centerline.

Example 9-2 On a certain 3.8 liter engine, dimension C is measured to be 3.472 in. while dimension D is measured to be 3.466 in. What is the out-of-round measurement?

The difference between these two measurements is $3.472 - 3.466 = 0.006$ in.

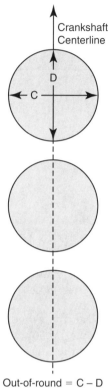

Out-of-round = C − D

FIGURE 9.2 Cylinder bore wear is usually uneven and should be measured along the crankshaft centerline (D) and perpendicular to the crankshaft centerline (C).

The distance any valve opens while the engine is running is determined by the size of the camshaft lobe and the type of rocker arm. The size of the camshaft lobe is found by finding the difference between measurements E and F in Figure 9.3.

Example 9-3 For a certain camshaft, dimension E is measured to be 2.103 in., while dimension F is taken to be 1.807 in. What is the lobe size for this valve? The lobe size is also referred to as "lift at the cam."

The difference between E and F here is 2.103 − 1.807 = 0.296 in.

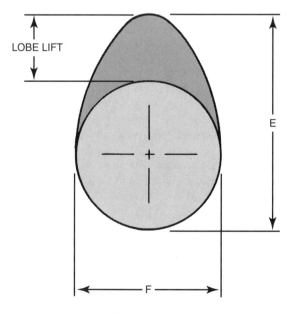

LOBE LIFT = E − F

FIGURE 9.3 The valve lift is determined by the camshaft lobe height.

Source: James D. Halderman and Chase D. Mitchell, Jr., *Automotive Technology: Principles, Diagnosis, and Service,* Second Edition © 2003. Reprinted by permission of Pearson Education, Inc., Upper Saddle River, NJ.

The **rocker arm ratio** is found by dividing dimension "A" by dimension "B" in the diagram shown in Figure 9.4.

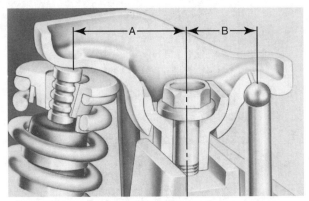

FIGURE 9.4 The rocker arm ratio is the ratio of A to B.

Source: James D. Halderman and Chase D. Mitchell, Jr., *Automotive Technology: Principles, Diagnosis, and Service,* Second Edition © 2003. Reprinted by permission of Pearson Education, Inc., Upper Saddle River, NJ.

Example 9-4 Suppose for the rocker arm in Figure 9.4, dimension A is 1.712 in., and dimension B is 1.070 in. Find the rocker arm ratio.

The rocker arm ratio here will be $\frac{1.712 \text{ in.}}{1.070 \text{ in.}} = 1.6{:}1$.

The **overall valve lift** is found by multiplying the size of the camshaft lobe by the rocker arm ratio.

$$\text{Lift at valve} = (\text{Lift at cam}) \times (\text{Rocker arm ratio})$$

Example 9-5 Find the overall valve lift for the rocker arm in Figure 9.4.

In this case, our lift at the cam, or lobe size, was 0.296 in. Our rocker arm ratio is 1.6:1, so

$$\text{Lift at valve} = (0.296 \text{ in.}) \times (1.6) = 0.4736 \text{ in.}$$

DISPLACEMENT

Engines are usually identified by their displacement. To calculate displacement, a few terms must be reviewed. The **bore** of an engine is the diameter of each cylinder. The **stroke** of an engine is the distance the piston travels from top-dead-center (TDC) to bottom-dead-center (BDC). **Swept volume** is the amount of space displaced by the piston in traveling from BDC to TDC. See Figure 9.5 for an illustration of these measurements.

Since the swept volume for one cylinder is in the geometric shape of a cylinder, the volume can be calculated as we've done earlier, by using the formula

$$\text{Volume of a cylinder} = \pi \times \text{radius}^2 \times \text{height}$$

Since the bore is the diameter of each cylinder, the radius will be 1/2 of the bore. The height of each cylinder is simply the stroke. Then we can say that

$$\text{Swept volume of each cylinder} = \pi \times \left(\frac{1}{2} \times \text{bore}\right)^2 \times \text{stroke}$$

Displacement is simply the total swept volume of an engine. Assuming all cylinders are the same size,

Engine displacement = (Number of cylinders) × (Swept volume of each cylinder)

166 CHAPTER 9

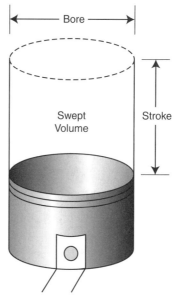

FIGURE 9.5 The displacement of an engine depends on the bore and the stroke of the cylinder.

Or, more specifically,

$$\text{Engine displacement} = (\text{Number of cylinders}) \times \pi \times \left(\frac{1}{2} \times \text{bore}\right)^2 \times \text{stroke}$$

Because π and $\left(\frac{1}{2}\right)^2$ are always present, this formula is often written as

$$\text{Engine displacement} = 0.7854 \times (\text{Number of cylinders}) \times (\text{bore})^2 \times (\text{stroke})$$

Note that these formulas are used regardless of whether the dimensions are given in inches, centimeters, or millimeters. Just remember that the units used in the calculation must all be the same, and will determine the units of displacement.

Example 9-6 Calculate the swept volume of one cylinder with a 3.500-in. bore and 3.800-in. stroke.

We'll use the formula from above,

$$\text{Swept volume of each cylinder} = \pi \times \left(\frac{1}{2} \times \text{bore}\right)^2 \times \text{stroke}$$

Filling in the values we're given,

$$\text{Swept volume of each cylinder} = \pi \times \left(\frac{1}{2} \times 3.500 \text{ in.}\right)^2 \times 3.800 \text{ in.}$$
$$= 36.56 \text{ in.}^3$$

Example 9-7 Calculate the swept volume of one cylinder with a 10.6-cm bore and a 9.4-cm stroke.

Again,

$$\text{Swept volume of each cylinder} = \pi \times \left(\frac{1}{2} \times \text{bore}\right)^2 \times \text{stroke}$$

With the values from above,

$$\text{Swept volume of each cylinder} = \pi \times \left(\frac{1}{2} \times 10.6 \text{ cm}\right)^2 \times 9.4 \text{ cm}$$
$$= 829.53 \text{ cm}^3 \text{ or } 829.53 \text{ cc}$$

Example 9-8 Find the displacement of a 4-cylinder engine with a 9.0-cm bore and 8.2-cm stroke.

We'll answer this question without relying on the formula for displacement. As in the previous two examples, the swept volume will be given by

$$\text{Swept volume of each cylinder} = \pi \times \left(\frac{1}{2} \times 9.0 \text{ cm}\right)^2 \times 8.2 \text{ cm} = 521.66 \text{ cm}^3$$

Since there are four cylinders,

$$\text{Engine displacement} = 4 \times 521.66 \text{ cm}^3 = 2086.64 \text{ cm}^3$$

This engine could be referred to as a 2087 cc engine. Usually engine sizes are rounded to the nearest 100 cm³, we'd call this a 2100 cc engine.

$$\text{We can also convert this to liters: } \frac{2100 \text{ cc}}{1} \times \frac{1 \text{ L}}{1000 \text{ cc}} = \frac{2100 \text{ L}}{1000} = 2.1 \text{ L}$$

Example 9-9 Starting in 1969, the Ford Motor Company produced an 8-cylinder engine with a 4.000-in. bore and a 3.500 in. stroke, and advertised it as a 351-in.³ (and later 5.8 L) engine. Is this correct?

This time we'll just use the formula for engine displacement.

$$\text{Engine displacement} = (8 \text{ cylinders}) \times \pi \times \left(\frac{1}{2} \times 4.000 \text{ in.}\right)^2 \times 3.500 \text{ in.} = 351.86 \text{ in.}^3$$

The displacement actually rounds to 352 in.³.
To check the displacement in liters, we'll do the conversion.

$$\frac{351.68 \text{ in.}^3}{1} \times \frac{1 \text{ L}}{60.98 \text{ in.}^3} = 5.77 \text{ L}$$

The displacement would round to the advertised 5.8 L.

Note this information about the previous example. Ford advertised this engine as a 351 to avoid confusion with an older engine with the same bore and stroke already using the 352 label. This was not an uncommon practice during the 1960s, and other examples exist.

COMPRESSION RATIO

On the compression stroke, the air and fuel get hotter as they are squeezed into a small volume. Engines must be designed so that these increasing temperatures are not hot enough to ignite the air and fuel too early (this condition is called preignition). This is done primarily by controlling the compression ratio. **Compression ratio** is a comparison of the volume of the air and fuel when the piston is at BDC to the volume of air and fuel when the piston is at TDC. An illustration of BDC and TDC is shown in Figure 9.6.

Most gasoline engines have compression ratios around 9.0:1. Newer engines with sensors that detect preignition can have compression ratios up to about 10.0:1, while turbocharged engines have compression ratios that can be as low as 8.0:1. Higher compression ratios result in greater power output by the engine. The octane rating of the fuel being used, however, can limit how high the compression ratio can be. Engines with higher compression ratios require fuel with higher octane ratings.

To measure compression ratio, the volume of every space that contains air and fuel within a cylinder needs to be known. We'll start by considering all the spaces that exist when the engine is at TDC, and how we can find the volume of each space. See Figure 9.7.

When the piston is at the top of the stroke, air and fuel can exist in the combustion chamber of the cylinder head, in the space created by the head gasket, in the flat space above the piston but below the deck of the block (called deck height volume), and if it

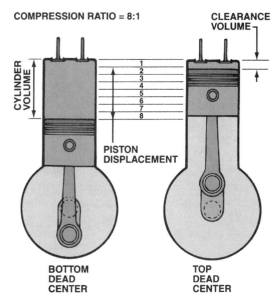

FIGURE 9.6 Compression ratio greatly affects how hot the air and fuel get as the piston approaches TDC.

Source: James D. Halderman and Chase D. Mitchell, Jr., *Automotive Technology: Principles, Diagnosis, and Service,* Second Edition © 2003. Reprinted by permission of Pearson Education, Inc., Upper Saddle River, NJ.

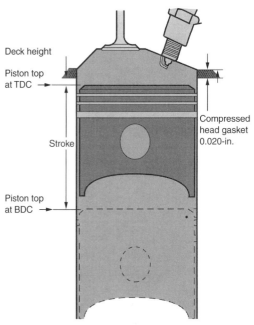

FIGURE 9.7 The combustion chamber, head gasket, deck height, and piston shape all affect the total volume at TDC.

Source: James D. Halderman and Chase D. Mitchell, Jr., *Automotive Technology: Principles, Diagnosis, and Service,* Second Edition © 2003. Reprinted by permission of Pearson Education, Inc., Upper Saddle River, NJ.

exists, the piston relief. All of these volumes combined make up what is called the **clearance volume** of a cylinder.

> Clearance volume = Combustion chamber volume + Head gasket volume
> + Deck height volume + Piston relief volume

Combustion Chamber Volume is determined by the type of cylinder head. Usually, this volume can be looked up if the casting number of the cylinder head is known. If not, it can be measured using a burette. Usually, combustion chamber volume is given in cm^3, and must be converted to cubic inches. Figures 9.8 through 9.12 show the equipment needed and the steps taken to measure the combustion chamber volume.

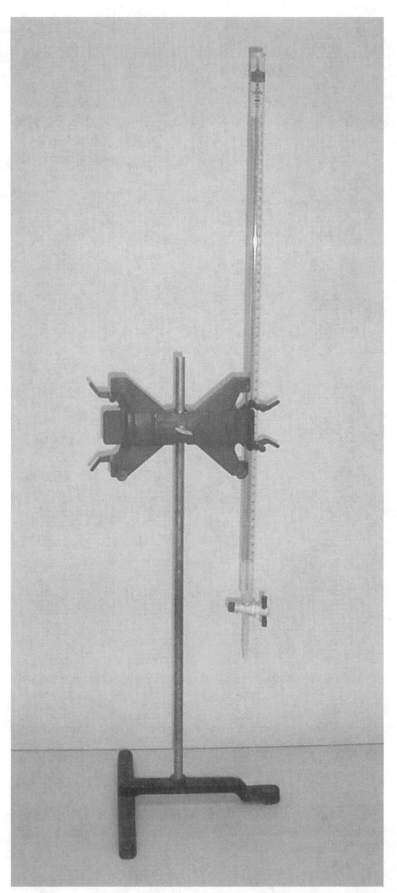

FIGURE 9.8 A burette is a tall, thin glass cylinder with a small valve on the bottom. It is used to measure volume in cubic centimeters. A 100 mL burette such as this works well for measuring cylinder head volume.

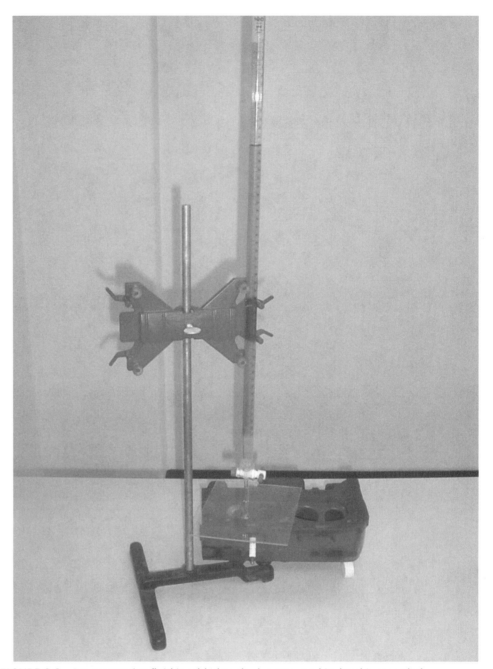

FIGURE 9.9 A noncorrosive fluid is added to the burette, and its level is recorded.

Example 9-10 Find the volume, in cubic inches, of a 64 cm³ combustion chamber.
Use the conversion process we've used before.

$$\frac{64 \text{ cm}^3}{1} \times \frac{1 \text{ in.}^3}{16.39 \text{ cm}^3} = 3.90 \text{ in.}^3$$

Head Gasket Volume is found by using the formula for the volume of a cylinder. The height of the cylinder is the thickness of the gasket, and the radius of the cylinder is approximately 1/2 × bore. As before,

$$\text{Volume of a cylinder} = \pi \times \text{radius}^2 \times \text{height}$$

So we have

$$\boxed{\text{Head gasket volume} = \pi \times \left(\frac{1}{2} \times \text{bore}\right)^2 \times \text{gasket thickness}}$$

FIGURE 9.10 A plastic cover with a small hole is sealed on top of the combustion chamber.

FIGURE 9.11 The fluid is allowed to fill the combustion chamber, taking care to avoid bubbles.

Example 9-14 A cylinder has a bore of 3.940 in., a deck height of 0.010 in., and uses 14.7 cm^3 relief pistons, 58 cm^3 combustion chambers, and a gasket that is 0.035 in. thick. What is the clearance volume?

We need to find the volume in cubic inches for each portion, then add them all together. For the combustion chambers and piston relief, we must convert:

$$\text{Combustion chambers: } \frac{58 \text{ cm}^3}{1} \times \frac{1 \text{ in.}^3}{16.39 \text{ cm}^3} = 3.54 \text{ in.}^3$$

$$\text{Piston relief volume: } \frac{14.7 \text{ cm}^3}{1} \times \frac{1 \text{ in.}^3}{16.39 \text{ cm}^3} = 0.90 \text{ in.}^3$$

For the head gasket volume and deck height volume, we'll use the appropriate formulas.

$$\text{Head gasket volume} = \pi \times \left(\frac{1}{2} \times 3.940 \text{ in.}\right)^2 \times 0.035 \text{ in.} = 0.43 \text{ in.}^3$$

$$\text{Deck height volume} = \pi \times \left(\frac{1}{2} \times 3.940 \text{ in.}\right)^2 \times 0.010 \text{ in.} = 0.12 \text{ in.}^3$$

$$\text{Clearance volume} = 3.54 \text{ in.}^3 + 0.90 \text{ in.}^3 + 0.43 \text{ in.}^3 + 0.12 \text{ in.}^3 = 4.99 \text{ in.}^3$$

Now let's consider the volume when the piston is at BDC. The total volume of air and fuel now is all of the volume we've just considered plus the swept volume, which we already know how to find from the displacement examples. Then we can summarize that

$$\text{Compression ratio} = \frac{\text{Volume of air and fuel at BDC}}{\text{Volume of air and fuel at TDC}}$$

or, more simply,

$$\text{Compression ratio} = \frac{\text{Swept volume} + \text{Clearance volume}}{\text{Clearance volume}}$$

Example 9-15 A cylinder has a swept volume of 42.83 in.^{3}, and a clearance volume of 5.21 in.^{3}. Find the compression ratio.

Using these numbers in the formula above this example,

$$\text{Compression ratio} = \frac{42.83 \text{ in.}^3 + 5.21 \text{ in.}^3}{5.21 \text{ in.}^3} = \frac{48.04 \text{ in.}^3}{5.21 \text{ in.}^3} = 9.2:1$$

Example 9-16 An engine has the following specifications. Find the compression ratio.

Bore: 4.000 in. Combustion chamber volume: 58 cm^3
Stroke 2.870 in. Piston relief volume: 9.2 cm^3
Deck height: 0.008 in. Head gasket thickness: 0.045 in.

First, find the clearance volume.

$$\text{Combustion chamber volume} = \frac{58 \text{ cm}^3}{1} \times \frac{1 \text{ in.}^3}{16.39 \text{ cm}^3} = 3.54 \text{ in.}^3$$

$$\text{Piston relief volume} = \frac{9.2 \text{ cm}^3}{1} \times \frac{1 \text{ in.}^3}{16.39 \text{ cm}^3} = 0.56 \text{ in.}^3$$

$$\text{Head gasket volume} = \pi \times \left(\frac{1}{2} \times 4.000 \text{ in.}\right)^2 \times 0.045 \text{ in.} = 0.57 \text{ in.}^3$$

$$\text{Deck height volume} = \pi \times \left(\frac{1}{2} \times 4.000 \text{ in.}\right)^2 \times 0.008 \text{ in.} = 0.10 \text{ in.}^3$$

$$\text{Clearance volume} = 3.54 \text{ in.}^3 + 0.56 \text{ in.}^3 + 0.57 \text{ in.}^3 + 0.10 \text{ in.}^3 = 4.77 \text{ in.}^3$$

ENGINE

Now, find the swept volume.

$$\text{Swept volume of each cylinder} = \pi \times \left(\frac{1}{2} \times 4.000 \text{ in.}\right)^2 \times 2.870 \text{ in.} = 36.07 \text{ in.}^3$$

Finally, since we know these two volumes,

$$\text{Compression ratio} = \frac{36.07 \text{ in.}^3 + 4.77 \text{ in.}^3}{4.77 \text{ in.}^3} = \frac{40.84 \text{ in.}^3}{4.77 \text{ in.}^3} = 8.6:1.$$

CHANGING COMPRESSION RATIO

When an engine is machined in preparation for a rebuild, there are a number of changes to each of the volumes we've been discussing. If the deck of a block is warped, it is machined. This removal of deck material reduces the deck height volume. Usually, though, that change is not enough to change the compression ratio. When a block is bored to accept larger pistons, though, the change in swept volume can increase the compression ratio measurably.

Example 9-17 An engine has an original bore of 3.730 in., and a stroke of 3.200 in. The clearance volume for this engine is 4.35 in.^{3}. If the engine is bored 0.030, what is the increase in compression ratio?

Using the original bore and stroke, we have

$$\text{Swept volume} = \pi \times \left(\frac{1}{2} \times 3.730 \text{ in.}\right)^2 \times 3.200 \text{ in.} = 34.97 \text{ in.}^3$$

Using our clearance volume of 4.35 in.^{3}, we find

$$\text{Compression ratio} = \frac{34.97 \text{ in.}^3 + 4.35 \text{ in.}^3}{4.35 \text{ in.}^3} = \frac{39.32 \text{ in.}^3}{4.35 \text{ in.}^3} = 9.0:1$$

Now, if the engine is bored 0.030 in. oversize, we have a new bore of 3.760 in. Recalculating with this new value,

$$\text{New swept volume} = \pi \times \left(\frac{1}{2} \times 3.760 \text{ in.}\right)^2 \times 3.200 \text{ in.} = 35.53 \text{ in.}^3$$

The head gasket volume and deck volume need not be recalculated with the new bore because their change in volume would not be enough to change the compression ratio.

$$\text{New compression ratio} = \frac{35.53 \text{ in.}^3 + 4.35 \text{ in.}^3}{4.35 \text{ in.}^3} = \frac{39.88 \text{ in.}^3}{4.35 \text{ in.}^3} = 9.2:1$$

Boring the engine 0.030 in. oversize increased the compression ratio from 9.0:1 to 9.2:1. This modification did not greatly affect the compression ratio.

Another common engine modification is the changing of cylinder heads. If a cylinder head is replaced, it must be replaced with another head with the same combustion chamber volume. If not, the compression ratio could become too high or too low.

Example 9-18 An engine with a 4.000-in. bore and 3.500-in. stroke has flat-top pistons and a deck height of 0 in. Originally, cylinder heads with 78-cc combustion chambers were used with a gasket that was 0.030 in. thick. The owner replaces the cylinder head with one using 64-cc combustion chambers, and a similar gasket. What is the change in compression ratio?

First, find the compression ratio when the original 78-cc heads were used.

$$\text{Swept volume} = \pi \times \left(\frac{1}{2} \times 4.000 \text{ in.}\right)^2 \times 3.500 \text{ in.} = 43.98 \text{ in.}^3$$

$$\text{Head gasket volume} = \pi \times \left(\frac{1}{2} \times 4.000 \text{ in.}\right)^2 \times 0.030 \text{ in.} = 0.38 \text{ in.}^3$$

$$\text{Combustion chamber volume} = \frac{78 \text{ cm}^3}{1} \times \frac{1 \text{ in.}^3}{16.39 \text{ cm}^3} = 4.76 \text{ in.}^3$$

$$\text{Clearance volume} = 4.76 \text{ in.}^3 + 0.38 \text{ in.}^3 = 5.14 \text{ in.}^3$$

$$\text{Compression ratio} = \frac{43.98 \text{ in.}^3 + 5.14 \text{ in.}^3}{5.14 \text{ in.}^3} = \frac{49.12 \text{ in.}^3}{5.14 \text{ in.}^3} = 9.6:1$$

Now, we'll recalculate using the new combustion chamber size of 64 cc. Note that the swept volume will be the same, as will the head gasket volume.

$$\text{New combustion chamber volume} = \frac{64 \text{ cm}^3}{1} \times \frac{1 \text{ in.}^3}{16.39 \text{ cm}^3} = 3.90 \text{ in.}^3$$

$$\text{New clearance volume} = 3.90 \text{ in.}^3 + 0.38 \text{ in.}^3 = 4.28 \text{ in.}^3$$

$$\text{New compression ratio} = \frac{43.98 \text{ in.}^3 + 4.28 \text{ in.}^3}{4.28 \text{ in.}^3} = \frac{48.26 \text{ in.}^3}{4.28 \text{ in.}^3} = 11.3:1.$$

With the changed cylinder heads, the engine would not run well on standard unleaded fuel.

DIESEL APPLICATIONS

Compression ratios are also very important in the operation of diesel engines, such as the one in the machinery in Figure 9.14. Since the diesel fuel ignites due to the high temperatures created by compression, an incorrect ratio could cause the combustion to begin at the wrong time. Typical compression ratios for diesel engines range from about 16.0:1 to 19.0:1, which is much higher than for gasoline engines.

FIGURE 9.14 Engines powered by diesel fuel operate with higher compression ratios than those powered by gasoline.

ENGINE

FIGURE 9.15 The combustion chambers on many diesel cylinder heads are almost flat. This means the combustion chamber volume is typically very small.

Although the basic principles are the same, there are some minor differences to consider when calculating the compression ratio for a diesel engine. First, many diesel engines do not have combustion chambers in the cylinder head. The head and valves are nearly flat, while the pistons are dished heavily to create the desired clearance volume. See Figure 9.15. This means that when calculating compression ratio, there is often no combustion chamber volume to consider. See Figure 9.16. The piston relief volume, however, will be fairly large.

Also, most diesel engines are not rebored in a rebuild. Typically, the cylinder sleeves are replaced, maintaining the original swept volume for the engine.

Example 9-19 One cylinder of a certain I-6 diesel engine has a swept volume of 148.86 in.^{3}, and a clearance volume of 9.39 in.^{3}. What is the compression ratio?

$$\text{Compression ratio} = \frac{148.86 \text{ in.}^3 + 9.39 \text{ in.}^3}{9.39 \text{ in.}^3} = \frac{158.25 \text{ in.}^3}{9.39 \text{ in.}^3} = 16.9{:}1.$$

Example 9-20 A diesel engine with a bore of 4.500 in. and a stroke of 5.000 in. is being rebuilt. The engine builder can use pistons that (along with the deck height and gasket thickness) have a clearance volume of 76 cc. He can also use pistons with a larger dish, creating a total clearance volume of 84 cc. What will the compression ratio be in each case?

In either case, we need to know the swept volume. With our dimensions,

$$\text{Swept volume} = \pi \times \left(\frac{1}{2} \times 4.500 \text{ in.}\right)^2 \times 5.000 \text{ in.} = 79.52 \text{ in.}^3$$

Measure the height of each camshaft lobe shown below.

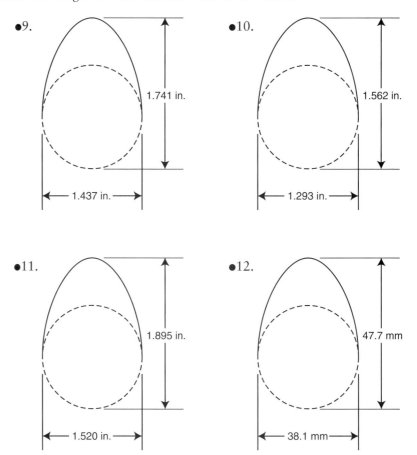

Determine the ratio for each rocker arm shown below.

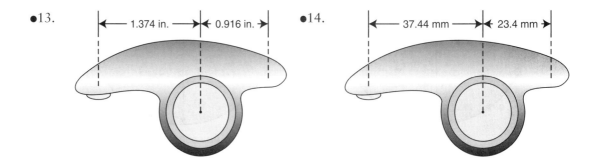

●●15. The intake lobe for a certain camshaft has 0.280 in. of lift at the cam. If this cam is used with a rocker arm ratio of 1.6:1, what is the lift at the valve?

●●16. The exhaust lobe for a performance camshaft has 0.328 in. of lift. When used with 1.5:1 rocker arms, what is the valve lift?

●●17. A 1.7:1 rocker arm is installed on an engine where both the intake and exhaust have 0.72 cm of lift at the camshaft. What is the lift at the valve?

●●●18. Exhaust valve lift is 0.464 in. when used with 1.6:1 rocker arms. What is the lift at the cam?

●●●19. A camshaft advertises 0.345 in. of lift at the cam, and 0.552 in. of lift at the valve. What rocker arm ratio was used when taking these measurements?

Displacement

Round your answers to the nearest tenth unless otherwise noted.

●1. An engine has a 9.2 cm bore and 8.7 cm stroke. What is the radius of each cylinder?

●2. The bore of an engine is 3.762 in., while the stroke is 3.340 in. What is the radius of each cylinder? Do not round this answer.

●3. In a certain 4-cylinder engine, the swept volume of each cylinder is 478.31 cm^3. What is the displacement of this engine, in cubic centimeters?

●4. An 8-cylinder engine contains cylinders with a swept volume of 36.28 $in.^3$. What is the displacement of this engine?

●5. Each cylinder in an inline 6-cylinder diesel engine has a swept volume of 68.53 $in.^3$. What is this engine's displacement?

●6. A 2-cylinder motorcycle engine is designed so that each cylinder has a swept volume of 524.47 cm^3. What is the displacement of this engine, in cubic centimeters?

●●7. A cylinder is designed with a 3.640-in. bore and 3.450-in. stroke. What is the swept volume of this cylinder?

●●8. What is the swept volume of a cylinder with a 3.500-in. bore and 3.900-in. stroke?

●●9. The bore of a certain cylinder is 9.2 cm, and the stroke is 9.0 cm. What is the swept volume, in cubic centimeters?

●●10. A cylinder has an 89 mm bore and an 82 mm stroke. What is the swept volume, in cubic centimeters?

●●11. The displacement of an engine is 241.73 $in.^3$.
   a. What is the displacement in cubic centimeters (cc)?
   b. What is the displacement in liters?

●●12. The displacement of an engine is calculated to be 2192.61 cm^3. What is the displacement in cubic inches?

Find the engine displacement for each of the engines in the table below.

Problem	Engine configuration	Bore	Stroke	Displacement
●●13.	I-4	3.720 in.	3.250 in.	in.^{3}
●●14.	V-8	3.680 in.	3.000 in.	in.^{3}
●●15.	V-6	4.000 in.	3.480 in.	in.^{3}
●●16.	I-5	9.8 cm	7.6 cm	cm^3
●●17.	I-6	8.6 cm	7.9 cm	cm^3
●●18.	I-4	91 mm	84 mm	cm^3

●●19. The engine in problem 14 is bored 0.040-in. oversize. What is the new displacement?

●●20. The engine in problem 17 is bored 0.10 cm oversize. What is the new displacement?

●●●21. The bore of a certain V-6 engine is 3.940 in., and the stroke is 3.280 in. Find the displacement, in cubic centimeters and in liters.

●●●22. A V-8 engine has a bore of 97 mm and a stroke of 86 mm. Calculate the displacement in cubic inches.

Compression Ratio

Round your answers for problems 1–14 to the nearest hundredth.

●1. Find the volume, in cubic inches, of a 62 cm^3 combustion chamber.

●2. Convert the volume of a 57 cc (or cm^3) combustion chamber to cubic inches.

●3. A piston dish has a volume of 12.8 cm^3. What is this volume, in cubic inches?

●4. Valve cutouts on a certain piston have a total volume of 6.5 cm^3. Convert this to cubic inches.

●5. Find the head gasket volume for a 0.037-in. thick gasket with a bore of 4.000 in.

●6. A head gasket has a bore of 3.760 in., and is 0.025 in. thick. Find the volume.

●7. A head gasket is 0.11 cm thick, and has a bore of 8.80 cm. Find the volume, in cm^3.

●8. The deck height for a certain V-8 engine is 0.015 in., and the bore is 3.842 in. Find the deck height volume.

●9. The engine in the previous problem has 0.005 in. milled from the deck, so that the new deck height is 0.010 in. Find the new deck height volume.

ENGINE **183**

●●10. A 4-cylinder engine has a deck height of 0.8 mm. The bore of that engine is 92.0 mm. Find the deck height volume, in cm^3.

Find the clearance volume, in cubic inches, for each described below. Note that volumes are given in both $in.^3$ and cm^3.

	Combustion chamber volume	Head gasket volume	Deck height volume	Piston relief volume	Clearance volume ($in.^3$)
●11.	3.47 $in.^3$	0.50 $in.^3$	0.31 $in.^3$	0.34 $in.^3$	
●12.	3.27 $in.^3$	0.46 $in.^3$	0.28 $in.^3$	1.06 $in.^3$	
●●13.	56 cm^3	0.43 $in.^3$	0.29 $in.^3$	12.3 cm^3	
●●14.	68 cc	0.52 $in.^3$	0.32 $in.^3$	4.6 cm^3	

Round your answers for numbers 15–25 to the nearest tenth.

●●15. Find the compression ratio for a cylinder with a swept volume of 43.71 $in.^3$ and a clearance volume of 5.31 $in.^3$.

●●16. An engine has a clearance volume of 4.15 $in.^3$ and a swept volume of 37.48 $in.^3$ per cylinder. Calculate the compression ratio.

●●17. A racing engine has a swept volume of 537.4 cc, and a clearance volume of 53.9 cc. Find the compression ratio. Will you need to convert these volumes to $in.^3$?

●●18. A turbocharged engine has a clearance volume of 56.9 cm^3. The swept volume is 445.2 cm^3. What is the compression ratio of this engine?

Find the compression ratio for each engine described below.

	Combustion chamber volume	Head gasket volume	Deck height volume	Piston relief volume	Swept volume	Compression ratio
●●19.	3.90 $in.^3$	0.51 $in.^3$	0.30 $in.^3$	0 $in.^3$	45.2 $in.^3$	
●●20.	3.68 $in.^3$	0.48 $in.^3$	0.27 $in.^3$	0.94 $in.^3$	46.1 $in.^3$	
●●●21.	57 cc	0.45 $in.^3$	0.25 $in.^3$	8.7 cc	38.6 $in.^3$	

Find the compression ratio for each engine given the specifications below.

	Bore	Stroke	Head gasket thickness	Deck height	Combustion chamber volume	Piston relief volume	Compression ratio
●●●22.	3.750 in.	3.000 in.	0.030 in.	0.012 in.	63 cc	0 cc	
●●●23.	3.800 in.	3.100 in.	0.035 in.	0.015 in.	61 cc	5.7 cc	
●●●24.	9.20 cm	8.00 cm	0.10 cm	0 cm	49 cc	9.0 cc	
●●●25.	90.0 mm	81.2 mm	1.0 mm	0.5 mm	50 cc	0 cc	

Changing Compression Ratio

Round all compression ratios to the nearest tenth.

●●1. Originally, an engine has a swept volume of 34.02 in.^{3} and a clearance volume of 4.25 in.^{3}. The engine is later bored 0.020 in. oversize, and has a new swept volume of 34.74 in.^{3}.

   a. What was the original compression ratio?
   b. What is the compression ratio after it is bored oversize?

●●2. Before it is bored 0.020 in. oversize, an engine has a swept volume of 36.06 in.^{3}. After it is rebored, it has a swept volume of 36.42 in.^{3}. The clearance volume is 4.41 in.^{3}.

   a. What is the compression ratio before the rebore?
   b. What is the compression ratio after the rebore?

●●3. An engine with a swept volume of 28.50 in.^{3} originally has a clearance volume of 3.47 in.^{3}. Later, an incorrect cylinder head is installed, creating a new clearance volume of 3.78 in.^{3}.

   a. Will the compression ratio go up or down with the incorrect head?
   b. What is the original compression ratio?
   c. What is the compression ratio with the incorrect head?

●●4. The swept volume of a certain cylinder is 468.3 cm^3. With the proper head gasket, pistons and cylinder head, it has a clearance volume of 58.5 cm^3. To raise the compression ratio for racing, a cylinder head with smaller combustion chambers is installed, so that the clearance volume is decreased to 49.7 cm^3.

   a. What was the compression ratio before this modification?
   b. What is the compression ratio after the modification?

●●●5. In problem 22 of the previous section, you calculated the compression ratio for the engine with the following specifications:

Bore	Stroke	Head gasket thickness	Deck height	Combustion chamber volume	Piston relief volume	Compression ratio
3.750 in.	3.000 in.	0.030 in.	0.012 in.	63 cc	0 cc	8.7:1

Now, recompute the compression ratio, assuming the engine is bored 0.040 in. oversize.

●●●6. In problem 23 of the previous section, you calculated the compression ratio for an engine with these specifications:

Bore	Stroke	Head gasket thickness	Deck height	Combustion chamber volume	Piston relief volume	Compression ratio
3.800 in.	3.100 in.	0.035 in.	0.015 in.	61 cc	5.7 cc	8.6:1

What is the new compression ratio if the engine were bored 0.030 in. oversize?

ENGINE

•••7. Refer again to problem 22 from the previous section.

Bore	Stroke	Head gasket thickness	Deck height	Combustion chamber volume	Piston relief volume	Compression ratio
3.750 in.	3.000 in.	0.030 in.	0.012 in.	63 cc	0 cc	8.7:1

What would be the new compression ratio if the cylinder heads were replaced with heads having 70-cc combustion chambers?

•••8. Here again are the specifications for problem 23 in the previous section:

Bore	Stroke	Head gasket thickness	Deck height	Combustion chamber volume	Piston relief volume	Compression ratio
3.800 in.	3.100 in.	0.035 in.	0.015 in.	61 cc	5.7 cc	8.6:1

If 55-cc combustion chambers were used, what would be the new compression ratio?

Diesel Applications

Round all compression ratios to the nearest tenth.

•1. Each cylinder in a diesel I-4 has a swept volume of 107.99 in.^{3}. Dished pistons creating a total clearance volume of 6.58 in.^{3} are used. What is the compression ratio of this engine?

•2. The swept volume of each cylinder in a small diesel engine is 824.3 cm^3. Pistons are installed so that the total clearance volume will be 49.5 cm^3. Find the compression ratio for this engine.

Find the compression ratio given the specifications for each engine below.

	Bore	Stroke	Clearance volume	Compression ratio
••3.	4.750 in.	6.000 in.	6.44 in.^{3}	
••4.	5.300 in.	6.200 in.	8.71 in.^{3}	
•••5.	5.640 in.	6.000 in.	143.8 cc	
•••6.	5.000 in.	5.800 in.	120.5 cc	

CHAPTER 10

Crankshafts and Camshafts

ENGINE BALANCING

Because of the high rotational speeds the crankshaft can reach, it is important that the moving components in an engine be balanced. There are three primary reasons for this. First, and most obvious, is for smooth engine operation. Anyone who has driven a car with a wheel out of balance knows that vehicle vibration is annoying and physically tiring. Second, the engine must be balanced to prolong bearing life. A crankshaft assembly that is out of balance will wear out main bearings much more quickly than a balanced assembly would. The third reason to maintain engine balance is to increase fuel economy. It takes extra energy to spin an object that is not balanced. In racing, an unbalanced engine means wasted horsepower.

What does it mean to "balance an engine?" Actually, balancing an engine means not only balancing the crankshaft, but the connecting rods and pistons as well. Some engines use balance shafts or auxiliary shafts to maintain balance. We will consider only basic inline crankshaft, connecting rod, and piston assemblies, and outline the process for balancing this basic configuration. These components are shown in Figure 10.1. It should be noted, however, that an engine is really never balanced. Balancing an engine really means that we're adjusting the masses inside the engine to come as close to being truly balanced as possible.

Let's start with the engine components. There are three primary components to focus on when balancing an engine: the crankshaft, the connecting rods, and the piston assembly, which includes the piston, wrist pin, and rings. Each of these components is doing one of two things while the engine is running: It is either rotating or reciprocating. For example, the crankshaft simply turns in a circle, so it is called a rotating component. The piston assembly, however, does

(a)

FIGURE 10.1 The crankshaft, connecting rods, and piston assemblies are the components that affect engine balance. (a) Crankshaft; (b) (next page) Connecting rod and piston.

CRANKSHAFTS AND CAMSHAFTS

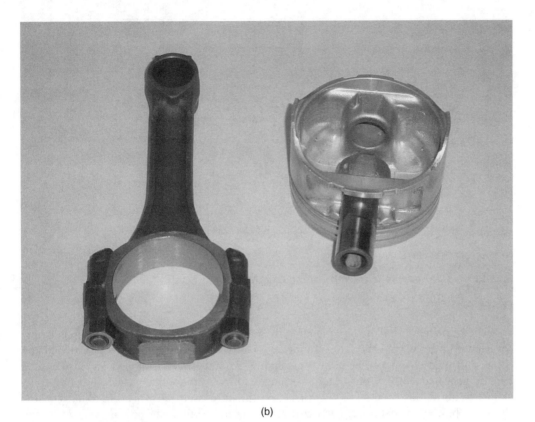

(b)

FIGURE 10.1 (*Continued*)

not rotate; it is continually going up and down in the cylinder, so we call it a reciprocating component. The connecting rod, however, is split; the large end is attached to the crankshaft and rotates, while the small end is attached to the piston and reciprocates. See Figure 10.2.

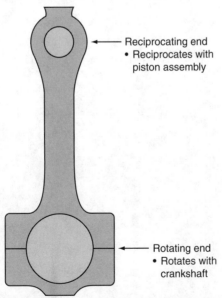

FIGURE 10.2 The two ends of the connecting rod have different motions. The small end reciprocates, and the large end rotates.

The **rotating assembly** consists of the crankshaft, the large end of the connecting rod, and the connecting rod bearings. The **reciprocating assembly** consists of the piston assembly and the small end of the connecting rod. See Figure 10.3.

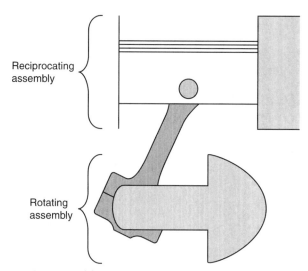

FIGURE 10.3 The rotating assembly and the reciprocating assembly.

The first step in balancing an engine is to make sure that the reciprocating assembly for each cylinder weighs the same. To do this, each piston is assembled with wrist pin and rings, and then weighed. The lightest assembly is noted, and metal is removed from the heavier piston assemblies until each piston assembly has the same weight. Pistons are designed to allow this process to take place; on the bottom of most pistons, usually below the wrist pin, extra metal has been left, assuming that some may be removed when balancing the pistons. This extra metal can be seen in Figure 10.4.

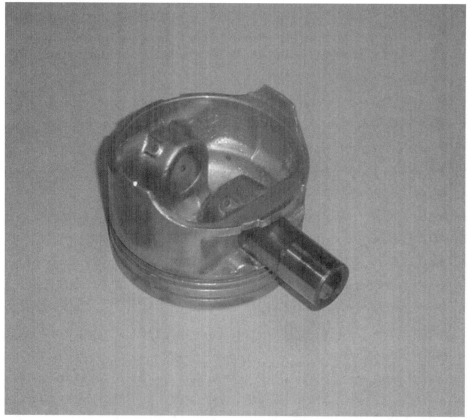

FIGURE 10.4 Most pistons have extra material on the bottom to allow for balancing.

CRANKSHAFTS AND CAMSHAFTS

Example 10-1 The piston assemblies for an inline 6-cylinder engine are each weighed. Their weights are shown on the table below. Find the standard weight that all pistons should be after balancing.

Piston number	Weight
1	487 g
2	498 g
3	492 g
4	485 g
5	490 g
6	499 g

Since the heavier pistons should have metal removed to match the lightest piston, all pistons should weigh 485 g when they are balanced.

The next step in balancing these components is to balance the connecting rods. Since the large end and small end must be considered separately, they must be balanced separately. This is done while supporting one end of the connecting rod while placing the other end on a scale. Weighing the large end of the connecting rod is shown in Figure 10.5. Weighing the small end of the connecting rod is shown in Figure 10.6.

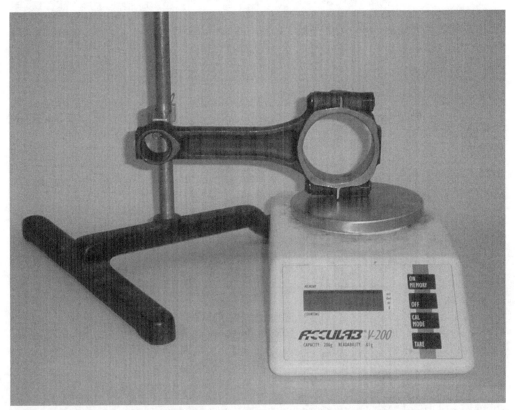

FIGURE 10.5 The large end of the connecting rod is weighed by supporting the small end, and placing the large end on a scale.

Like pistons, connecting rods have extra material called **balance pads** on each end to allow for balancing. Balance pads are shown in Figure 10.7.

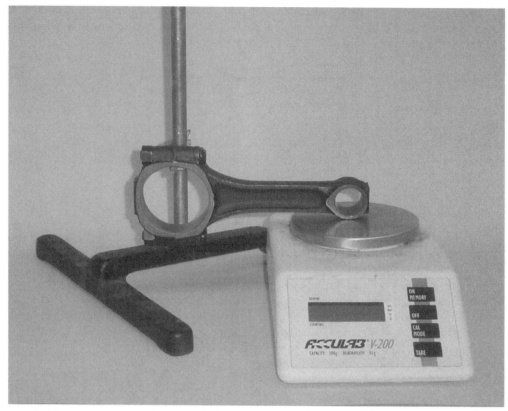

FIGURE 10.6 Weighing the small end of a connecting rod is done in the same way as the large end.

Example 10-2 The connecting rods for the same engine have been weighed, and the values recorded on the table below. Find the desired weight for both the small end and the large end.

Connecting rod number	Small end weight	Large end weight
1	211 g	529 g
2	205 g	524 g
3	214 g	533 g
4	211 g	530 g
5	208 g	522 g
6	210 g	525 g

The lightest small end weighs 205 g, so metal should be removed from the heavier small ends until all six weigh 205 g. The lightest large end is 522 g, so materials should be removed from the heavier rods until they all weigh 522 g on the large end. This means that every connecting rod should have a total weight of 727 g.

Finally, the crankshaft must be balanced. Before doing this, however, a basic principle should be clarified. As you know, the crankshaft has counterweights opposite each connecting rod journal. These counterweights make the engine as balanced as possible. The question is: What should those counterweights actually balance?

Since the large end of the connecting rod is always attached to the connecting rod journal, the counterweights must *always* balance the weight of the large end, called the **rotating weight**. The total weight of the reciprocating components (piston assembly and small end of rod) is called the **reciprocating weight**. Does the crankshaft counterweight need to balance this also? It does, but only half the time. Let's see why.

CRANKSHAFTS AND CAMSHAFTS

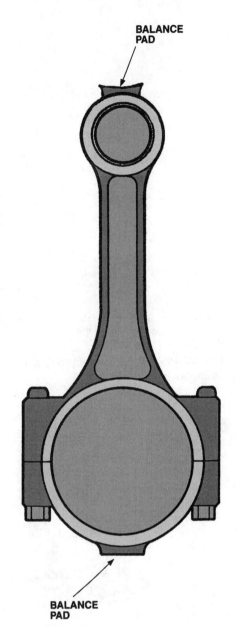

FIGURE 10.7 Balance pads allow each end of a connecting rod to be balanced.

Source: James D. Halderman and Chase D. Mitchell, Jr., *Automotive Technology: Principles, Diagnosis, and Service,* Second Edition © 2003. Reprinted by permission of Pearson Education, Inc., Upper Saddle River, NJ.

When the piston is at TDC or BDC, the piston and small end of rod are (momentarily) not moving. Since the crankshaft at that exact moment is moving but the reciprocating weight is not, the counterweight does not need to balance the reciprocating weight. See Figure 10.8.

When the piston is exactly halfway between TDC and BDC, the piston and small end of the rod are moving, so the crankshaft counterweight must balance the reciprocating weight. See Figure 10.9.

The crankshaft is balanced in the same way as a wheel. **Bobweights** are used to simulate the weight of the rod and piston assemblies, and the crank is spun on a balancing machine that indicates where metal must be removed from counterweights. Bobweights are shown in Figure 10.10. Since the bobweights should be chosen to best balance the engine, they need to simulate the rotating weight all the time, and the reciprocating weight half of the time. Then the formula for calculating the bobweight is given by

$$\text{Bobweight} = \text{Rotating weight} + \frac{\text{Reciprocating weight}}{2}$$

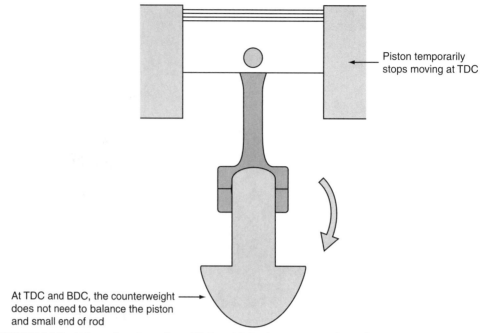

FIGURE 10.8 When the piston is at TDC or BDC, the piston and small end of the rod are momentarily at rest.

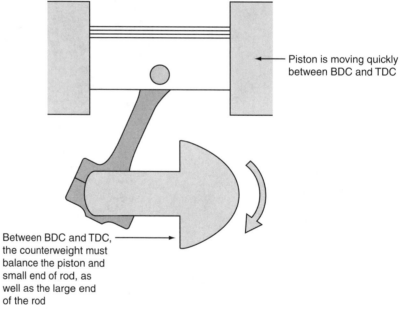

FIGURE 10.9 The movement of the piston, connecting rod, and counterweight halfway between TDC and BDC.

where

Rotating weight = Large end of connecting rod with bearings

Reciprocating weight = Small end of rod + Piston assembly

and all of the small ends, large ends, and piston assemblies weigh the same.

Example 10-3 The piston assemblies and connecting rods for an inline 4-cylinder have all been balanced. Each piston assembly weighs 438 g, each connecting rod small end weighs 204 g, and each large end (with bearings) weighs 510 g. Calculate the bobweight used for each cylinder of the engine.

CRANKSHAFTS AND CAMSHAFTS

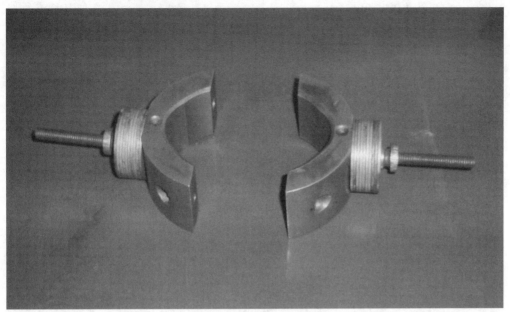

FIGURE 10.10 Bobweights are attached to the connecting rod journals of the crankshaft while it is being balanced. The correct bobweight must be used to correctly simulate the weight of the piston and connecting rod.

From the information given, we know the following:

Rotating weight = Large end of connecting rod with bearings

Rotating weight = 510 g

Reciprocating weight = Small end of rod + Piston assembly

Reciprocating weight = 438 g + 204 g = 642 g

$$\text{Bobweight} = \text{Rotating weight} + \frac{\text{Reciprocating weight}}{2}$$

$$\text{Bobweight} = 510\,\text{g} + \frac{642\,\text{g}}{2}$$

$$\text{Bobweight} = 831\,\text{g}$$

Example 10-4 The components of an I-4 engine are weighed, and the values recorded below. Calculate the bobweight to be used for each cylinder when balancing this crankshaft.

Cylinder number	Piston assembly	Rod, small end	Rod, large end
1	501 g	213 g	467 g
2	505 g	216 g	475 g
3	493 g	218 g	480 g
4	497 g	213 g	471 g

Each piston assembly should be made to weigh 493 g.
Each connecting rod small end should be made to weigh 213 g.
Each connecting rod large end should be made to weigh 467 g.
The rotating weight is then 467 g. The reciprocating weight must be

$$493\,\text{g} + 213\,\text{g} = 706\,\text{g}$$

Finally, using this information,

$$\text{Bobweight} = 467\,\text{g} + \frac{706\,\text{g}}{2}$$

$$\text{Bobweight} = 820\,\text{g}$$

The crankshaft is then balanced by removing metal from the counterweights on the crankshaft until the entire assembly is in balance. Holes drilled for balancing are shown in Figure 10.11.

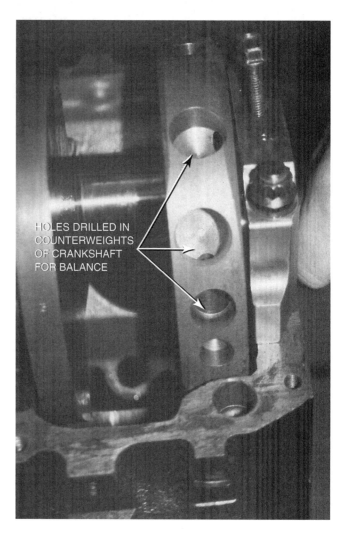

FIGURE 10.11 Metal is removed from the counterweight of a crankshaft until the bobweight is balanced when the crankshaft is rotating.

Source: James D. Halderman and Chase D. Mitchell, Jr., *Automotive Technology: Principles, Diagnosis, and Service,* Second Edition © 2003. Reprinted by permission of Pearson Education, Inc., Upper Saddle River, NJ.

TORQUE AND HORSEPOWER

When a crankshaft (or drive shaft, or wheel for that matter) turns, it needs to transmit a certain amount of twisting force to move the vehicle. The same idea applies to tightening or loosening a lug nut; if a rusted lug nut does not loosen, greater twisting force must be applied to break it free. **Torque** is a measure of twisting force. Think for a moment about that rusted lug nut. Besides using tricks like oil or heat, there are two ways to break that nut free. You can either apply more force to the wrench (use more muscle) or you can use a longer wrench (but use the same amount of muscle). Either way, the torque will be increased. Since the force applied and the length of the lever arm are both factors in creating torque, we have the following:

$$\text{Torque} = \text{Force} \times \text{Length of lever arm}$$

For example; suppose you apply 10 lb of force to a wrench that is 1 ft long. You then create 10 lb · ft (or 10 ft · lb) of torque. See Figure 10.12.

The length of the lever arm is the distance from the center of the rotating component to the point where the force is applied. The force must be applied at a right angle to the lever arm. While the customary unit of torque is the pound · foot (lb · ft), the metric system uses the Newton · meter (N · m). Pound · feet are often called foot · pounds, and 1 lb · ft = 1 ft · lb.

CRANKSHAFTS AND CAMSHAFTS

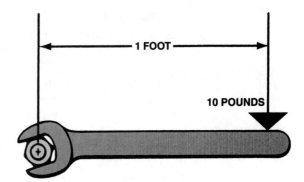

FIGURE 10.12 Torque is found by multiplying the force by the length of the lever arm through which it is applied.

Source: James D. Halderman and Chase D. Mitchell, Jr., *Automotive Technology: Principles, Diagnosis, and Service*, Second Edition © 2003. Reprinted by permission of Pearson Education, Inc., Upper Saddle River, NJ.

Example 10-5 When tightening a head bolt, a technician applies 50 lb of force to the end of a 2-ft torque wrench. What will be the torque reading on the wrench?

Here, the length of the lever arm is 2 feet. The amount of force is 50 lb.

$$\text{Torque} = \text{Force} \times \text{Length of lever arm}$$
$$\text{Torque} = 50 \text{ lb} \times 2 \text{ ft} = 100 \text{ lb} \cdot \text{ft of torque}$$

The equation for torque can be rearranged to find the necessary force if the torque and length of lever arm are known, or to find the length of the necessary lever arm if torque and force are fixed.

$$\text{Force} = \frac{\text{Torque}}{\text{Length of lever arm}}$$
$$\text{Length of lever arm} = \frac{\text{Torque}}{\text{Force}}$$

Example 10-6 How much force must be applied to a 1.5-ft wrench to create 45 lb · ft. of torque?

$$\text{Force} = \frac{\text{Torque}}{\text{Length of lever arm}}$$
$$\text{Force} = \frac{45 \text{ lb} \cdot \text{ft}}{1.5 \text{ ft}} = 30 \text{ lb}$$

Example 10-7 A winch creating 125 lb · ft of torque needs to lift 250 lb. What is the length of the lever arm needed to do this?

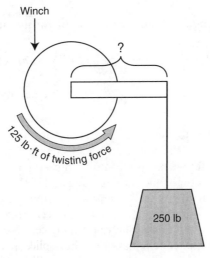

$$\text{Length of lever arm} = \frac{125 \text{ lb} \cdot \text{ft}}{250 \text{ ft}} = 0.5 \text{ ft} = 6 \text{ in.}$$

When we hear of an engine that produces, say, 300 lb·ft of torque at 5000 RPM, it tells us how much twisting force the engine is producing at a certain engine speed. Gears are used to increase or decrease torque before it reaches the rear wheels. We'll work with gear ratios in Chapter 11.

Torque, though, is not the best way to measure an engine's performance. **Horsepower** is a way of measuring how much work an engine can do in a certain amount of time. After all, if you used a very long wrench, you could create more torque with the muscles in your arm than could the engine in a pickup truck, but the truck will be able to pull a trailer up a hill in less time. In addition, changing gear ratios will change the amount of torque available at the wheels, but different gear ratios do not change horsepower. We'll discuss this more later.

Horsepower, then, is a better way to measure the power output of an engine. Since the amount of twisting force and the speed of the engine are taken into account, the following formula should make sense.

$$\text{Horsepower} = \frac{\text{Torque} \times \text{RPM}}{5252}$$

Although we will not create that formula from scratch, it is based on the fact that the definition of 1 HP is the ability to lift 550 lb 1 ft in 1 second. If we were to multiply that by 60, we'd have 33,000 lb·ft in 1 minute. The number 5252 in the denominator comes from the fact that a horsepower is 33,000 lb·ft per minute rotating in a circle with a circumference of $2 \times \pi \times$ radius. After dividing, we see that $33,000 \div (2 \times \pi) = 5252$.

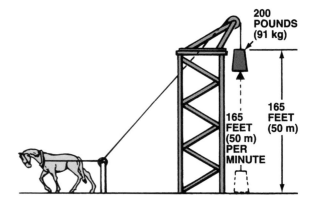

FIGURE 10.13 This horse produces one horsepower by lifting 200 lb 165 ft in 1 minute. This is because 200 lb × 165 ft in 1 minute = 33,000 lb·ft per minute.

Source: James D. Halderman and Chase D. Mitchell, Jr., *Automotive Technology: Principles, Diagnosis, and Service,* Second Edition © 2003. Reprinted by permission of Pearson Education, Inc., Upper Saddle River, NJ.

Example 10-8 An engine creates 240 lb·ft of torque at 3800 RPM. How many horsepower is this?

Using a torque value of 240 lb·ft and an RPM value of 3800, we have

$$\text{Horsepower} = \frac{240 \times 3800}{5252} = 173.6 \text{ HP @3800 RPM}$$

A dynamometer is used to measure the power output of an engine. A chassis dynamometer measures power at the wheels, while an engine dynamometer measures power at the engine. A dynamometer works by measuring the torque output of an engine and noting the RPM. The "dyno" actually measures torque and RPM, and calculates horsepower. The engine is actually producing more horsepower than any dynamometer will measure. Much of the horsepower created, though, is used to overcome the internal friction of the engine. The horsepower that is left after internal friction has been overcome is called **brake horsepower** or **bhp**. Brake horsepower is the power that is created by the engine that can be used to power the vehicle.

Example 10-9 A dynamometer measures the torque produced by an engine at various engine speeds. The values are recorded below.

Engine speed	Torque
3000 RPM	178 lb·ft
3500 RPM	198 lb·ft
4000 RPM	205 lb·ft
4500 RPM	212 lb·ft
5000 RPM	223 lb·ft
5500 RPM	215 lb·ft

Calculate the horsepower at each RPM and make a horsepower vs. RPM graph.

Since we know

$$\text{Horsepower} = \frac{\text{Torque} \times \text{RPM}}{5252}$$

we can fill in the horsepower value for each engine speed.

Engine speed	Torque	Horsepower
3000 RPM	178 lb·ft	101.7 HP
3500 RPM	198 lb·ft	131.9 HP
4000 RPM	205 lb·ft	156.1 HP
4500 RPM	212 lb·ft	181.6 HP
5000 RPM	223 lb·ft	212.3 HP
5500 RPM	215 lb·ft	225.1 HP

Now, graphing the horsepower value at each engine speed, we have a graph that looks like this:

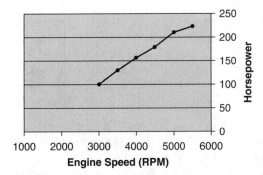

Notice that the torque output begins to drop after an engine speed of 5000 RPM, but horsepower continues to rise.

CAMSHAFT EVENT TIMING

The camshaft is the biggest factor in determining whether an engine will be a high-revving race engine or a high-torque truck engine. The camshaft also determines some very important properties like idle quality and emission levels. The amount of time the intake and exhaust valves stay open, the distance they open, and when they open relative to each other will determine an engine's characteristics. All of these things are controlled by the camshaft.

Every camshaft has its own "personality." A camshaft diagram is a way to visually compare the timing of camshaft events. Let's look at an example camshaft diagram in Figure 10.14 and determine what we can learn from the information shown.

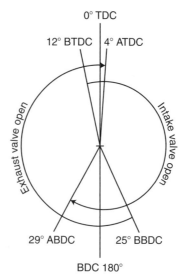

FIGURE 10.14 A camshaft diagram allows us to see the timing of camshaft events relative to crankshaft position.

First, we need to recognize that camshafts are typically used in 4-cycle engines. That means the four strokes listed below must be taking place:

1. **Intake stroke** The piston is going down in the bore, drawing in air and fuel through the open intake valve.
2. **Compression stroke** The piston is going back up in the bore, both valves are closed, and the air and fuel mixture is being compressed.
3. **Power stroke** The fuel is ignited, and with both valves closed, the burning gas expands while pushing the piston down.
4. **Exhaust stroke** The piston goes back up, forcing the burned fuel out of the open exhaust valve.

The crankshaft angles given as the open and closing points of the intake and exhaust valves are actually the points at which the camshaft is providing 0.050 in. of lift or more. Duration values may also be based on 0.050 in. of camshaft lift, but are frequently based on 0.006 in. of camshaft lift. When duration values assume 0.006 in. of lift, it is referred to as advertised duration.

Note the following events from the camshaft diagram above in Figure 10.14.

1. **Intake valve opens** Our camshaft diagram shows us that the intake valve actually begins opening 12° BTDC, or 12° before top dead center. This is usually done to give the fuel a head start in filling the cylinder.
2. **Intake valve closes** The arrow marking the intake valve duration shows that the intake valve stays open until 29° ABDC, or 29° after bottom dead center. That means the intake valve is still open for the first part of the compression stroke. That's because the momentum of the fuel mixture is still helping the cylinder to become filled.
3. **Both valves temporarily closed** After the intake valve closes, both valves stay closed for the remainder of the compression stroke and most of the power stroke. Notice that the diagram indicates the exhaust valve is open, but that's not until the second time around our diagram. Remember, it takes two crankshaft revolutions to make a complete cycle.
4. **Exhaust valve opens** On the second revolution of the crankshaft, the arrow marking the exhaust valve duration shows the exhaust valve opening 25° BBDC, or 25° before bottom dead center. That means the exhaust valve actually opens just before the end of the power stroke. That's done to allow the burned fuel at the top to start leaving the cylinder.
5. **Exhaust valve closes** To finish up the second revolution, the exhaust valve closes, but not until 4° ATDC, or 4° after top dead center. You should notice that this is the end of the cycle, and the next one has already begun. In other words, the exhaust valve closes a few degrees *after* the intake valve has already opened.

CRANKSHAFTS AND CAMSHAFTS

This is done to create **scavenging** in the cylinder, which means using the momentum of exiting exhaust to start pulling in air and fuel for the next cycle.

That brings us to the next topic: cam characteristics. From the diagram in Figure 10.14, we can find the valve overlap, the intake valve duration, the exhaust valve duration, and the intake valve centerline.

1. **Valve overlap**   In the diagram in Figure 10.14, the exhaust valve is still open at 12° BTDC, when the intake valve opens. Both valves are open until 4° ATDC, when the exhaust valve closes. That means both valves are open for 12° of crankshaft rotation before TDC, and another 4° of crankshaft rotation after TDC. Then both valves are open for a total of 16°, and we have what is called the valve overlap.

$$12° + 4° = 16° \text{ valve overlap}$$

2. **Intake valve duration**   The intake opens 12° BTDC. Then, it stays open for the entire 180° of crankshaft rotation while the piston is going down on the intake stroke. It stays open another 29° after the piston hits BDC. That means the intake valve is open a total of 221°, which is called the intake duration.

$$12° + 180° + 29° = 221° \text{ intake duration}$$

3. **Exhaust valve duration**   The exhaust valve opens 25° before BDC. It stays open the entire time the piston is going up during the exhaust stroke, which is another 180° of crankshaft rotation. It remains open until 4° ATDC. That gives us a total of 209° of crankshaft revolution, which is called the exhaust duration.

$$25° + 180° + 4° = 209° \text{ exhaust duration}$$

4. **Intake centerline**   This is the moment exactly halfway between the point at which the intake valve opens and the point at which it closes. For many camshafts, this is also the point when the intake valve is open the farthest, before it starts closing again. We know that the intake valve duration is 221°. Since half of this time was spent opening and half was spent closing (usually), we can see that the intake valve takes $221° \div 2 = 110.5°$ of crankshaft rotation to fully open. Then the intake centerline is located 110.5° after the intake valve opens. The intake opens 12° before TDC, so the intake centerline is located $110.5° - 12° = 98.5°$ after TDC, which is during the intake stroke.

The intake centerline is usually not indicated on the cam diagram.

Example 10-10   For the camshaft diagram shown, find:

a. The intake duration.
b. The exhaust duration.
c. The valve overlap.
d. The intake centerline.

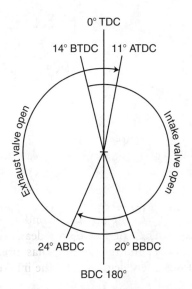

a. The intake valve opens 14° before TDC, then stays open the full 180° of crankshaft rotation while the piston goes down on the intake stroke. It stays open another 24° after the piston reaches BDC, so that gives us a total of 218°.

$$14° + 180° + 24° = 218° \text{ intake duration}$$

b. The exhaust valve opens 20° before BDC, and remains open 180° of crankshaft rotation during the exhaust stroke. It does not close until 11° after TDC, which adds up to 211°.

$$11° + 180° + 20° = 211° \text{ exhaust duration}$$

c. Both valves are open from 14° before TDC on the exhaust stroke until 11° after TDC on the intake stroke, for a total of 25°.

$$14° + 11° = 25° \text{ valve overlap}$$

d. The intake duration was found to be 218°. Then the intake valve takes $218° \div 2 = 109°$ of crankshaft rotation to get fully opened. Since it began 14° before top dead center, its centerline occurs at $109° - 14° = 95°$ after top dead center.

$$109° - 14° = 95° \text{ ATDC intake centerline}$$

Example 10-11 Create a camshaft event diagram from the following information:

	Open	Close
Intake valve	5° ATDC	29° ABDC
Exhaust valve	44° BBDC	10° BTDC

Begin by making a blank diagram.

Now, use a protractor to mark the lines where the intake opens and closes. Then, make a circular arrow to show the intake duration. Notice that the intake opens 5° ATDC and closes at 29° ABDC. Label this portion "Intake valve open."

CRANKSHAFTS AND CAMSHAFTS 201

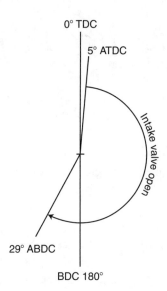

Next, mark the lines where the exhaust opens and closes. Make the circular arrow that shows the exhaust duration. Here, the exhaust valve opens 44° BBDC and closes 10° BTDC. Label this part "Exhaust valve open."

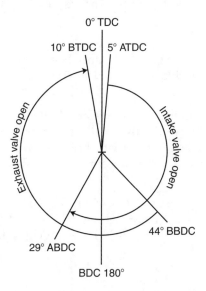

Notice that there is no valve overlap for this camshaft. The engine using this cam would probably have lower emissions, since no unburned fuel can escape through the exhaust valve before it closes.

The intake duration would be 204° for this cam. Verify by noting that the intake opens a bit after TDC, and is open for 180° − 5° = 175° crankshaft degrees of the intake stroke, and another 29° after BDC.

$$175° + 29° = 204° \text{ intake duration}$$

The exhaust duration would be 214° for this camshaft. The exhaust opens 44° BBDC, but closes 10° BTDC of the exhaust stroke. That means the valve is only open for 170° of crankshaft rotation during the exhaust stroke.

$$44° + 170° = 214° \text{ exhaust duration}$$

The intake centerline would be located at 107°. The intake duration was 204°, so it takes 204° ÷ 2 = 102° to fully open. Since it does not start opening until 5° ATDC, it is fully open at 107° ATDC.

$$5° + 102° = 107° \text{ ATDC intake centerline}$$

CHAPTER 10 *Practice Problems*

Engine Balancing

●1. A bare piston weighs 366 g, the wrist pin weighs 88 g, and the rings weigh a total of 39 g. What is the total weight of this piston assembly?

●2. A forged piston weighs 358 g. The wrist pin for that piston weighs 77 g. Each of the two compression rings weigh 18 g, and the oil ring weighs 13 g. What is the total weight of this piston assembly?

●3. The large end of a connecting rod weighs 493 g, and the rod bearing weighs 17 g. What is the rotating weight for this assembly?

●4. The small end of a connecting rod weighs 169 g, and the piston assembly weighs 462 g. What is the reciprocating weight of this assembly?

●5. A balanced piston and rod assembly has a rotating weight of 462 g and a reciprocating weight of 631 g. What bobweight should be used when balancing this engine?

●6. The reciprocating weight of a balanced piston and rod assembly is 598 g, while the rotating weight is 396 g. How much weight should be used for a bobweight?

●●7. The piston assemblies of a 4-cylinder engine are balanced so that they all weigh 415 g. The connecting rods are also balanced so that the small ends each weigh 206 g, and the large ends (with bearings) weigh 392 g. Calculate the proper bobweight to use when balancing the crankshaft.

●●8. Each of the components of a 6-cylinder engine are balanced. The piston assemblies all weigh 385 g. The connecting rods each weigh 179 g on the small end, and 359 g on the large end (including the bearings). What is the proper bobweight for balancing the crankshaft?

●●9. The weights of the components of a 4-cylinder engine are given below.

Cylinder number	Piston assembly	Rod, small end	Rod, large end (including bearings)
1	385 g	197 g	411 g
2	396 g	194 g	409 g
3	392 g	202 g	406 g
4	388 g	200 g	410 g

a. What will be the weight of all the piston assemblies when they are balanced?
b. What will be the weight of all the small connecting rod ends when they are balanced?
c. What will be the weight of all the large connecting rod ends when they are balanced?
d. What will be the reciprocating weight?
e. What will be the rotating weight?
f. What bobweight should be used when balancing the crankshaft?

CRANKSHAFTS AND CAMSHAFTS

●●10. The weights of the components of an inline 6-cylinder engine are given below.

Cylinder number	Piston assembly	Rod, small end	Rod, large end (including bearings)
1	435 g	202 g	426 g
2	438 g	208 g	431 g
3	431 g	209 g	434 g
4	430 g	208 g	428 g
5	434 g	210 g	423 g
6	438 g	203 g	427 g

a. What will be the weight of all the piston assemblies when they are balanced?
b. What will be the weight of all the small connecting rod ends when they are balanced?
c. What will be the weight of all the large connecting rod ends when they are balanced?
d. What will be the reciprocating weight?
e. What will be the rotating weight?
f. What bobweight should be used when balancing the crankshaft?

●●11. The weights of the components of an inline 4-cylinder engine are given below.

Cylinder number	Piston assembly	Rod, small end	Rod, large end (including bearings)
1	447 g	199 g	436 g
2	442 g	199 g	432 g
3	450 g	208 g	438 g
4	444 g	203 g	437 g

a. What will be the weight of all the piston assemblies when they are balanced?
b. What will be the weight of all the small connecting rod ends when they are balanced?
c. What will be the weight of all the large connecting rod ends when they are balanced?
d. What will be the reciprocating weight?
e. What will be the rotating weight?
f. What bobweight should be used when balancing the crankshaft?

●●12. The weights of the components of an inline 6-cylinder engine are given below.

Cylinder number	Piston assembly	Rod, small end	Rod, large end (including bearings)
1	417 g	187 g	397 g
2	419 g	192 g	395 g
3	425 g	191 g	393 g
4	421 g	189 g	401 g
5	418 g	196 g	396 g
6	424 g	189 g	399 g

a. What will be the weight of all the piston assemblies when they are balanced?
b. What will be the weight of all the small connecting rod ends when they are balanced?

    c. What will be the weight of all the large connecting rod ends when they are balanced?
    d. What will be the reciprocating weight?
    e. What will be the rotating weight?
    f. What bobweight should be used when balancing the crankshaft?

Torque and Horsepower

●1. A technician applies 62 lb to the end of a torque wrench that is 2 ft long. How much torque is being applied to the bolt?

●2. Although he knows it's a bad idea, a technician tries to free a corroded bolt by slipping a 4-ft long pipe over the end of his ratchet to create a "cheater-bar." If he applies 76 lb of force to the pipe, how much torque is being applied to the bolt?

●●3. An 18-in. breaker bar is used to free a nut. The tech applied 32 lb of force to the end of that bar. How much torque (in lb · ft) did it take to break the nut free?

●●●4. A 9-in. wrench is used to tighten a head bolt. If the technician applied 60 lb of force to the end of the wrench, how much torque is applied to the bolt?

●5. What is the length of the wrench needed to tighten a bolt to 135 lb · ft if a technician can supply 90 lb of force?

●6. What is the maximum radius of the spool of a winch that must lift 400 lb if the winch turns with 150 lb · ft of torque?

●7. A winch turning with a torque of 240 lb · ft uses a spool with a radius of 3 in. to wind a cable. How much weight can the winch lift?

●●8. How much force is needed to create 160 lb · ft of torque using a wrench that is 18 in. long?

Round your answers to problems 9–14 to the nearest tenth.

●9. A diesel engine produces 430 lb · ft of torque at 2700 RPM. How many horsepower is this?

●10. A racing engine produces 345 lb · ft of torque at 7300 RPM. What is the horsepower rating at this speed?

●11. A lawnmower engine produces a maximum of 8.4 lb · ft of torque at 2000 RPM.
    a. How many horsepower does this engine create at this speed?
    b. Is this necessarily the maximum horsepower the engine can produce? Why?

- 12. A truck produces 350 lb·ft of torque at 3800 RPM. How many horsepower is this?

- 13. A dynamometer collects the measurements below for a certain V-8 engine.
    a. Complete the table below by calculating the horsepower at each engine speed.
    b. Make a graph of horsepower vs. RPM on the chart below.

Engine speed (RPM)	Torque (lb·ft)	Horsepower
2500	137	
3000	158	
3500	179	
4000	215	
4500	243	
5000	261	
5500	250	

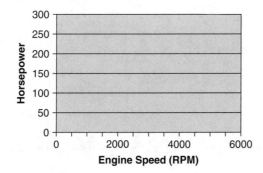

- 14. A dynamometer collects the following data for a turbocharged 4-cylinder engine.
    a. Complete the table by calculating the horsepower at each engine speed.
    b. Make a graph of horsepower vs. RPM on the chart below.

Engine speed (RPM)	Torque (lb·ft)	Horsepower
4000	138	
4500	145	
5000	153	
5500	165	
6000	172	
6500	185	
7000	167	

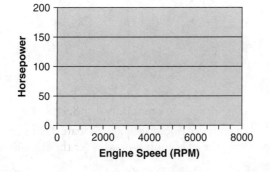

Camshaft Event Timing

●1. Describe, in your own words, what "intake duration" means.

●2. Describe, again in your own words, what "exhaust duration" means.

●3. Does every camshaft have a valve-overlap interval? Why or why not?

Use the camshaft event diagram below to answer questions 4–7.

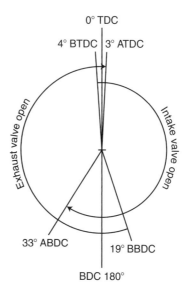

●●4. How many total degrees of crankshaft rotation is the intake valve open?

●●5. What is the exhaust duration for this camshaft?

●●6. What is the valve overlap for this camshaft?

●●●7. Where (in degrees ATDC) is the intake centerline for this camshaft?

Use the following camshaft event diagram to answer questions 8–11.

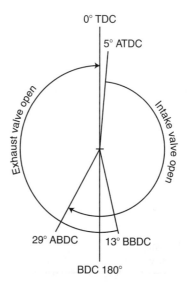

CRANKSHAFTS AND CAMSHAFTS   207

●●8. How many total degrees of crankshaft rotation is the exhaust valve open?

●●9. What is the intake duration for this camshaft?

●●10. What is the valve overlap for this camshaft?

●●●11. Where (in degrees ATDC) is the intake centerline for this camshaft?

●●12. Use the following cam event information to create a camshaft diagram and answer problems 13–15.

	Open	Close
Intake valve	2° ATDC	20° ABDC
Exhaust valve	29° BBDC	1° ATDC

●●13. Find the intake duration for the camshaft in problem 12.

●●14. Find the exhaust duration for the camshaft in problem 12.

●●15. Calculate the valve overlap for the camshaft in problem 12.

●●16. Use the following cam event information to create a camshaft diagram and answer problems 17–20.

	Open	Close
Intake valve	6° BTDC	26° ABDC
Exhaust valve	40° BBDC	2° ATDC

0° TDC

BDC 180°

●●17. Find the intake duration for this camshaft.

●●18. Find the exhaust duration for this camshaft.

●●19. Calculate the valve overlap for this camshaft.

●●●20. Find the intake centerline (in degrees ATDC) for this camshaft.

CHAPTER 11

Thermodynamics

AIR/FUEL RATIOS AND VOLUMETRIC EFFICIENCY

The amount of air and fuel used in combustion affects engine speed, power output, fuel efficiency, emissions, and more. An understanding of the mathematics behind these ideas makes diagnosis and repair easier.

An **air:fuel ratio** describes how many pounds of air are mixed with 1 lb of fuel for the combustion process. Under most conditions, air and gasoline burn most efficiently when they are mixed in a ratio of 14.7:1. That means that in perfect conditions, the engine will draw 14.7 lb of air for every 1 lb of fuel that it consumes. An engine which uses more fuel (or less air) than this is said to be running **rich**. For example, if an engine is running rich, it might use only 13 lb of air for every pound of fuel. Likewise, if an engine uses less fuel (or more air) it is said to be running **lean**. A lean-running engine might use 16 lb of air for every 1 lb of fuel (16:1).

Example 11-1 A gallon of gasoline weighs about 6 lb. How many pounds of air are used in the complete burning of 1 gallon of gasoline, assuming a 14.7:1 air/fuel ratio?

We can answer this question by using a proportion.

$$\frac{14.7 \text{ lb of air}}{1 \text{ lb of gasoline}} = \frac{x \text{ lb of air}}{6 \text{ lb of gasoline}}$$

Cross-multiplying, we get

$$14.7 \times 6 = 1 \times x$$
$$88.2 = x$$

It will take 88.2 lb of air to completely burn 6 lb (1 gallon) of gasoline.

For diesel engines, the air fuel ratio varies. By changing the amount of fuel injected into the cylinders of a diesel engine, the speed of the engine is controlled. The amount of air drawn by any cylinder in a diesel engine does not depend greatly on engine speed. That means that the ratio of air to diesel fuel will be different at low speeds than at high speeds.

Example 11-2 A diesel engine draws 160 lb of air in burning 5 lb of fuel. What is the air/fuel ratio at this rate?

Here, the air/fuel ratio can be written as 160:5, or

$$\frac{160 \text{ lb of air}}{5 \text{ lb of fuel}}$$

Reducing this ratio, we have

$$\frac{32 \text{ lb of air}}{1 \text{ lb of fuel}}$$

Which means the air/fuel ratio is 32:1.

These air/fuel ratios can be changed by restricting either the air or fuel supplied to the engine. To make a gasoline engine efficient, engineers try to make the engine "breathe" as easily as possible. Since the engine is actually pumping air as it runs, it is ideal to waste as little energy as possible on this pumping process.

Ideally, a cylinder would fill completely with air and fuel on the intake stroke. That is, if the cylinder has a displacement of 40 in.^{3}, it would draw 40 in.^{3} of air and fuel. We could also say that the cylinder is 100% filled with air and fuel. The intake system of an engine has an air filter, throttle body, and many other parts that resist airflow. As a result, the cylinders are not able to draw all of the air that they should. One cylinder that displaces 40 in.^{3} may only draw 30 in.^{3} of air and fuel. We could also say that the cylinder only draws 30/40 or 75% of its capacity. **Volumetric efficiency** is the percentage of each cylinder that fills with air and fuel as the engine is running. Before we can calculate volumetric efficiency, we need to calculate how much air and fuel an engine could ideally consume, called the **theoretic airflow**.

Example 11-3 Calculate the theoretic airflow of a 305-in.^{3} engine operating at 4000 RPM.

Ideally, the engine should draw 305 in.^{3} for each revolution, but only half of the cylinders have an intake stroke during any revolution.

$$\frac{305 \text{ in.}^3}{1 \text{ revolution}} \times \frac{1}{2} \times \frac{4000 \text{ rev}}{1 \text{ minute}} = \frac{1{,}220{,}000 \text{ in.}^3}{2 \text{ minutes}} = 610{,}000 \text{ in.}^3/\text{minute}$$

So the engine should consume 610,000 in.^{3} per minute at an engine speed of 4000 RPM.

Usually, airflow is given in CFM or cubic feet per minute. Since 1728 in.^{3} = 1 ft^3, we can repeat the problem and see that

$$\frac{305 \text{ in.}^3}{1 \text{ revolution}} \times \frac{1}{2} \times \frac{4000 \text{ rev}}{1 \text{ minute}} \times \frac{1 \text{ ft}^3}{1728 \text{ in.}^3} = \frac{1{,}220{,}000 \text{ ft}^3}{3456 \text{ minutes}} = 353.0 \text{ ft}^3/\text{minute}$$

Then engine has a theoretic airflow of 353.0 CFM at 4000 RPM.

The computation above can be simplified since the only variables are the engine's displacement and the engine's speed. Multiplying all the other factors together, we can find theoretic airflow by using the following:

$$\text{Theoretic airflow} = \frac{\text{Displacement} \times \text{Engine speed}}{3456}$$

Notice that the units of displacement *must* be in cubic inches (in.^{3}). Also, engine speed must have units of RPM, and the theoretic airflow will be in CFM.

Example 11-4 Find the theoretic airflow of a 170-in.^{3} engine operating at 3500 RPM.

We'll simply use the formula above:

$$\text{Theoretic airflow} = \frac{170 \text{ in.}^3 \times 3500 \text{ RPM}}{3456} = 172.2 \text{ CFM}$$

Example 11-5 Find the theoretic airflow of a 2.5 L engine operating at 4500 RPM.

First, convert 2.5 L to cubic inches.

$$\frac{2.5 \text{ L}}{1} \times \frac{61.0 \text{ in.}^3}{1 \text{ L}} = 152.5 \text{ in.}^3 \text{ or about } 153 \text{ in.}^3$$

Now, use the formula for theoretic airflow.

$$\text{Theoretic airflow} = \frac{153 \text{ in.}^3 \times 4500 \text{ RPM}}{3456} = 199.2 \text{ CFM}$$

Theoretic airflow, however, is just that: theoretic. Most engines do not draw the same amount of air as their theoretic airflow. Most engines draw less. Some racing engines and

THERMODYNAMICS

turbocharged or supercharged engines actually draw more. The amount of air actually consumed by an engine (in CFM) is measured by an airflow meter attached to the engine. This amount of air actually drawn by the engine is called the **actual airflow**. By measuring the actual airflow and comparing it to the theoretic airflow, the volumetric efficiency can be calculated.

$$\text{Volumetric efficiency} = \frac{\text{Actual airflow}}{\text{Theoretic airflow}}$$

Example 11-6 The theoretic airflow of an engine is calculated to be 217.4 CFM at 3000 RPM. The actual airflow is measured to be 196.3 CFM. Calculate the volumetric efficiency.

From above, we have

$$\text{Volumetric efficiency} = \frac{196.3 \text{ CFM}}{217.4 \text{ CFM}} = 0.903 = 90.3\%$$

Example 11-7 A 289-in.^{3} engine operates at 4000 RPM. An airflow meter finds that the engine consumes 287.2 CFM. Find the volumetric efficiency of this engine.

First, find the theoretic airflow:

$$\text{Theoretic airflow} = \frac{289 \text{ in.}^3 \times 4000 \text{ RPM}}{3456} = 334.5 \text{ CFM}$$

The actual airflow is given to be 287.2 CFM, so we have

$$\text{Volumetric efficiency} = \frac{287.2 \text{ CFM}}{334.5 \text{ CFM}} = 0.859 = 85.9\%$$

Example 11-8 A motorcycle engine with a displacement of 1049 cc consumes air at a rate of 49.0 CFM at 2500 RPM. Find the volumetric efficiency.

First, convert 1049 cc to in.^{3}.

$$\frac{1049 \text{ cc}}{1} \times \frac{1 \text{ in.}^3}{16.39 \text{ cc}} = 64.0 \text{ in.}^3$$

Now find the theoretic airflow.

$$\text{Theoretic airflow} = \frac{64 \text{ in.}^3 \times 2500 \text{ RPM}}{3456} = 46.3 \text{ CFM}$$

Finally, calculate the volumetric efficiency.

$$\text{Volumetric efficiency} = \frac{46.3 \text{ CFM}}{49.0 \text{ CFM}} = 0.945 = 94.5\%.$$

INDUCTION SYSTEM AND CARBURETOR SIZING

Knowing the theoretic airflow for a given engine allows us to choose a carburetor or fuel injection system that is sized appropriately for a given engine. Using a carburetor that is too small will limit horsepower at high speeds, while using one that is too large usually decreases throttle response. We'll refer to carburetors and fuel injection systems as induction systems, but note that all of our calculations refer to both types of fuel delivery systems. A sample carburetor can be seen in Figure 11.1.

While volumetric efficiency changes with RPM, it is common for a naturally aspirated (not turbocharged or supercharged) street engine to have a volumetric efficiency of about 80% at high RPM. Some older engines are less, and many high performance engines are much higher. That means that at around 6000 RPM, the engine actually draws about 80%

FIGURE 11.1 The carburetor size must be matched to the engine's airflow to maximize performance and fuel economy.

of its theoretic airflow. Turbocharged and supercharged engines can have volumetric efficiencies around 110%. That means they actually draw air at a rate greater than their theoretic airflow.

Based on that idea, to choose the correct size for an induction system, we can simply multiply the volumetric efficiency by the theoretic airflow.

$$\text{Induction system airflow} = \text{Volumetric efficiency} \times \text{Theoretic airflow}$$

It is important to note that the theoretic airflow must be calculated at the highest speed an engine will reach during use.

Example 11-9 A naturally aspirated engine has a volumetric efficiency of 80%. It has a theoretic airflow of 432.0 CFM at the peak RPM of 6500. What size fuel injection system should be used on this engine?

From the formula for induction system airflow above,

$$\text{Induction system airflow} = 0.80 \times 432.0 \text{ CFM} = 345.6 \text{ CFM}$$

Example 11-10 A turbocharged engine with a volumetric efficiency of 110% has a theoretic airflow of 492.3 CFM at the peak speed of 7000 RPM. What size carburetor should be placed on this engine?

Again,

$$\text{Induction system airflow} = 1.10 \times 492.3 \text{ CFM} = 541.5 \text{ CFM}$$

Usually, carburetors are sold in 50 CFM intervals, and a carburetor flow rating must always have a value larger than the maximum airflow of the engine. A 550 CFM carburetor should be used.

THERMODYNAMICS

Example 11-11 A 350-in.^{3} engine operates up to 5500 RPM. Assume it is built for racing, and has a volumetric efficiency of 90%. Find the proper carburetor size for this engine.

First, find the theoretic airflow as we did earlier.

$$\text{Theoretic airflow} = \frac{350 \text{ in.}^3 \times 5500 \text{ RPM}}{3456} = 557.0 \text{ CFM}$$

$$\text{Induction system airflow} = 0.90 \times 557.0 \text{ CFM} = 501.3 \text{ CFM}$$

A 550-CFM carburetor would be about right for this engine.

Example 11-12 A 2.1 L engine has a turbocharger and will operate up to 7500 RPM. Find the proper maximum flow rate needed for a fuel injection system assuming 115% volumetric efficiency.

As we did earlier, find the theoretic airflow by first converting the displacement to in.^{3}.

$$\frac{2.1 \text{ L}}{1} \times \frac{61.0 \text{ in.}^3}{1 \text{ L}} = 128.1 \text{ in.}^3 \text{ or about } 128 \text{ in.}^3$$

$$\text{Theoretic airflow} = \frac{128 \text{ in.}^3 \times 7500 \text{ RPM}}{3456} = 277.8 \text{ CFM}$$

$$\text{Induction system airflow} = 1.15 \times 277.8 \text{ CFM} = 319.5 \text{ CFM}$$

The maximum airflow is about 319.5 CFM, and a 350 CFM system would be large enough.

INDICATED HORSEPOWER AND TORQUE

As we've discussed earlier, much of the power generated by an engine is consumed in overcoming internal friction. That means that when an engine is placed on the dynamometer and brake horsepower is measured, only a fraction of the engine's actual horsepower is being observed. **Indicated horsepower (ihp)** is a measure of the overall power generated by the burning of gasoline within the cylinder, before friction can steal some of that useful power. **Indicated torque**, then, is a measure of the overall torque generated by the burning fuel before frictional losses. There is no way a dynamometer can measure these values, but we can calculate them. Later, we'll compare the indicated horsepower to the brake horsepower to find how much power is wasted in simply turning the engine.

There are two parts to calculating indicated horsepower. The first part is to find the mean effective pressure of a typical cylinder. The **mean effective pressure (MEP)** is the pressure generated inside a cylinder by the burning (and expanding) fuel and air. The MEP is found by using equipment that measures the pressure inside the cylinders while the engine is running. This pressure changes as the piston goes down in the cylinder, which is why the mean (average) pressure is used. Figure 11.2 shows how MEP acts on the piston.

Once the MEP for a given engine has been measured, we can perform the needed calculation. We'll work more with the idea in Chapter 12, but it's a basic fact that the pressure on the piston is determined by the MEP and the area of the piston. See Figure 11.3 for a diagram of this calculation.

$$\text{Force} = \text{Pressure} \times \text{Area}$$

$$\text{Downward force on piston} = \text{MEP} \times \text{Piston face area}$$

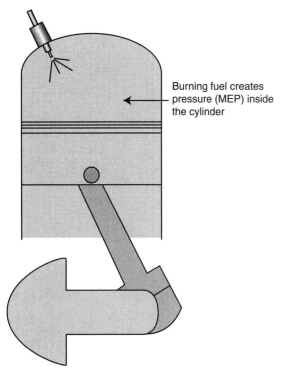

FIGURE 11.2 This figure shows a typical cylinder in which MEP is creating a downward force on the piston.

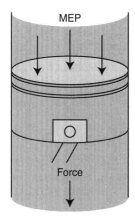

MEP × Piston face area = Downward force on piston

FIGURE 11.3 The mean effective pressure (MEP) can be used to find the force on the piston.

The distance through which this force is acting is simply the length of the stroke. We'll divide the stroke by 12 to convert it to feet as required by the definition of horsepower. Also, on any revolution, half of the cylinders are producing power.

$$\text{Work} = \text{Force} \times \text{Distance}$$

$$\text{Work done per revolution} = \text{MEP} \times \text{Piston face area} \times \frac{\text{Stroke}}{12} \times \frac{\text{Number of cylinders}}{2}$$

Finally, multiplying by the speed of the engine (and dividing by 33,000 to convert the units to horsepower), we have

$$\text{Indicated horsepower} = \text{MEP} \times \text{Piston area} \times \frac{\text{Stroke}}{12} \times \frac{\text{Number of cylinders}}{2} \times \frac{\text{RPM}}{33,000}$$

From Chapter 8, however, we saw that

$$\text{Engine displacement} = \text{Piston area} \times \text{Stroke} \times \text{Number of cylinders}$$

Since all measurements are in inches, *the engine's displacement must be in cubic inches* for this formula to work.

Then the formula above can be simplified to give

$$\text{Indicated horsepower} = \frac{\text{MEP} \times \text{Engine displacement} \times \text{RPM}}{792,000}$$

THERMODYNAMICS

Apparently, then, all we need to compute the theoretical horsepower created by an engine is the MEP, the engine's displacement, and the speed (RPM) of the engine.

Example 11-13 The MEP of a 260-in.^{3} engine operating at 3500 RPM is 180 lb/in.^{2} (or 180 psi). What is the indicated horsepower of this engine?

We can simply use the formula we just developed

$$\text{Indicated horsepower} = \frac{\text{MEP} \times \text{Engine displacement} \times \text{RPM}}{792{,}000}$$

with the information we were given:

$$\text{Indicated horsepower} = \frac{180 \times 260 \times 3500}{792{,}000} = 206.8 \text{ ihp}$$

Example 11-14 A supercharged 3.8 L engine has an MEP value of 215 psi while operating at 4500 RPM. What is the indicated horsepower for this engine?

We must first convert the displacement to cubic inches.

$$\frac{3.8 \text{ L}}{1} \times \frac{61.0 \text{ in.}^3}{1 \text{ L}} = 231.8 \text{ in.}^3 \text{ or about } 232 \text{ in.}^3$$

Now, use the formula above with the values we're given.

$$\text{Indicated horsepower} = \frac{215 \times 232 \times 4500}{792{,}000} = 283.4 \text{ ihp}$$

Although not as useful, we can also calculate the indicated torque for an engine. Using the fact that

$$\text{Torque} = \frac{\text{Horsepower} \times 5252}{\text{RPM}}$$

we can see that

$$\text{Indicated torque} = \frac{\text{Indicated horsepower} \times 5252}{\text{RPM}}$$

$$\text{Indicated torque} = \frac{\text{MEP} \times \text{Engine displacement} \times \text{RPM}}{792{,}000} \times \frac{5252}{\text{RPM}}$$

This formula can be simplified to

$$\text{Indicated torque} = \frac{\text{MEP} \times \text{Engine displacement}}{150.8}$$

Example 11-15 Find the indicated torque of a 183-in.^{3} engine if the MEP is measured to be 155 psi at 2500 RPM.

Notice that the speed of the engine does not matter in this case. Using the engine displacement and MEP,

$$\text{Indicated torque} = \frac{155 \times 183}{150.8} = 188.1 \text{ lb} \cdot \text{ft of torque}$$

MECHANICAL EFFICIENCY

Indicated horsepower and torque values are not typically used to rate engine power because some part of the indicated values are used to overcome the friction involved with engine operation. Recall that brake horsepower (bhp) is the useful horsepower that can be used to do work outside of the engine. Brake horsepower is always less than indicated horsepower.

Ideally, an engine could be designed that had no internal friction. That is, it took no effort to pump the oil, move the pistons, turn the water pump, and so on. Then none of the indicated horsepower would be wasted, and the brake horsepower would equal the indicated horsepower. We could say that the engine was efficient, because it wasted very little energy. Mechanical efficiency is the percentage of indicated horsepower that becomes brake horsepower. Another way to say this is that the mechanical efficiency is the percentage of theoretic horsepower that can actually be used to move the vehicle.

$$\text{Mechanical efficiency} = \frac{\text{Brake horsepower}}{\text{Indicated horsepower}}$$

Note that the indicated horsepower is always calculated, while the brake horsepower is always measured on a dynamometer.

Example 11-16 The 260-in.^{3} engine in Example 11-13 had an indicated horsepower value of 206.8 horsepower at 3500 RPM. On a dynamometer, the engine produces 166 brake horsepower at 3500 RPM. What is the mechanical efficiency at this engine speed?

The indicated horsepower of this engine was calculated to be 206.8 hp at 3500 RPM. The brake horsepower was measured to be 166 hp at 3500 RPM.

$$\text{Mechanical efficiency} = \frac{\text{Brake horsepower}}{\text{Indicated horsepower}}$$

$$\text{Mechanical efficiency} = \frac{166 \text{ bhp}}{206.8 \text{ ihp}} = 0.803 = 80.3\%$$

Apparently, the remaining power (19.7%) was used to overcome the friction within the engine (internal friction).

Example 11-17 A 4-cylinder engine with a displacement of 156 in.^{3} is placed on a dynamometer while the MEP is measured. At 5,500 RPM, the MEP is measured and found to be 175 lb/in.^{2}, while the dynamometer measures the brake horsepower to be 148 bhp. What is the mechanical efficiency of this engine at 5,500 RPM?

First, calculate the indicated horsepower.

$$\text{Indicated horsepower} = \frac{175 \times 156 \times 5{,}500}{792{,}000} = 189.6 \text{ ihp}$$

Since we know the brake horsepower to be 148 bhp, we have

$$\text{Mechanical efficiency} = \frac{148 \text{ bhp}}{189.6 \text{ ihp}} = 0.781 = 78.1\%$$

ALTITUDE COMPENSATION

Anyone who has ever driven a vehicle in the high-elevation mountain passes of Wyoming or Colorado knows that a vehicle has less power at high altitudes. The reason for this is quite simple. Gravity pulls our atmosphere downward just as it does anything else. As a result, there is more air (and greater pressure) near sea level than there is at an altitude high in the mountains. Since there is less pressure, engines operating at high altitudes draw in less air (and less fuel) on each intake stroke, meaning they produce less power.

At sea level, the pressure due to the atmosphere around us is about 14.7 psi. This drops about 0.5 psi for every 1000 ft above sea level (for reasonably low altitudes). Since 0.5 psi is about 3.4% of 14.7 psi, we can say that the amount of oxygen available for combustion drops by about 3.4% for every 1000 ft above sea level.

This value of 3.4% decreases as we go higher, though. As a general rule of thumb, an internal combustion engine loses 3% of its power for ever 1000 ft in elevation, or about 1% for every 100 m.

THERMODYNAMICS

Example 11-18 An engine produces 160 bhp at sea level. How much power can it produce on a mountain road, where the elevation is 5000 ft?

There are several approaches to solve this problem. Let's set up a ratio to find what percentage of the engine's power is being lost.

$$\frac{3\% \text{ of total power}}{1000 \text{ ft}} = \frac{x}{5000 \text{ ft}}$$

Cross-multiplying, we have

$$x \times 1000 \text{ ft} = 0.03 \times 5000 \text{ ft}$$

$$x = 0.15 \text{ or } 15\% \text{ of the engine's total power}$$

Since the engine loses 15% of its total power, we can multiply to find the total power lost.

$$15\% \text{ of } 160 \text{ bhp is lost}$$

$$0.15 \times 160 \text{ bhp} = 24 \text{ bhp lost}$$

Then at this higher elevation the engine produces

$$160 \text{ bhp} - 24 \text{ bhp} = 136 \text{ bhp}$$

There's really no need to go through all that trouble. Again, we can develop a formula since the computation is always the same.

> Brake horsepower lost in higher altitude
> $$= \text{Horsepower at sea level} \times \frac{\text{Elevation in feet}}{1000} \times 0.03$$

Keep in mind that this is the amount of power lost, and must be subtracted from the engine's horsepower at sea level to find the amount of power being produced.

Example 11-19 A truck producing 155 bhp at sea level is driven through a mountain pass in Wyoming. The elevation of the pass is 9600 feet ASL (above sea level). How much brake horsepower is the engine producing?

Use the formula above to calculate the loss in horsepower.

Brake horsepower lost in higher altitude

$$= \text{Horsepower at sea level} \times \frac{\text{Elevation in feet}}{1000} \times 0.03$$

$$= 155 \text{ bhp} \times \frac{9600}{1000} \times 0.03 = 44.64 \text{ bhp}$$

Then the engine produces

$$155 \text{ bhp} - 44.64 \text{ bhp} = 110.36 \text{ or about } 110 \text{ bhp}.$$

The engine here lost almost 30% of its power. That's noticeable to the driver!

Example 11-20 For a little adventure, you decide to ride your moped, which produces 3.2 bhp at sea level to the top of Pike's Peak in Colorado (4,300 meters ASL). How much horsepower will the moped produce at this altitude?

First, convert 4,300 m to feet.

$$\frac{4,300 \text{ m}}{1} \times \frac{3.28 \text{ ft}}{1 \text{ m}} = 14,108 \text{ ft}$$

Now, apply our formula for power loss.

Brake horsepower lost in higher altitude $= 3.2 \text{ bhp} \times \dfrac{14,108}{1000} \times 0.03 = 1.35 \text{ bhp}$

Then the moped produces

$$3.2 \text{ bhp} - 1.35 \text{ bhp} = 1.85 \text{ bhp}$$

Looks like you'll need to push that moped to the top.

CHAPTER 11 *Practice Problems*

Air/Fuel Ratios and Volumetric Efficiency

●1. A certain engine operates at a 14.7:1 air/fuel ratio. How many pounds of air are drawn in as the engine uses 4 lb of gasoline?

●2. At idle, an engine is set to operate at an air/fuel ratio of 13.5:1. How many pounds of air are needed to completely burn 8 lb of gasoline?

●3. A "lean-burn" engine operates with an air/fuel ratio of 17:1. How many pounds of fuel are needed while the engine consumes 340 lb of air?

●4. Using a 14.7:1 air/fuel ratio, an engine draws 441 lb of air. How much gasoline does it use in that time?

●5. A certain engine is tested and uses 75.0 lb of air while burning 5.0 lb of fuel. What is the air/fuel ratio for this engine?

●6. An engine consumes 36.5 lb of air and 2.5 lb of fuel during a test period at wide-open throttle. What is the air/fuel ratio?

●7. A diesel engine idles and consumes 75 lb of air and 2.5 lb of fuel. What is the air/fuel ratio under these conditions?

●8. Under a heavy load, a diesel engine consumes 6.2 lb of fuel and 111.6 lb of air. What is the air/fuel ratio during this period?

●9. Find the theoretic airflow of a 150-in.^{3} engine operating at 3700 RPM.

●10. What is the theoretic airflow of a 327-in.^{3} engine at a speed of 4200 RPM?

●11. A 2.2-L engine operates at 3700 RPM. Find the theoretic airflow.

●12. Calculate the theoretic airflow of a 4.1-L engine operating at 5000 RPM.

●13. Find the volumetric efficiency of an engine having a theoretic airflow of 213.9 CFM and an actual airflow of 202.4 CFM.

●14. A V-8 engine operating at 6000 RPM has a theoretic airflow of 482.7 CFM. An airflow meter shows an actual airflow of 421.0 CFM. Find the volumetric efficiency.

●●15. Find the volumetric efficiency of a 225-in.^{3} engine operating at 2800 RPM and having an actual airflow of 172.3 CFM.

●●16. A 140-in.^{3} engine operating at 4200 RPM has an actual airflow of 158.3 CFM. Find the volumetric efficiency.

●●17. A turbocharged 122-in.^{3} engine operating at 6000 RPM has an actual airflow of 223.7 CFM. Calculate the volumetric efficiency.

THERMODYNAMICS 219

••18. A turbocharged 232-in.³ engine draws air at a rate of 382.4 CFM at 5500 RPM. Find the volumetric efficiency.

•••19. A 2.4-L engine operating at 4800 RPM draws air at a rate of 192.0 CFM. Find the volumetric efficiency.

•••20. An 8.0-L engine operating at 4000 RPM has an actual airflow of 513 CFM. Calculate the volumetric efficiency.

Induction System and Carburetor Sizing

•1. A truck engine has a maximum theoretic airflow of about 445 CFM, and 80% volumetric efficiency. What is the maximum airflow through the induction system?

•2. A motorcycle has a theoretic airflow of 68.3 CFM and 90% volumetric efficiency. What is the maximum airflow through the throttle body (induction system) of this motorcycle?

•3. A turbocharged 4-cylinder engine has a maximum theoretic airflow of 265.0 CFM. Assuming it has been built for racing and has 120% volumetric efficiency, what is the maximum airflow through the carburetor? What size carburetor should be used?

•4. A V-8 engine with a high-rise manifold has 95% volumetric efficiency at the maximum engine speed of 7000 RPM. If the theoretic airflow at this speed is 662 CFM, what is the maximum airflow through the manifold? What size carburetor should be used?

••5. A 280-in.³ engine has 90% volumetric efficiency at the peak speed of 6500 RPM. What is the maximum airflow through the throttle body?

••6. At the peak speed of 3000 RPM a 210-in.³ tractor engine has 75% volumetric efficiency. Find the maximum airflow needed by this engine, and determine the proper carburetor size.

••7. A turbocharged 140-in.³ engine has 110% volumetric efficiency at the redline (peak) engine speed of 7000 RPM. Find the maximum airflow for the fuel injection system on this engine.

••8. Refer to the engine in problem 7, but assume the intake system is modified to increase the volumetric efficiency to 115%. Calculate the new maximum airflow.

•••9. A 2.9-L engine has 80% volumetric efficiency at the redline engine speed of 6000 RPM. Calculate the maximum airflow through the induction system.

•••10. A 4.6-L engine having 100% volumetric efficiency at the peak speed of 6250 RPM is fitted with a new throttle body. What is the minimum flow rating for the new throttle body?

Indicated Horsepower and Torque

•1. A 176-in.³ engine has an MEP value of 190 lb/in.² while the engine is operating at 3800 RPM. Find the indicated horsepower for this engine.

•2. The mean effective pressure inside the cylinders of a 318-in.³ V-8 engine is 205 psi while the engine is operating at 4000 RPM. Find the indicated horsepower of this engine.

CHAPTER 12

Transmission and Gear Ratios

GEAR RATIOS

The engine produces plenty of power needed to move the vehicle, but only within a certain speed range. The transmission and axle ratio must change from the engine speed to a speed that is appropriate for the wheels. At the same time, the gears in the transmission must change the torque output of the engine to an amount that is enough to accelerate the vehicle.

To understand gear ratios, we need to first understand levers. Imagine trying to lift a heavy boulder by using a long pry bar and a fulcrum, as in Figure 12.1.

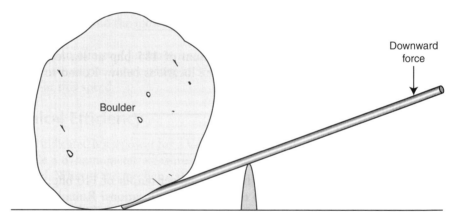

FIGURE 12.1 The distance from the downward force to the fulcrum is the lever arm.

The distance from the boulder to the fulcrum is much less than the distance from the fulcrum to the point where you'd apply a downward force. From your own experience, you know that even if you only weigh 100 lb, you could provide a lifting force of much more than that (maybe 500 lb) due to your ingenuity in using a lever.

Of course, there's no free lunch when it comes to doing work, and the increase in lifting force comes at a price. You might need to push your end of the board downward several inches (maybe up to 5 in.) to lift the boulder even 1 in. This tradeoff of distance for force is exactly what happens in a gear ratio.

Imagine if a similar lever were used to turn a gear instead of lifting a boulder, as in Figure 12.2.

The input force was applied and moved the lever 12 in., while the output distance was only 4 in. Since the output side of the lever only moved 1/3 of the input distance, torque was increased by a factor of 3.

Gears work on the same principle. When two gears mesh, one will be the gear providing power, and is called the **drive** gear. The other, which gets turned, is called the **driven** gear.

When the drive gear is smaller than the driven gear, the gears increase the amount of torque, but decrease the speed. Usually it is easier to count the teeth on a gear to determine

TRANSMISSION AND GEAR RATIOS

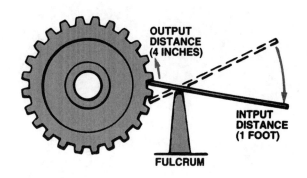

FIGURE 12.2 A longer input distance is used to achieve a shorter output distance while torque is increased.

Source: James D. Halderman and Chase D. Mitchell, Jr., *Automotive Technology: Principles, Diagnosis, and Service,* Second Edition © 2003. Reprinted by permission of Pearson Education, Inc., Upper Saddle River, NJ.

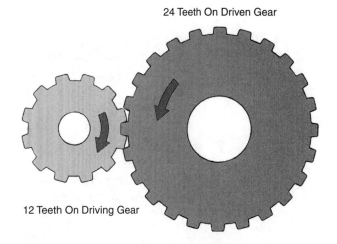

FIGURE 12.3 The size of the drive gear and driven gear determines the gear ratio.

Source: James D. Halderman and Chase D. Mitchell, Jr., *Automotive Technology: Principles, Diagnosis, and Service,* Second Edition © 2003. Reprinted by permission of Pearson Education, Inc., Upper Saddle River, NJ.

its size than it is to measure the radius, although either method will give the same result. Two gears with teeth counted are shown in Figure 12.3. Using a proportion, we can say the following:

$$\frac{\text{Driven gear teeth}}{\text{Drive gear teeth}} = \frac{\text{Output torque}}{\text{Input torque}}$$

$$\frac{\text{Driven gear teeth}}{\text{Drive gear teeth}} = \frac{\text{Input speed}}{\text{Output speed}}$$

Another example of meshing gears is the ring and pinion gearset, shown in Figure 12.4. The following example looks at this type of gear pair.

Example 12-1 A ring gear with 37 teeth meshes with a pinion gear having 12 teeth. If the drive shaft of the vehicle turns with 370 lb · ft of torque at 3000 RPM, how much torque is present at the rear wheels, and what is their speed?

First, find the torque at the rear wheels.

The ring gear is the driven gear, and has 37 teeth.
The pinion gear is the drive gear, and has 12 teeth.
The input torque is the drive shaft torque, and is 370 lb · ft.

Our proportion can be written using these values and solved.

$$\frac{\text{Driven gear teeth}}{\text{Drive gear teeth}} = \frac{\text{Output torque}}{\text{Input torque}}$$

$$\frac{37 \text{ teeth}}{12 \text{ teeth}} = \frac{n}{370 \text{ lb} \cdot \text{ft}}$$

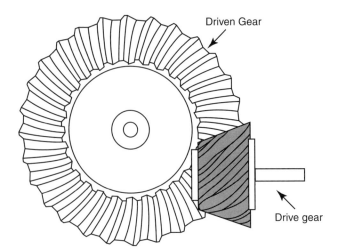

FIGURE 12.4 The number of teeth on the ring and pinion gears determines the speed and amount of torque at the drive wheels.

$$12 \times n = 37 \times 370$$
$$n = 1140.833 \text{ lb} \cdot \text{ft}$$

The torque is increased to about 1140.8 lb·ft at the rear wheels.

Repeat this idea to find the speed of the rear wheels.

Again, the driven gear has 37 teeth and the drive gear has 12. The input speed is the drive shaft speed, which is 3000 RPM.

$$\frac{\text{Driven gear teeth}}{\text{Drive gear teeth}} = \frac{\text{Input speed}}{\text{Output speed}}$$

$$\frac{37 \text{ teeth}}{12 \text{ teeth}} = \frac{3000 \text{ RPM}}{n}$$

$$37 \times n = 3000 \times 12$$
$$n = 972.97 \text{ RPM}$$

The wheels will turn at a rate of about 973.0 RPM.

Each pair of gears that mesh together will always change torque and speed by the same factor. The change in torque and speed is determined by the gear ratio of the two gears. The **gear ratio** is the ratio of the driven gear to the drive gear, usually expressed as a decimal number rounded to the hundredth place.

$$\text{Gear ratio} = \frac{\text{Driven gear teeth}}{\text{Drive gear teeth}}$$

Example 12-2 Find the gear ratio of the differential described in Example 12-1.

Recall that the driven gear is the ring gear, and has 37 teeth, and that the drive gear is the pinion gear, which has 12 teeth. Then we can compute the gear ratio.

$$\text{Gear ratio} = \frac{\text{Driven gear teeth}}{\text{Drive gear teeth}} = \frac{37 \text{ teeth}}{12 \text{ teeth}} = 3.083$$

Rounding, and writing as a ratio we can say that the axle ratio here is 3.08:1.

This means the drive shaft must turn 3.08 times to make the drive wheels turn 1 time.

We can also use the gear ratio for two gears to determine the output speed and torque instead of using ratios as we did earlier. Rearranging the speed and torque formulas above we have

TRANSMISSION AND GEAR RATIOS

> Output torque = Input torque × Gear ratio

> Output speed = Input speed ÷ Gear ratio

Since an increase in numeric gear ratio will result in an increase in output torque we say that we have a **direct relationship** between those quantities. Increasing the numeric ratio will decrease the output speed, however, and we call this an **inverse relationship.**

Example 12-3 Use the gear ratio of 3.08:1 to find the torque output and speed of the rear wheels assuming the same conditions as Example 12.1.

Recall that the drive shaft is turning at 3000 RPM with 370 lb·ft of torque. Since our gear ratio is 3.08:1, we have

Output torque = Input torque × Gear ratio
Output torque = 370 lb·ft × 3.08 = 1139.6 lb·ft
Output speed = Input speed ÷ Gear ratio
Output speed = 3000 RPM ÷ 3.08 = 974.0 RPM

These values are slightly different than those found in Example 12-1 using the proportion method. The rounding of the gear ratio to 3.08 introduced a bit of error into our calculation.

These three examples featured a driven gear that was larger than the drive gear, which resulted in a gear ratio numerically larger than 1. A **gear reduction** occurs when a gear ratio is greater than 1, such as in Figure 12.5. A gear reduction increases torque but reduces speed.

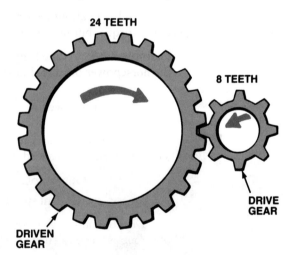

FIGURE 12.5 These gears will have a ratio of exactly (24 ÷ 8) or 3.00:1.

Source: James D. Halderman and Chase D. Mitchell, Jr., *Automotive Technology: Principles, Diagnosis, and Service,* Second Edition © 2003. Reprinted by permission of Pearson Education, Inc., Upper Saddle River, NJ.

It is possible for the driven gear to be smaller than the drive gear. A **gear overdrive** or **overdrive** results when a gear ratio is less than 1. An overdrive reduces torque, but increases speed. This is exactly what the overdrive gear on a transmission does. The increase in output speed increases fuel economy by speeding up the rear wheels (and slowing the engine). This reduction in torque means the vehicle will not accelerate well in an overdrive gear. See Figure 12.6 for an example of an overdrive.

If two gears mesh and have the same number of teeth, their gear ratio will be 1:1. This is called a **direct drive** gear set. Speed and torque are not changed in a 1:1 ratio.

Example 12-4 An overdrive unit having a ratio of 0.87:1 is installed on the back of a transmission. The input shaft of this overdrive unit turns with 176 lb·ft of torque at 1800 RPM. What will be the torque and speed of the output shaft of this overdrive unit?

Since this gear ratio is less than 1, we can expect the torque to decrease and the speed to increase. From above, we have

To find the overall first gear ratio, multiply these two gear ratios together.

$$\text{First gear ratio} = 1.273 \times 3.000 = 3.819.$$

We would say that the first gear ratio here is 3.82:1.

It would take 3.82 turns of the input shaft to make the output shaft turn one time.

Example 12-7 Find the gear ratio for fourth gear on the transmission shown below.

Fourth gear

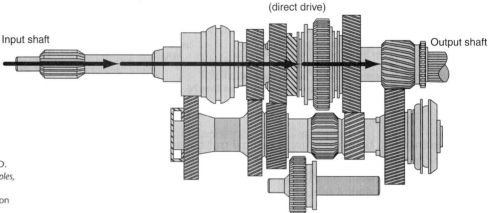

Source: James D. Halderman and Chase D. Mitchell, Jr., *Automotive Technology: Principles, Diagnosis, and Service,* Second Edition © 2003. Reprinted by permission of Pearson Education, Inc., Upper Saddle River, NJ.

Most manual transmissions have a direct drive gear before entering the overdrive gears. In this diagram, the input shaft is connected directly to the output shaft. Since the input and output shafts turn at the same rate, the ratio is 1:1.

The overall drive ratio is often referred to in automotive magazines that feature the technical specifications of a vehicle. The **overall drive ratio** is the overall gear ratio for any gear of the transmission when used with a certain axle ratio.

Example 12-8 Find the overall drive ratio for first gear when the transmission in Example 12-6 is connected to a differential having a 3.500:1 axle ratio.

Again, to find the overall drive ratio, we can multiply the gear ratios together.

The ratio for first gear above (before rounding) is 3.819:1.
The rear axle ratio has a ratio of 3.500:1.
Since $3.500 \times 3.819 = 13.367$, the overall drive ratio for first gear is 13.367:1.

Using the conventional 2 decimals, our overall drive ratio is 13.37:1.

Example 12-9 Suppose the transmission and rear axle we've been discussing are in a vehicle with an engine that produces 165 lb · ft of torque at 3400 RPM. Find the amount of torque and speed at the drive wheels in first gear.

From the previous section, we know the following:

$$\text{Output torque} = \text{Input torque} \times \text{Gear ratio}$$
$$\text{Output torque} = 165 \text{ lb} \cdot \text{ft} \times 13.37 = \text{about } 2206 \text{ lb} \cdot \text{ft}$$
$$\text{Output speed} = \text{Input speed} \div \text{Gear ratio}$$
$$\text{Output speed} = 3400 \text{ RPM} \div 13.37 = \text{about } 254 \text{ RPM}$$

PLANETARY GEAR RATIOS

Automatic transmission gear ratios are obtained by the use of planetary gear sets. A planetary gear set consists of three primary components: A sun gear, planetary gears (and a carrier), and a ring gear. See Figure 12.8. Since there are three components to a planetary gear set, gear ratios are calculated a bit differently.

TRANSMISSION AND GEAR RATIOS

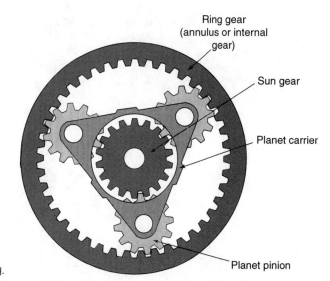

FIGURE 12.8 The three components of a planetary gear set are shown here.

Source: James D. Halderman and Chase D. Mitchell, Jr., *Automotive Technology: Principles, Diagnosis, and Service*, Second Edition © 2003. Reprinted by permission of Pearson Education, Inc., Upper Saddle River, NJ.

Holding one of the three components of a planetary gear set fixed allows a power transfer to occur. Of the two remaining components, one will become the input component while the other becomes the output component. Each of these is controlled by a series of clutches and bands within the transmission.

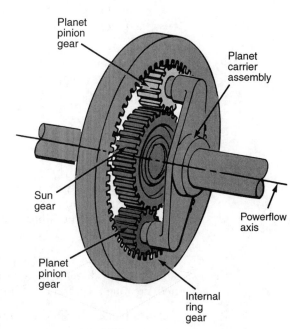

FIGURE 12.9 Controlling which component is held fixed will determine the gear ratio of the assembly.

Source: James D. Halderman and Chase D. Mitchell, Jr., *Automotive Technology: Principles, Diagnosis, and Service*, Second Edition © 2003. Reprinted by permission of Pearson Education, Inc., Upper Saddle River, NJ.

In Figure 12.9, suppose the sun gear were held fixed. Turning the planet gear carrier would cause the planetary gears to spin, causing the ring gear to turn as well. The ring gear would turn faster than the planet gear carrier, resulting in an overdrive. The gear ratio for this scenario can be calculated as follows:

$$\text{Gear ratio} \begin{cases} \text{Sun gear fixed,} \\ \text{Planet gear carrier is input} \\ \text{Ring gear is output} \end{cases} = \frac{\text{Ring gear teeth}}{\text{Sun gear teeth} + \text{Ring gear teeth}} :1$$

Also, the input shaft and output shaft will turn in the same direction.

Example 12-10 Find the gear ratio for the planetary gear set shown below.

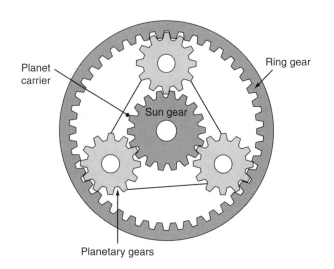

Input shaft: Connected to the planet carrier
Output shaft: Connected to the ring gear
Fixed: Sun gear

Ring gear: 80 teeth
Sun gear: 30 teeth
Planetary gears: 25 teeth each

From the information we're given, we have the scenario described above, and can use the following ratio formula.

$$\text{Gear ratio} \begin{cases} \text{Sun gear fixed,} \\ \text{Planet gear carrier is input} \\ \text{Ring gear is output} \end{cases} = \frac{\text{Ring gear teeth}}{\text{Sun gear teeth + Ring gear teeth}} : 1$$

$$\text{Gear ratio} = \frac{80}{80 + 30} = \frac{80}{110} = 0.73 : 1$$

By changing which of the three components is held fixed, and which becomes the input and output, various gear ratios can be created. Also, by changing which component is held fixed, the direction of rotation can be changed. Table 12.1 describes the eight possible cases, and the formula for finding the gear ratio for each case.

Notice that if any two components are locked together, all three will turn as a unit, creating a direct drive. If none of the components is held fixed, no power will be transmitted, creating a neutral condition.

Example 12-11 Find the gear ratio for a planetary gear set used in an automatic transmission that has reverse rotation and the following conditions.

Input shaft: Connected to the sun gear
Output shaft: Connected to the ring gear
Fixed: Planet carrier

Ring gear: 60 teeth
Sun gear: 24 teeth
Planetary gears: 18 teeth each

Use Table 12.1 on the next page to find the gear ratio for this condition.

Fixed	Input	Output	Input/Output Rotation	Gear Ratio
Planet carrier	Sun gear	Ring gear	Reversed	$\frac{\text{Ring}}{\text{Sun}}:1$

TABLE 12.1 Gear Ratio Formulas for the Eight Gear Combinations*

Fixed	Input	Output	Input/Output Rotation	Gear Ratio
Sun gear	Planet carrier	Ring gear	Same	$\frac{Ring}{Ring + Sun}:1$
Sun gear	Ring gear	Planet carrier	Same	$\frac{Ring + Sun}{Ring}:1$
Planet carrier	Sun gear	Ring gear	Reversed	$\frac{Ring}{Sun}:1$
Planet carrier	Ring gear	Sun gear	Reversed	$\frac{Sun}{Ring}:1$
Ring gear	Planet carrier	Sun gear	Same	$\frac{Sun}{Ring + Sun}:1$
Ring gear	Sun gear	Planet carrier	Same	$\frac{Ring + Sun}{Sun}:1$
Any two components locked together	Any	Any	Same	1:1 Direct drive
None	Any	Any	None	Neutral

*For computing gear ratios, Ring = ring gear tooth count; Sun = sun gear tooth count.

The gear ratio is then $\frac{Ring}{Sun}:1$. With our values, we have

$$\frac{60}{24}:1 = 2.5:1$$

Example 12-12 First gear of the same transmission holds the ring gear fixed, and locks the input shaft to the sun gear. Calculate the gear ratio created by this combination.

Again, use Table 12.1 to find the information for this condition.

Fixed	Input	Output	Input/Output Rotation	Gear Ratio
Ring gear	Sun gear	Planet gear	Reversed	$\frac{Ring + Sun}{Sun}:1$

The gear ratio is then $\frac{Ring + Sun}{Sun}:1$. With the values from Example 12-11, we have

$$\frac{60 + 24}{24}:1 = 3.5:1.$$

TIRE SIZING

You may know that if the tire size installed on a vehicle is changed, it can affect not only the speedometer accuracy but the odometer and computer on the vehicle. For this reason, it is best to replace tires with new ones of the same size. Should a change take place, however, an understanding of the system used to describe tire size is useful.

The standard system used to describe tire size is based on metric dimensions. The type of tire, the width, the sidewall height, and the rim size can all be determined based on the size given on the sidewall of newer tires, as shown in Figure 12.10.

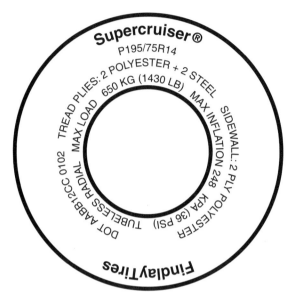

FIGURE 12.10 The metric size system on this tire gives the size as P195/75R14.

The tire shown has the size shown on the sidewall. Each part of this labeling system tells us something about the tire. The following markings are taken from Figure 12.10.

- P   Designates the type of tire. Common markings are P for passenger cars, and LT for light truck applications.
- 195   The overall width of the tire, in millimeters. Typical widths range from 155 for small vehicles to 325 or more for high performance vehicles. The width is measured at the widest point, as shown in Figure 12.11.
- 75   The **aspect ratio.** This means the sidewall height of the tire is 75% of the width. Typical aspect ratios vary from 80% for smaller tires to as low as 35% for performance tires.
- R   This designates that the tire was built using radial construction.
- 14   This is the diameter of the rim, in inches, that the tire must be mounted on. Most rim sizes vary from 13 in. to 17 in. See Figure 12.12.

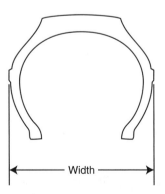

FIGURE 12.11 The width of the tire, in millimeters, is measured at the widest place.

TRANSMISSION AND GEAR RATIOS

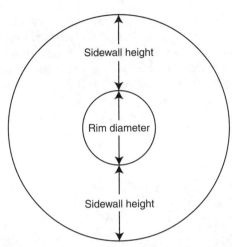

FIGURE 12.12 The overall height of the tire is equal to the rim diameter plus two times the sidewall height.

Example 12-13 What is the overall width, in inches, of a P215/70R15 tire?

The width in this case is 215 mm. Converting this to inches, we have

$$\frac{215 \text{ mm}}{1} \times \frac{1 \text{ in.}}{25.4 \text{ mm}} = 8.5 \text{ in.}$$

Example 12-14 What is the sidewall height, in inches, of a P245/65R16 tire?

The width is 245 mm. The sidewall height is 65% of the width, so we can multiply to find the sidewall height.

Sidewall height = 65% of 245 mm = 0.65 × 245 mm = 159.25 mm.

Now, convert this to inches.

$$\frac{159.25 \text{ mm}}{1} \times \frac{1 \text{ in.}}{25.4 \text{ mm}} = 6.3 \text{ in.}$$

Example 12-15 Find the overall diameter of a truck tire with size LT265/75R16.

First, find the sidewall height, which is 75% of the width for this tire. Refer to Figure 12.12 for a diagram of rim diameter and sidewall height.

$$75\% \text{ of } 265 \text{ mm} = 0.75 \times 265 \text{ mm} = 198.75 \text{ mm}$$

$$\frac{198.75 \text{ mm}}{1} \times \frac{1 \text{ in.}}{25.4 \text{ mm}} = 7.8 \text{ in.}$$

The overall tire diameter is found by adding the sidewall height twice to the rim diameter.

Overall tire diameter = 7.8 in. + 16 in. + 7.8 in. = 31.6 in.

SPEEDOMETER CALIBRATION

Changing the size of the tires installed on a vehicle will create error in the speedometer calibration. This is because the speedometer simply measures how fast the drive wheels are turning. Placing tires of a different size on a vehicle means that for each revolution, the car is now going to cover more (if larger tires are installed) or less (if smaller tires are used) distance for each revolution.

Suppose tires with a larger diameter are installed on a vehicle. The car will now travel farther for each revolution of the wheels than it did with the original wheels. That is, the car is moving farther and faster than it used to. That means that placing larger wheels on a vehicle causes the speedometer to give a reading that is too slow.

To correct this, the gear that drives the (mechanical) speedometer must be changed. To make the speedometer give a reading that is faster, to reflect the fact that larger wheels have been installed, a smaller speedometer gear must be used. This means that a gear with fewer teeth should be installed. See Figure 12.13.

FIGURE 12.13 Mechanical speedometers can be calibrated by changing the size of the speedometer drive gear.

Likewise, if smaller wheels are installed on a vehicle, it will travel slower than what the speedometer reflects. To make the speedometer reading "slow down," a larger speedometer gear should be used. That is, we should use a gear with more teeth.

A proportion can be used to find the new speedometer gear teeth. Since larger tires mean a smaller gear should be used, we have what is called an **inverse proportion**.

$$\frac{\text{Old tire diameter}}{\text{New tire diameter}} = \frac{\text{New gear tooth count}}{\text{Old gear tooth count}}$$

Example 12-16 The stock 26 in. tires on a car are replaced with 28 in. tires. If the old speedometer gear has 17 teeth, how many teeth should the new gear have?

We'll simply place the values we have here into the proportion above. Then, cross-multiply to find the new tooth count.

$$\frac{\text{Old tire diameter}}{\text{New tire diameter}} = \frac{\text{New gear tooth count}}{\text{Old gear tooth count}}$$

$$\frac{26}{28} = \frac{n}{17}$$

$$28 \times n = 26 \times 17$$

$$n = 15.8$$

Since a gear can't really have 15.8 teeth, the best gear to use would have 16 teeth.

Example 12-17 A vehicle originally has P205/75R15 tires. The owner replaces those tires with P225/60R14 tires. The old speedometer gear had 20 teeth. How many teeth should a new speedometer gear have?

Begin by finding the outside diameter of the old tires and the new tires.

Old tires: Sidewall height = 75% of 205 mm

$$\text{Sidewall height} = 0.75 \times 205 \text{ mm} \times \frac{1 \text{ in.}}{25.4 \text{ mm}} = 6.1 \text{ in.}$$

Overall tire diameter = 6.1 in. + 15 in. + 6.1 in. = 27.2 in.

New tires: Sidewall height = 60% of 225 mm

$$\text{Sidewall height} = 0.60 \times 225 \text{ mm} \times \frac{1 \text{ in.}}{25.4 \text{ mm}} = 5.3 \text{ in.}$$

Overall tire diameter = 5.3 in. + 14 in. + 5.3 in. = 24.6 in.

Now, use the formula above to calculate the new speedometer gear tooth count, knowing that the old tire diameter was 27.2 in., and the new tire diameter is 24.6 in.

$$\frac{\text{Old tire diameter}}{\text{New tire diameter}} = \frac{\text{New gear tooth count}}{\text{Old gear tooth count}}$$

$$\frac{27.2}{24.6} = \frac{n}{20}$$

$$24.6 \times n = 27.2 \times 20$$

$$n = 22.1$$

Again, a gear can't have 22.1 teeth, so the best gear to use would have 22 teeth.

Changing tire size is not the only way to create speedometer error. It's a common performance modification to change the rear axle ratio on a vehicle. This has the same effect as changing tire size in that it causes the car to move at a different rate than the speedometer gear is calibrated for.

Suppose an economical axle ratio (numerically lower) is replaced with a higher performance gear (numerically higher). This will increase acceleration, but will cause the vehicle to move less distance with each revolution of the engine. This means that the vehicle will move slower, and that the speedometer will give a reading that is too high. To correct this, a gear with fewer teeth must be used.

If a numerically higher gear were replaced with a numerically lower gear, a smaller speedometer gear would be needed. This direct relationship between numeric gear ratio and tooth count gives a direct proportion.

$$\frac{\text{New axle ratio}}{\text{Old axle ratio}} = \frac{\text{New gear tooth count}}{\text{Old gear tooth count}}$$

Example 12-18 To increase acceleration, the 2.73:1 differential gears are replaced with 3.55:1 gears. The old speedometer gear had 18 teeth. How many teeth should a new speedometer gear have?

We can use the proportion above by substituting our values directly, then solving.

$$\frac{3.55}{2.73} = \frac{n}{18}$$

$$2.73 \times n = 18 \times 3.55$$

$$n = 23.4$$

A speedometer gear with 23 teeth would be the best choice.

A proportion can also be used to correct a speedometer by doing a freeway test. If a speedometer is thought to be in error, it can be tested by driving at a speed of *exactly* 60 MPH on a freeway. At 60 MPH, the vehicle should take exactly one minute to travel

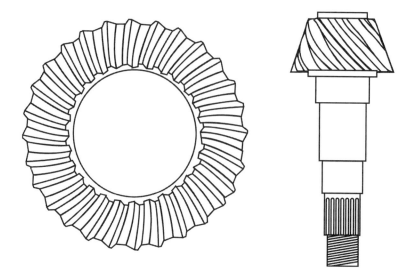

●6. An unmarked differential has a ring gear with 38 teeth and a pinion gear having 13 teeth. What is the gear ratio of this ring and pinion set?

●●7. A 4.10:1 rear differential could be created using _____ ring gear teeth and _____ pinion gear teeth.

●●8. How many teeth on the ring gear and pinion gear would be needed to create a 2.56:1 rear axle ratio? Ring gear _____ Pinion gear _____

Use proportions to find the missing values for each problem below.

●●9. Input torque: 120 lb · ft
Input speed: 500 RPM

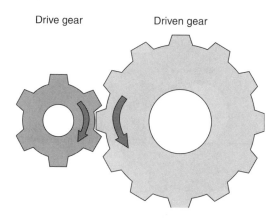

Output torque _____
Output speed _____

TRANSMISSION AND GEAR RATIOS

●●10. Input torque: 190 lb · ft
Input speed: 1200 RPM

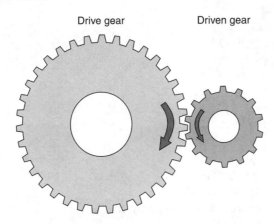

Output torque _____
Output speed _____

●●11. Input torque: 250 lb · ft
Input speed: 1350 RPM

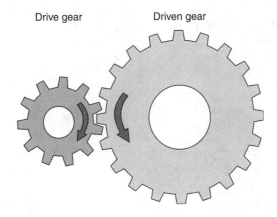

Output torque _____
Output speed _____

●●12. Input torque: 90 lb · ft
Input speed: 4600 RPM

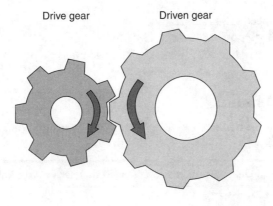

Output torque _____
Output speed _____

Use the gear ratios you calculated in problems 1–4 (rounded to two decimal places) to find the values for problems 13–16.

●●13. Input torque: 170 lb · ft
Input speed: 2000 RPM
Use the gear set shown in problem 1.

Gear ratio _____
Output torque _____
Output speed _____

●●14. Input torque: 205 lb · ft
Input speed: 2800 RPM
Use the gear set shown in problem 2.

Gear ratio _____
Output torque _____
Output speed _____

●●●15. Round your answers to one decimal place where necessary.
Input torque: 315 lb · ft
Input speed: 1900 RPM
Use the gear set shown in problem 3.

Input horsepower _____
Gear ratio _____
Output torque _____
Output speed _____
Output horsepower _____

●●●16. Round your answers to one decimal place where necessary.
Input torque: 113 lb · ft
Input speed: 3655 RPM
Use the gear set shown in problem 4.

Input horsepower _____
Gear ratio _____
Output torque _____
Output speed _____
Output horsepower _____

Overall Drive Ratios

Calculate the overall drive ratio for each gear of the transmission described below. Round to two decimal places.

●1. Transmission Gear	Transmission Gear Ratio	Drive Axle Ratio	Overall Gear Ratio
First	3.684:1	2.727:1	
Second	2.432:1	2.727:1	
Third	1.555:1	2.727:1	
Fourth	1:1	2.727:1	
Fifth	0.830:1	2.727:1	

TRANSMISSION AND GEAR RATIOS

●2.

Transmission Gear	Transmission Gear Ratio	Drive Axle Ratio	Overall Gear Ratio
First	4.092:1	3.454:1	
Second	2.622:1	3.454:1	
Third	1.624:1	3.454:1	
Fourth	1:1	3.454:1	
Reverse	4.222:1	3.454:1	

Calculate the gear ratio for each transmission speed described below. Round to two decimal places.

●3.

Transmission Gear	Input Shaft to Countershaft	Countershaft to Output Shaft	Transmission Gear Ratio
First	1.242:1	2.860:1	
Second	1.242:1	1.961:1	
Third	1.242:1	1.372:1	
Fourth	Direct drive	Direct drive	

●4.

Transmission Gear	Input Shaft to Countershaft	Countershaft to Output Shaft	Transmission Gear Ratio
First	1.195:1	3.000:1	
Second	1.195:1	2.13:1	
Third	1.195:1	1.453:1	
Fourth	Direct drive	Direct drive	

●●5. Second gear

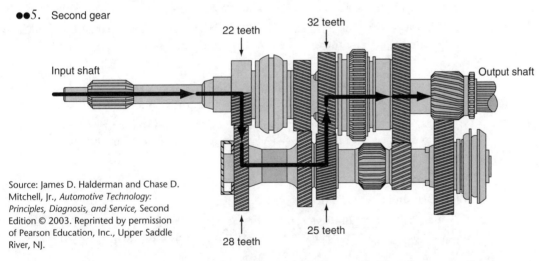

Source: James D. Halderman and Chase D. Mitchell, Jr., *Automotive Technology: Principles, Diagnosis, and Service*, Second Edition © 2003. Reprinted by permission of Pearson Education, Inc., Upper Saddle River, NJ.

●●6. Third gear

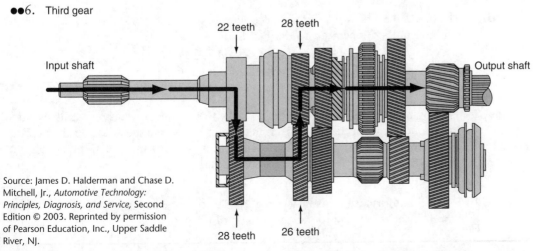

Source: James D. Halderman and Chase D. Mitchell, Jr., *Automotive Technology: Principles, Diagnosis, and Service*, Second Edition © 2003. Reprinted by permission of Pearson Education, Inc., Upper Saddle River, NJ.

●●7. Fourth gear

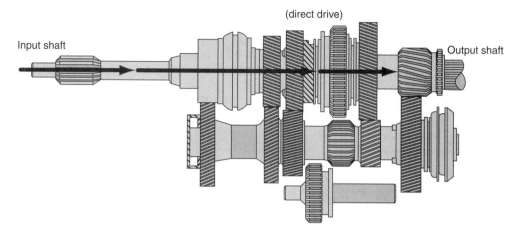

Source: James D. Halderman and Chase D. Mitchell, Jr., *Automotive Technology: Principles, Diagnosis, and Service,* Second Edition © 2003. Reprinted by permission of Pearson Education, Inc., Upper Saddle River, NJ.

●●8. Assume the transmission shown in problems 5–7 is used with a drive axle having a ratio of 3.100:1.
   a. Find the overall drive ratio for second gear.
   b. Find the overall drive ratio for third gear.
   c. Find the overall drive ratio for fourth gear.

●●9. An engine operates at 2300 RPM on the interstate in overdrive. The transmission's overdrive gear has a ratio of 0.91:1, and the drive axle has a ratio of 3.08:1.
   a. What is the overall drive ratio in overdrive? Round to the nearest hundredth.
   b. At what speed are the drive wheels turning? Round to the nearest tenth.

●●10. In fourth gear, a vehicle is cruising on a highway at 55 MPH. The engine is turning at 2800 RPM. Fourth gear has a ratio of 1:1, and the rear axle has a 3.45:1 ratio.
   a. What is the overall drive ratio in fourth gear? Round to the nearest hundredth.
   b. At what speed are the drive wheels turning? Round to the nearest tenth.

●●11. In first gear, a tow vehicle's engine is operating at 3400 RPM. First gear has a ratio of 4.73:1, and the rear axle has a 4.11:1 ratio.
   a. What is the overall drive ratio in first gear? Round to the nearest hundredth.
   b. At what speed are the drive wheels turning? Round to the nearest tenth.

●●12. The truck in problem 11 shifts to second gear, which has a ratio of 2.52:1. The engine continues to operate at 3400 RPM.
   a. What is the overall drive ratio in second gear?
   b. At what speed are the drive wheels turning in this gear?

●●13. The engine in problem 9 produces 110 lb·ft of torque at cruising speed. What is the torque output at the rear wheels in overdrive? Use the specifications from problem 9 to answer this question. All output values of torque and speed should use gear ratios rounded to the nearest hundredth.

●●14. Using the information given in problem 10, find the torque output at the drive wheels if the engine is producing 95 lb·ft of torque.

TRANSMISSION AND GEAR RATIOS

●●15. The tow vehicle in problem 11 is powered by a diesel engine producing 325 lb·ft of torque at 3400 RPM. What is the torque output at the rear wheels in first gear?

●●16. Using the information from problems 11 and 12, find the torque output at the drive wheels in second gear. Assume that the engine produces 325 lb·ft of torque.

Planetary Gear Ratios

Using Table 12.1, calculate the gear ratio created in the planetary gear shown below by each case as described.

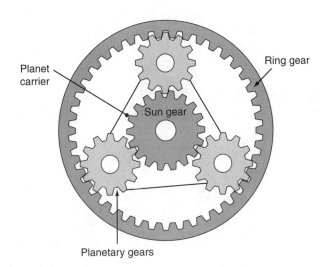

●●1. Input shaft: connected to the sun gear
Output shaft: connected to the ring gear
Fixed: planet carrier
Ring gear: 50 teeth
Sun gear: 30 teeth
Planetary gears: 15 teeth each

●●2. Input shaft: connected to the sun gear
Output shaft: connected to the ring gear
Fixed: planet carrier
Ring gear: 68 teeth
Sun gear: 34 teeth
Planetary gears: 21 teeth each

●●3. Input shaft: connected to the sun gear
Output shaft: connected to the planet carrier
Fixed: ring gear
Ring gear: 56 teeth
Sun gear: 24 teeth
Planetary gears: 19 teeth each

●●4. Input shaft: connected to the sun gear
Output shaft: connected to the planet carrier
Fixed: ring gear
Ring gear: 64 teeth
Sun gear: 28 teeth
Planetary gears: 16 teeth each

●●5. Input shaft: connected to the ring gear
Output shaft: connected to the planet carrier
Fixed: sun gear
Ring gear: 48 teeth
Sun gear: 22 teeth
Planetary gears: 15 teeth each

●●●6. Input shaft: connected to the ring gear
Output shaft: connected to the planet carrier
Fixed: sun gear
Ring gear: 51 teeth
Sun gear: 26 teeth
Planetary gears: 13 teeth each

●●●7. Input shaft: connected to the ring gear
Output shaft: connected to the sun gear
Fixed: planet carrier
Ring gear: 49 teeth
Sun gear: 23 teeth
Planetary gears: 15 teeth each

●●8. Input shaft: connected to the ring gear
Output shaft: connected to the sun gear
Fixed: planet carrier

Ring gear: 57 teeth
Sun gear: 29 teeth
Planetary gears: 20 teeth each

●●9. Input shaft: connected to the planet carrier
Output shaft: connected to the sun gear
Fixed: ring gear

Ring gear: 36 teeth
Sun gear: 18 teeth
Planetary gears: 12 teeth each

●●10. Input shaft: connected to the planet carrier
Output shaft: connected to the sun gear
Fixed: ring gear

Ring gear: 40 teeth
Sun gear: 16 teeth
Planetary gears: 13 teeth each

●●11. Input shaft: connected to the planet carrier
Output shaft: connected to the ring gear
Fixed: sun gear

Ring gear: 39 teeth
Sun gear: 17 teeth
Planetary gears: 14 teeth each

●●12. Input shaft: connected to the planet carrier
Output shaft: connected to the ring gear
Fixed: sun gear

Ring gear: 44 teeth
Sun gear: 18 teeth
Planetary gears: 16 teeth each

Tire Sizing

For each of the tire sizes below, find the overall width in millimeters, centimeters, and inches. Round to one decimal place.

1. P205/75R15   Overall width: _____ millimeters
                                      _____ centimeters
                                      _____ inches

2. P185/70R14   Overall width: _____ millimeters
                                      _____ centimeters
                                      _____ inches

3. LT235/65R15   Overall width: _____ millimeters
                                      _____ centimeters
                                      _____ inches

4. P205/75R14   Overall width: _____ millimeters
                                      _____ centimeters
                                      _____ inches

Find the sidewall height for each of the tire sizes given below. Report values in millimeters, centimeters, and inches. Round to one decimal place.

5. P315/55R16   Sidewall height: _____ millimeters
                                      _____ centimeters
                                      _____ inches

TRANSMISSION AND GEAR RATIOS

6. P215/70R14   Sidewall height:
   - _____ millimeters
   - _____ centimeters
   - _____ inches

7. LT295/60R16   Sidewall height:
   - _____ millimeters
   - _____ centimeters
   - _____ inches

8. LT275/65R16   Sidewall height:
   - _____ millimeters
   - _____ centimeters
   - _____ inches

Find the overall diameter, in inches, for each of the tire sizes given below.

9. P225/55R15   Overall diameter: _____ inches

10. P295/65R16   Overall diameter: _____ inches

11. P315/35R17   Overall diameter: _____ inches

12. LT225/75R16   Overall diameter: _____ inches

Speedometer Calibration

●1. A truck has 17 teeth on the speedometer drive. The stock 26 in. tires are replaced with 33 in. tires. How many teeth should the new speedometer gear have?

●2. By putting lower profile tires on a vehicle, you change the speedometer calibration. If the old tires were 25 in. tall, and the new tires on your car are 22 in. tall, how many teeth should the correction gear have? Assume the old gear has 23 teeth.

●3. The 24 in. tires are replaced with 25 in. tires on an economy car. The old speedometer gear has 22 teeth. How many teeth should a replacement gear have?

●●4. A vehicle is originally equipped with P215/75R15 tires. They are replaced with P235/65R15 tires. The old speedometer gear has 19 teeth. Is a replacement gear necessary?

●●5. A set of P185/75R14 tires is replaced with wider P225/65R15 tires. If the old speedometer gear has 21 teeth, how many should a new speedometer gear have?

●●6. The P235/75R15 tires on a truck are replaced with 31 × 10.5R15 tires, meaning the tires have a 31 in. diameter. The stock speedometer gear has 22 teeth. How many teeth should a replacement gear have?

- 7. In a muscle car, the 3.08 differential is replaced with a 3.73 differential. If the old speedometer gear had 22 teeth, how many should the new gear have?

- 8. To get better fuel economy, a customer has the 3.55 gear ratio in his 2WD truck replaced with a 2.73 gear set. The old speedometer gear has 18 teeth. How many should the new gear have?

- 9. The stock differential gear ratio of 2.56:1 is replaced with a new limited-slip differential and gear set having a ratio of 3.27:1. The old speedometer gear has 20 teeth. How many teeth should there be on a new gear?

- 10. A customer suspects that his speedometer is off. You take the car on the freeway, drive at exactly 60 MPH by the speedometer, and find that it takes 61 seconds to drive between mile markers. You look up his car in a manual, and find that his speedometer drive gear has 13 teeth. How many should a new gear have?

- 11. After making several modifications to a muscle car, a driver measures the time it takes to travel exactly 1 mile on the freeway while his speedometer reports 60 MPH. It takes 64 seconds to travel 1 mile in this test. The old speedometer gear has 18 teeth. How many teeth should a replacement gear have to correct the speedometer?

- 12. After getting pulled over for speeding, you suspect that your speedometer is not correct. On the freeway (where the speed limit is 65 MPH) you perform the speedometer test described in this section, and find that it takes 52 seconds to travel exactly 1 mile. (That's why you got the speeding ticket.) You remove the old speedometer gear and count 22 teeth. How many teeth should be on the gear that corrects the speedometer?

- 13. In building a performance car, you replace the 25.5-in. tall wheels with 24-in. tall wheels. At the same time, you replace the 3.08:1 differential with a 3.55:1 differential. If the old speedometer gear has 20 teeth, how many should the gear that corrects the speedometer have?

- 14. Originally the speedometer gear in your vehicle has 19 teeth. You replace the 3.27:1 differential with a 3.08:1 unit, and replace the 25 1/4 in. tires with 23 3/4 in. tires. How many teeth should a replacement speedometer gear have?

CHAPTER 13

Hydraulic Systems

FORCE, PRESSURE, AND AREA

Hydraulic systems are used to deliver power. Like gears, hydraulic systems can be used to change the form in which power is delivered. Suppose you were to drive a nail through a board using a hammer. There's no hydraulics at work here, but the same principles are in use. When the hammer hits the head of the nail, a large amount of force is being applied to the point of the nail, which is reasonably sharp. That large force applied to such a small area creates a pressure that allows the nail to go through the board. Had the hammer been used to hit the board without a nail, the force of the hammer would not be focused on a small point. Since the face of the hammer has a larger area, less pressure is created, and the hammer does not go through the board.

Another example of this relationship is exhibited in a woman wearing high-heeled shoes. The weight of the woman creates a downward force which does not change. If her heels have very small tips, the force is applied to a small area, creating a very large pressure underneath her heel. If she is wearing heels with a larger tip, the same force is spread over a larger area, and lower pressure is created. **Pressure** describes how much force is present over a given area.

The relationship between force, pressure, and area provide the foundation for all hydraulic systems. These three items are related by the following relationship.

$$\text{Force} = \text{Pressure} \times \text{Area}$$

This relationship can be written two other ways as well:

$$\text{Pressure} = \frac{\text{Force}}{\text{Area}}$$

$$\text{Area} = \frac{\text{Force}}{\text{Pressure}}$$

These formulas can be easily remembered using the following "F-P-A" circle shown in Figure 13.1. If any two of the three values in a force (F), pressure (P), and area (A) relationship are known, the third can be found using this circle.

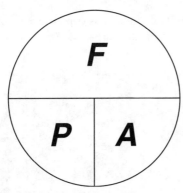

FIGURE 13.1 The relationship of the values of force (F), pressure (P), and area (A).

Example 13-1 Use the *F-P-A* circle to find the formula for pressure.

Since we want to find pressure, find P on the circle. The other two values are F and A. Since F is above A, we can conclude that

$$P = \frac{F}{A} \quad \text{or Pressure} = \frac{\text{Force}}{\text{Area}}$$

Example 13-2 If pressure and area are known, how is force found?

We need to find the value for force. P and A are the other values, and they are side by side, which means we should multiply. Then we know that

$$F = P \times A \quad \text{or Force} = \text{Pressure} \times \text{Area}$$

In the customary system, force is usually measured in pounds, and area is usually measured in square inches. Since

$$\text{Pressure} = \frac{\text{Force}}{\text{Area}}$$

it should make sense that the units of pressure are

$$\frac{\text{pounds}}{\text{in.}^2}$$

or pounds/in.^{2}. As we've done before, we'll abbreviate this as psi.

Example 13-3 A 175-lb man stands on a floor. The bottoms of his shoes have a total area of 70 in.^{2}. How much pressure is he applying to the floor underneath his shoes?

From the relationship above, we know that

$$\text{Pressure} = \frac{\text{Force}}{\text{Area}}$$

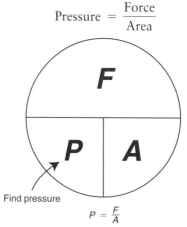

The weight of the man creates a 175-lb force spread over 70 in.^{2}, so

$$\text{Pressure} = \frac{175 \text{ lb}}{70 \text{ in.}^2} = 2.5 \text{ lb/in.}^2 = 2.5 \text{ psi}$$

Example 13-4 A hydraulic cylinder has a piston with an area of 12.56 in.^{2}. If the pressure inside the hydraulic cylinder is 400 lb/in.^{2}, how much extending force does the cylinder have?

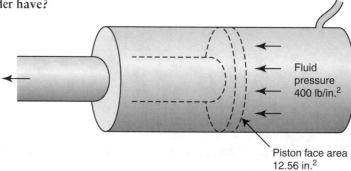

Here, we know the pressure and the area. From the *F-P-A* circle we can see that

$$\text{Force} = \text{Pressure} \times \text{Area}$$

so we have

$$\text{Force} = 400 \text{ lb/in.}^2 \times 12.56 \text{ in.}^2 = 5024 \text{ lb}$$

Note that if the correct formula is used, the force will have the correct unit of pounds.

Example 13-5 A hydraulic cylinder used to lift a snow plow must extend with a force of 1200 lb to lift the plow. The pump used to operate the cylinder is capable of creating pressures up to 350 lb/in.². How large should the piston (and cylinder) be to create the needed force?

We can see from the *F-P-A* circle that

$$\text{Area} = \frac{\text{Force}}{\text{Pressure}}$$

We need a force of 1200 lb using a pressure of 350 lb/in.².

$$\text{Area} = \frac{1200 \text{ lb}}{350 \text{ lb/in.}^2} = 3.43 \text{ in.}^2$$

The area of the piston must be 3.43 in.² to create 1200 lb of force.

Example 13-6 A round hydraulic cylinder has a piston diameter of 3.500 in. It is operated by a pump that generates a pressure of 900 lb/in². How much force will the extending cylinder create?

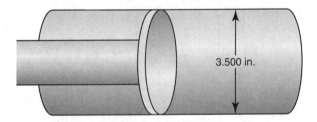

Since Force = Pressure × Area, we need to find the area of the cylinder's piston. The area of a circle is given by

$$\text{Area} = \pi \times \text{radius}^2$$

The radius in this case is 3.500 ÷ 2 = 1.750 in., so we have

$$\text{Area} = 3.14 \times (1.75 \text{ in.})^2 = 9.62 \text{ in.}^2$$

Then the force of the extending cylinder can be found.

$$\text{Force} = 900 \text{ lb/in.}^2 \times 9.62 \text{ in.}^2 = 8658 \text{ lb}$$

BRAKING SYSTEMS

Modern braking systems use hydraulic system concepts to convert a small amount of force exerted by your foot into a great amount of force applied to a disc brake rotor or drum. When the driver applies force to the brake pedal, the pedal lever itself increases the force applied to the input rod of the master cylinder. The vacuum brake booster also applies force to the input rod, and the rod applies all of this force to the piston inside the master cylinder, where brake system pressure is created.

This pressure, then, pushes on the piston or pistons inside the caliper of disc brakes and the wheel cylinder of drum brakes, which exert a force on the pads or shoes.

Figure 13.2 shows a diagram of hydraulic pressure used in a brake system.

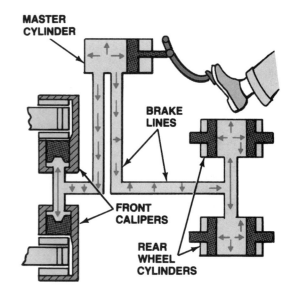

FIGURE 13.2 Typical brake systems are based on hydraulic principles.

Source: James D. Halderman and Chase D. Mitchell, Jr., *Automotive Technology: Principles, Diagnosis, and Service,* Second Edition © 2003. Reprinted by permission of Pearson Education, Inc., Upper Saddle River, NJ.

The mechanical advantage of a brake pedal lever increases the force applied to the master cylinder. The mechanical advantage is found by measuring the distance from the pivot or fulcrum to the brake pedal, and the distance from the fulcrum to the input rod of the master cylinder.

$$\text{Brake pedal mechanical advantage} = \frac{\text{Distance from fulcrum to pedal}}{\text{Distance from fulcrum to input rod}}$$

The division is usually performed for this ratio, so that it is written as a unit rate. We'll see this in Example 13-7.

The amount of force applied to the pedal by the driver is multiplied by the mechanical advantage of the lever as the force into the master cylinder is generated. See Figure 13.3.

Force produced by brake pedal = Force exerted by driver × Mechanical advantage

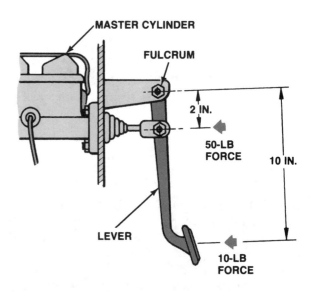

FIGURE 13.3 The brake pedal lever increases the force applied to the master cylinder.

HYDRAULIC SYSTEMS

Example 13-7 A driver applies 10 lb of force to a brake pedal. The distance from the pedal to the fulcrum is 10 in., and the distance from the fulcrum to the input rod of the master cylinder is 2 in. Find the force on the rod of the master cylinder.

The mechanical advantage is calculated using the formula above:

$$\text{Brake pedal mechanical advantage} = \frac{10 \text{ in.}}{2 \text{ in.}} = 5:1$$

The brake pedal lever, then, will multiply the driver's force by 5.

$$\text{Force produced by brake pedal} = 10 \text{ lb} \times 5 = 50 \text{ lb}$$

Suppose the brake pedal in Figure 13.3 is used with a master cylinder that uses a 1.000-in. diameter piston. If there is no power brake booster, how much pressure is created in the system when the driver applies 30 lb of force with her foot? See the diagram of this setup in Figure 13.4. From Example 13-7, the brake pedal has a mechanical advantage of 5:1. The driver applies 30 lb. This means the piston in the master cylinder will receive

$$30 \text{ lb} \times 5 = 150 \text{ lb}$$

We know that

$$\text{Pressure} = \frac{\text{Force}}{\text{Area}}$$

so we need to find the area of the piston in the master cylinder. The diameter is 1.000 in., so the radius is 0.500 in.

$$\text{Piston area} = \pi \times (0.500 \text{ in.})^2 = 0.785 \text{ in.}^2$$

Then

$$\text{Pressure} = \frac{150 \text{ lb}}{0.785 \text{ in.}^2} = 191.1 \text{ lb/in.}^2$$

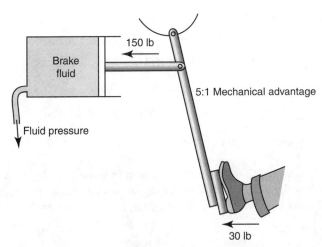

FIGURE 13.4 From this diagram, we can determine how much pressure is created when the driver applies 30 lb of force.

Example 13-8 If the braking system pressure is 250 lb/in.^{2}, find the force exerted by a caliper which uses a piston with a diameter of 2.200 in. Figure 13.5 shows an example of a caliper.

We know that

$$\text{Force} = \text{Pressure} \times \text{Area}$$

FIGURE 13.5 The caliper piston is usually round, which makes finding the area as simple as using our formula for the area of a circle: $A = \pi \times r^2$.

so we need to find the area of the caliper's piston. The diameter of the caliper piston is 2.200 in., so the radius is 1.100 in.

$$\text{Area} = \pi \times (1.100 \text{ in.})^2 = 3.80 \text{ in.}^2$$

The force exerted by the caliper can then be found.

$$\text{Force} = \text{Pressure} \times \text{Area} = 250 \text{ lb/in.}^2 \times 3.80 \text{ in.}^2 = 950 \text{ lb}$$

The caliper exerts 950 lb of pressure on the rotor.

Example 13-9 Assume the same brake system in Example 13-8 uses drum brakes in the rear, and that the wheel cylinders in the rear have a diameter of 1.300 in. Find the force exerted by the wheel cylinders. An example of a drum brake is shown in Figure 13.6.

Again, we must first find the area of the piston within the wheel cylinder. The cylinder has a diameter of 1.300 in., so the radius must be 0.650 in.

$$\text{Area} = \pi \times (0.650 \text{ in.})^2 = 1.327 \text{ in.}^2$$

As before, we can find the force produced by the wheel cylinder.

$$\text{Force} = \text{Pressure} \times \text{Area} = 250 \text{ lb/in.}^2 \times 1.327 \text{ in.}^2 = 331.75 \text{ lb}$$

Since the wheel cylinders are smaller than the caliper pistons, the force created is less.

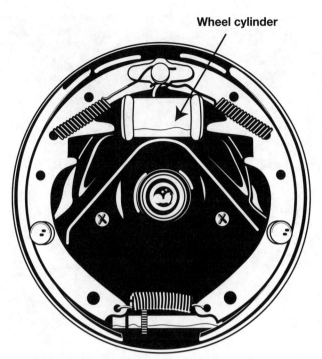

FIGURE 13.6 The force exerted by the wheel cylinder applies braking action within the drum.

HYDRAULIC PUMPS AND FLOW RATES

In order for any hydraulic system to work correctly, it must be powered by an appropriate pump. The boom on a backhoe, for example, must be operated by a pump that not only creates high enough pressure to move the bucket, but also pumps at a rate that makes the equipment move at a reasonable speed. The pump for a snow plow must lift the plow quickly, and work at high enough pressures to easily lift heavy snow.

There are several types of hydraulic pumps, including gear pumps, vane pumps, and piston pumps. Every hydraulic pump is rated according to the hydraulic pressure it can develop in lb/in.² and the rate of flow it provides, in **gallons per minute (GPM)**. Many pumps are designed so that their flow rate can be varied. A variable-flow pump is used to control the speed of the cylinder or hydraulic motor it powers. Two hydraulic pumps are shown in Figure 13.7.

We will focus on one particular type of hydraulic pump, a fixed-displacement piston pump, an example of which is shown in Figure 13.8. Every time a piston pump makes a complete revolution, each cylinder fills with fluid and pumps it out. Then to find the displacement per revolution of a hydraulic pump, we can use the same formula for displacement of an engine.

Piston pump displacement = (Number of cylinders) × (Swept volume of each cylinder)

Or,

$$\text{Piston pump displacement} = (\text{Number of cylinders}) \times \pi \times \left(\frac{1}{2} \times \text{bore}\right)^2 \times \text{stroke}$$

Example 13-10 Find the displacement of a 1-cylinder hydraulic pump having a bore of 0.800 in. and a stroke of 1.200 in.

From the formula directly above this problem, we have

$$\text{Piston pump displacement} = (1) \times \pi \times \left(\frac{1}{2} \times 0.800 \text{ in.}\right)^2 \times 1.200$$

$$= 0.603 \text{ in.}^3/\text{revolution}$$

FIGURE 13.7 The flow rating of a hydraulic pump depends on the amount of fluid it displaces per revolution and the speed at which it is turned.

FIGURE 13.8 A piston pump's displacement is found in the same way an engine's displacement is found.

HYDRAULIC SYSTEMS

This process should look familiar, since it's exactly the same process for finding engine displacement. Usually, though, flow rates are given in gallons per minute, or GPM. To calculate the flow rate in GPM, we simply take the pump displacement per revolution and multiply it by the number of revolutions per minute. We must also convert the displacement from cubic inches to gallons.

$$\text{Piston pump flow rate} = \text{Pump displacement (in.}^3\text{/revolution)} \times \text{RPM} \times \frac{1 \text{ gal}}{231 \text{ in.}^3}$$

This can be simplified to the following:

$$\text{Piston pump flow rate} = \frac{\text{Pump displacement} \times \text{RPM}}{231}$$

Example 13-11 Find the flow rate of a 2.850 in.³ displacement pump operating at 900 RPM.

From the formula above, we have

$$\text{Piston pump flow rate} = \frac{2.850 \times 900}{231} = 11.10 \text{ GPM}$$

Example 13-12 A 6-cylinder radial hydraulic pump has a bore of 0.750 in. and a stroke of 0.625 in. What will be the flow rate of this pump if it is operated at 1400 RPM?

First, we need to find the displacement of the pump.

$$\text{Piston pump displacement} = (6) \times \pi \times \left(\frac{1}{2} \times 0.750 \text{ in.}\right)^2 \times 0.625$$

$$= 1.657 \text{ in.}^3\text{/revolution}$$

Now, calculate the flow rate.

$$\text{Piston pump flow rate} = \frac{1.657 \times 1400}{231} = 10.04 \text{ GPM}$$

HYDRAULIC CYLINDERS, FORCE, AND ACTUATION SPEED

When a hydraulic pump operates a hydraulic cylinder, there are two factors to consider. The force and speed of the extending or contracting cylinder can be calculated if the size of the cylinder and the pump ratings are known. See Figure 13.9.

We've already used the *F-P-A* diagram to determine the force generated by an extending cylinder if the pump pressure is known. Many cylinders, called **double-acting cylinders**, are designed so that they generate force when extending or contracting. When contracting, the piston area is smaller because of the rod that is attached to the piston called the **ram**. The area of the ring-shaped back side of the piston, called the **annulus**, is smaller than the front face of the piston. This means that the extending force and contracting force will not be the same. This is shown in the diagram in Figure 13.10.

Example 13-13 A 6-in. diameter hydraulic piston has a 2-in. ram. What are the areas of the two piston faces used when extending and contracting?

When extending, the hydraulic pressure is applied to the entire circular piston having a 6-in. diameter (and 3-in. radius).

FIGURE 13.9 The size of the cylinder is a factor in how much force it can exert.

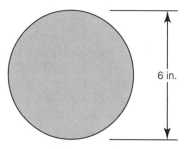

Then the area used to calculate extending force is

$$\text{Area (extending)} = \pi \times (3 \text{ in.})^2 = 28.27 \text{ in.}^2$$

HYDRAULIC SYSTEMS

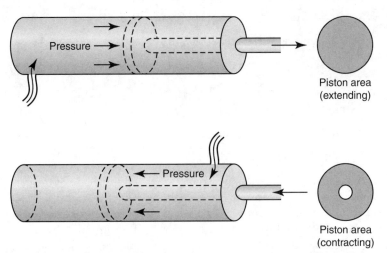

FIGURE 13.10 When extending (top), the hydraulic pressure is applied to the entire circular piston. When contracting (bottom), the hydraulic pressure is applied to a smaller area.

When contracting, the hydraulic pressure is applied to the annulus. We can calculate this by taking the area of the piston when extending and subtracting the area taken by the ram. The ram has a 2-in. diameter, and 1-in. radius.

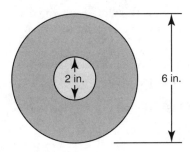

$$\text{Ram area} = \pi \times (1 \text{ in.})^2 = 3.14 \text{ in.}^2$$

Then the area of the annulus is

$$\text{Area (contracting)} = 28.27 \text{ in.}^2 - 3.14 \text{ in.}^2 = 25.13 \text{ in.}^2$$

Example 13-14 A pump operates a hydraulic cylinder at a pressure of 1200 lb/in.². The 3-in. hydraulic cylinder has a 1-in. ram. Calculate the force generated when the ram is extending, and the force generated when it is contracting.

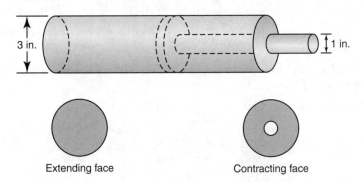

To find the extending force, begin by finding the area of the face of the piston.

$$\text{Face area (extending)} = \pi \times (1.5 \text{ in.})^2 = 7.07 \text{ in.}^2$$

Since we know Force = Pressure × Area,

Extending force = 1200 lb/in.² × 7.07 in.² = 8484 lb

To find the contracting force, first find the area of the annulus.

Ram area = π × (0.5 in.)² = 0.79 in.²

Annulus area (contracting) = 7.07 in.² − 0.79 in.² = 6.28 in.²

We can now find the contracting force.

Contracting force = 1200 lb/in.² × 6.28 in.² = 7536 lb

Extension or contraction speed can also be calculated if the size of the cylinder and the pump's flow rate are known. This speed, also known as ram speed, is usually given in inches per second or inches per minute. We'll use inches per second for our calculations. This formula is found by multiplying the flow rate (gallons/minute) by 231 so that it has units of in.³/minute, then divide by 60 to convert the units to in.³/second. Given the same flow rate, a large-diameter cylinder will extend more slowly than a small-diameter cylinder, so the area of the piston is also a factor.

$$\text{Ram speed (in./sec)} = \frac{\text{Pump flow rate} \times 231}{\text{Piston area} \times 60}$$

Note here that the pump's flow rate must be in GPM and the piston area should be in in.². Piston area can refer to the face or the annulus.

Example 13-15 A 3-in. hydraulic cylinder having a 1-in. ram (as in Example 13-14) is powered by a 5.0 GPM pump. Find the ram speed when the cylinder is extending, and when it is contracting.

To find the extension speed, we can use the formula above. In Example 13-14 we calculated the area used when extending.

Face area (extending) = π × (1.5 in.)² = 7.07 in.²

Now, we can use the formula for ram speed.

$$\text{Ram speed (in./sec)} = \frac{\text{Pump flow rate} \times 231}{\text{Piston area} \times 60}$$

$$\text{Ram speed (extending)} = \frac{5.0 \text{ GPM} \times 231}{7.07 \text{ in.}^2 \times 60} = 2.72 \text{ in./sec}$$

To find the contraction speed, we use the same process.
From Example 13-14,

Annulus area (contracting) = 7.07 in.² − 0.79 in.² = 6.28 in.²

$$\text{Ram speed (contracting)} = \frac{5.0 \text{ GPM} \times 231}{6.28 \text{ in.}^2 \times 60} = 3.07 \text{ in./sec}$$

We can now see that this type of cylinder contracts faster but with less force than it extends.

Example 13-16 Find the contraction speed of a 5-in. hydraulic cylinder with a 1.5 in. ram when powered by a 7.5 GPM pump.

First, we must find the area of the annulus.

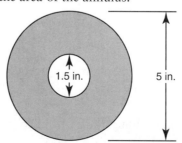

HYDRAULIC SYSTEMS

The radius of the outer piston is 2.5 in., and the radius of the ram is 0.75 in. Then

$$\text{Annulus area (contracting)} = [\pi \times (2.5 \text{ in.})^2] - [\pi \times (0.75 \text{ in.})^2] = 17.87 \text{ in.}^2$$

As before, we can now calculate the ram speed when contracting.

$$\text{Ram speed (contracting)} = \frac{7.5 \text{ GPM} \times 231}{17.87 \text{ in.}^2 \times 60} = 1.62 \text{ in./sec}$$

CHAPTER 13 *Practice Problems*

Force, Pressure, and Area

1. A hydraulic cylinder is designed so that the piston has an area of 7.07 in.^{2}. It operates at a system pressure of 1200 lb/in.^{2}. How much force can the cylinder produce when it is extending?

2. A hydraulic pump generating pressures up to 600 psi is used to actuate a cylinder having a piston area of 5.31 in.^{2}. What force can be created by the extending cylinder?

3. In a pump, 220 lb are applied to a piston having an area of 0.78 in.^{2}. What pressure is created by this pump? Round your answer to the nearest tenth.

4. A hydraulic cylinder has a piston with an area of 3.80 in.^{2}. How much pressure is needed to allow the cylinder to produce 520 lb of force? Round your answer to the nearest tenth.

5. In an automatic transmission, a servo (piston and cylinder assembly) must create a force of 1700 lb. If the operating pressure of the servo is 210 psi, what should be the area of the piston in the servo? Round your answer to the nearest tenth.

6. What should the area of a hydraulic cylinder piston be if a pressure of 400 lb/in.^{2} must create an extending force of 1400 lb?

7. What force is created by the servo in an automatic transmission if the piston area is 3.14 in.^{2} and the fluid pressure is 250 psi?

8. The pressure in problem 7 is increased to 280 psi. What force can that servo create using this increased pressure?

9. The slave cylinder in a hydraulic clutch has a piston area of 1.77 in.^{2}. If the slave cylinder must exert a force of 420 lb to operate the clutch, what pressure is needed? Round your answer to the nearest tenth.

10. Small electric air compressors that plug into the cigarette lighter of a vehicle can use very small motors to create high pressures. Suppose one of these small pumps is designed to apply just 5 lb of force to a piston having an area of 0.082 in.^{2}. How much pressure can be generated? Round your answer to the nearest hundredth.

11. The master cylinder in a braking system uses a piston within a cylinder to create brake line pressure. If the input rod of the master cylinder is subjected to 180 lb of force, and the braking system operates at 350 lb/in.^{2}, what must be the area of the piston inside the master cylinder? Round your answer to the nearest hundredth.

12. What should be the area of a hydraulic cylinder piston if 400 lb of force must be created using just 125 lb/in.^{2} of pressure?

●●13. Hydraulic cylinders are measured according to their length and piston diameter. While the length of the cylinder does not affect the force, the diameter does. If a cylinder has a 3-in. diameter piston, what extending force will be created by operating it at 750 psi? Use 3.14 for π. Round your answer to the nearest pound.

●●14. The servo in an automatic transmission has a 2 in. diameter. If the servo is activated at 300 psi, how much force is created by the servo? Round to the nearest pound.

●●15. How much pressure is needed to make a 2.5-in. diameter servo exert a 900-lb force? Round to the nearest pound.

●●16. How much pressure is needed to make a 2.8-in. diameter servo exert a 950-lb force? Use 3.14 for π. Round your answer to the nearest pound per square inch.

●●●17. The slave cylinder of a clutch must exert a force of 600 lb on the release lever. If the operating pressure of the clutch is 400 psi, what should be the diameter of the clutch? Use 3.14 for π.

●●●18. What diameter piston should be used in a master cylinder so that 400 psi are created in the braking system when the piston has 150 lb applied to it? Round to the nearest hundredth.

Braking Systems

A brake pedal lever is shown below. Find the mechanical advantage if the conditions in problems 1–4 are met. Where necessary, round to the nearest hundredth.

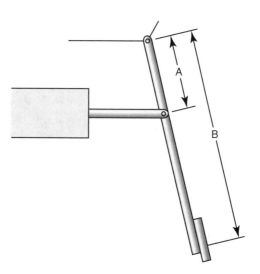

●1. Dimension A is 2 in. and dimension B is 9 in. _____

●2. Dimension A is 1.8 in. and dimension B is 8.1 in. _____

●3. Dimension A is 1.75 in. and dimension B is 8.34 in. _____

●4. Dimension A is 1.6 in. and dimension B is 6.2 in. _____

●5. A driver applies 25 lb of force to a brake pedal having a 4.5:1 mechanical advantage. How much force is transferred to the master cylinder?

●6. In a panic situation, a driver applies 105 lb of force to a brake pedal. If the lever has a 3.8:1 mechanical advantage, how much force is delivered to the master cylinder?

For problems 7–10, assume that the brake system pressure is 220 lb/in.².

- •7. How much force is generated by a caliper having a piston diameter of 1.800 in.? Use 3.14 as the approximation for π, and round to the nearest pound.

- •8. How much force is produced by a dual-piston caliper if each piston has a 1.200-in. diameter? Use 3.14 as the approximation for π, and round to the nearest pound.

- •9. How much force is produced by a wheel cylinder that has a bore of 1 in.? Round to the nearest pound.

- •10. How much force is created by a wheel cylinder with a bore of 1.150 in.? Use 3.14 as the approximation for π, and round to the nearest pound.

- ••11. A driver presses the brake pedal with 40 lb of force. The brake pedal lever has a 2.5:1 mechanical advantage, and transfers the force to a master cylinder having a piston diameter of 0.800 in. How much pressure is created in the braking system? Round to the nearest pound per square inch.

- ••12. Assuming a master cylinder uses a piston having a 0.780-in. diameter, how much pressure is created in the braking system during the panic stop described in problem 6? Round to the nearest pound per square inch.

- ••13. In a gradual stop, 10 lb of force is applied to a brake pedal having a mechanical advantage of 3.1:1. If the master cylinder it activates has a piston diameter of 0.750 in., how much pressure is created in the system? Round to the nearest tenth of a pound per square inch.

- ••14. The master cylinder in a vehicle uses a piston that has a 0.640-in. diameter. If it is designed to be used with a 4.0:1 mechanical advantage brake pedal lever, how

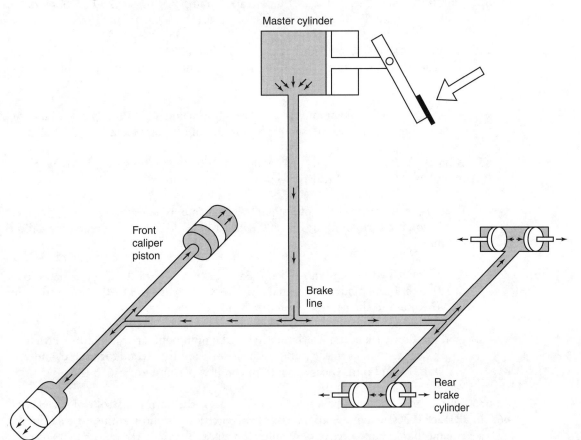

much pressure could be created if the driver applies 135 lb of force to the pedal? Round to the nearest pound per square inch.

●●●15. Suppose the master cylinder in the braking system shown on page 261 has a diameter of 3/4 in. Assume there is 80 lb of force acting on a brake pedal with a 5:1 ratio and answer the following questions. Use 3.14 as the approximation for π, and round to the nearest tenth where necessary.
   a. How many pounds of force are being applied to the master cylinder piston?
   b. How much pressure is being generated in the braking system?
   c. If a front wheel caliper has a piston diameter of 1.900 in., how many pounds of force are exerted by the caliper?
   d. How much force is produced by a rear wheel cylinder having a bore of 1.200 in.?

●●●16. A driver applies 115 lb of force to a brake pedal that is connected to the master cylinder through a brake pedal lever having a 4.4:1 ratio. The master cylinder piston has a diameter of 0.840 in. Use 3.14 as the approximation for π, and round answers to the nearest tenth where necessary.
   a. How many pounds of force are being applied to the master cylinder piston?
   b. How much pressure is being generated in the braking system?
   c. Find the force produced by a 2.000-in. caliper piston.
   d. Find the force produced by a 1.140-in. diameter wheel cylinder.

Hydraulic Pumps and Flow Rates

●1. A single-cylinder hydraulic pump has a 1.000-in. bore and a 1.000-in. stroke. What is the displacement of this pump? Round to the nearest hundredth of a square inch.

●2. Find the displacement of a 4-cylinder radial pump having a 0.780-in. bore and a 1.000-in. stroke. Round to the nearest hundredth of a square inch.

●3. Determine the displacement of an 8-cylinder radial hydraulic pump having a bore of 1.100 in. and a stroke of 0.950 in. Round to the nearest hundredth of a square inch.

●4. A 2-cylinder hydraulic pump has a 0.600-in. bore and a 0.750-in. stroke. Find the displacement. Round to the nearest hundredth of a square inch.

●5. A pump displaces 1.560 in.^{3} per revolution. What is the flow rate, in GPM, if the pump operates at 1500 RPM? Round to the nearest tenth of a gallon per minute.

●6. Find the flow rate, in GPM, of a small hydraulic pump that displaces 1.283 in.^{3} per revolution if it turns at a rate of 2000 RPM. Round to the nearest tenth of a gallon per minute.

●●7. A 4-cylinder hydraulic pump is designed to have a bore of 0.625 in. and a stroke of 1.000 in. If the pump is operated at 1800 RPM, what will be the flow rating, in GPM? Round to the nearest tenth of a gallon per minute.

●●8. Find the flow rating of a 2-cylinder hydraulic pump operating at 2200 RPM. The bore of each cylinder in the pump is 0.500 in., and the stroke for each cylinder is 0.800 in. Round to the nearest tenth of a gallon per minute.

●●9. An 8-cylinder pump is operated at 1500 RPM. If the bore and stroke of each cylinder are 0.700 in. and 0.650 in. respectively, find the flow rating of this pump. Round to the nearest tenth of a gallon per minute.

●●10. A single-cylinder pump having a 3/4-in. bore and 5/8-in. stroke is operated by an electric motor at 2000 RPM. What is the flow rating of this pump? Round to the nearest tenth of a gallon per minute.

●●●11. A 6-cylinder radial pump is designed to use a 20.0-mm bore and 15.0-mm stroke. If operated at 1000 RPM, find the flow rate in GPM. Round to the nearest tenth of a gallon per minute.

●●●12. Find the flow rating of a 2-cylinder hydraulic pump operating at 1700 RPM. The bore of each cylinder in the pump is 14.0 mm, and the stroke for each cylinder is 18.0 mm. Round to the nearest tenth of a gallon per minute.

●●●13. Many variable displacement pumps operate by changing the stroke of the individual cylinders. Suppose a 12-cylinder pump uses a 0.800-in. bore, and the stroke can be changed from 0.300 in. to 0.700 in. If operated at 1500 RPM, what is the *range* of flow ratings that this pump can produce? Round to the nearest tenth of a gallon per minute.

●●●14. A variable-displacement piston pump has 9 cylinders and is operated at 1300 RPM. If each cylinder has a bore of 0.500 in., and a stroke that is adjustable from 0.450 in. to 0.750 in., what are the minimum and maximum flow ratings for this pump? Round to the nearest tenth of a gallon per minute.

Hydraulic Cylinders, Force, and Actuation Speed

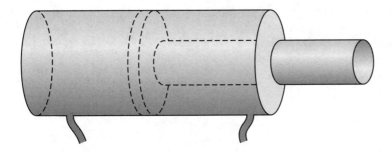

●1. A typical hydraulic cylinder such as the one above is 4 in. in diameter and has a 2-in. ram. Round answers to the nearest hundredth of a square inch.
   a. Find the area of the face of the piston used when the cylinder is extending.
   b. Find the area of the annulus used when the cylinder is contracting.

●2. A 3-in. hydraulic cylinder uses a 2-in. ram. Round answers to the nearest hundredth of a square inch.
   a. Find the area of the face of the piston used when the cylinder is extending.
   b. Find the area of the annulus used when the cylinder is contracting.

●3. The boom on a backhoe is actuated by a hydraulic cylinder that has a 12 in.-diameter with a 6-in. ram. Use 3.14 as the approximation for π, and round to the nearest tenth of a square inch.
   a. Find the area of the face of the piston used when the cylinder is extending.
   b. Find the area of the annulus used when the cylinder is contracting.

●4. The hydraulic cylinder on a dump truck is 10 in. in diameter and has a 5-in. ram. Round all answers to the nearest tenth of a square inch.
   a. Find the area of the face of the piston used when the cylinder is extending.
   b. Find the area of the annulus used when the cylinder is contracting.

●●5. Suppose the hydraulic cylinder in problem 1 was actuated by a pump producing 1300 psi at a flow rate of 4.0 GPM.
   a. What extending force is exerted by the cylinder?
   b. What is the contracting force exerted by the cylinder?
   c. What is the extension speed of the cylinder? Round to the nearest hundredth of an inch per second.
   d. What is the contraction speed of the cylinder? Round to the nearest hundredth of an inch per second.

●●6. If the cylinder in problem 2 is part of a hydraulic system powered by a 5.0 GPM pump rated at 800 lb/in.^{2}, answer the following questions.
   a. What extending force is exerted by the cylinder?
   b. What is the contracting force exerted by the cylinder?
   c. What is the extension speed of the cylinder? Round to the nearest hundredth of an inch per second.
   d. What is the contraction speed of the cylinder? Round to the nearest hundredth of an inch per second.

●●7. The backhoe described in problem 3 operates at hydraulic pressures up to 1500 psi. The pump for the system has a flow rating of 12.5 GPM.
   a. What extending force is exerted by the boom cylinder?
   b. What is the contracting force exerted by the boom cylinder?
   c. What is the extension speed of the cylinder? Round to the nearest hundredth of an inch per second.
   d. What is the contraction speed of the cylinder? Round to the nearest hundredth of an inch per second.

●●8. The dump truck in problem 4 has a hydraulic pump that creates pressures up to 1300 psi at 6.8 GPM.
   a. What extending force is exerted by the truck's lifting cylinder?
   b. What is the extension speed of the cylinder?

CHAPTER 14

Electrical Systems

CURRENT, VOLTAGE, RESISTANCE, AND OHM'S LAW

Many of the ideas that are associated with a hydraulic system are similar to the basic principles of an electrical circuit. While hydraulic systems pump fluid from one point to another, electrical circuits move electrons from one point to another. In a hydraulic circuit, the pressure of the fluid in the system and the rate at which it flows will impact the operation of the system. In the same way, the amount of electromotive force acting on electrons and the rate at which they flow will impact the operation of an electrical circuit. A simple circuit is shown in Figure 14.1.

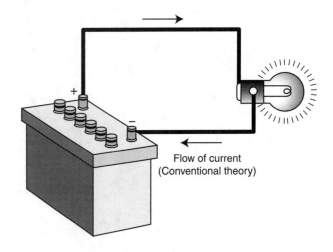

FIGURE 14.1 Electrons actually flow from the negative terminal of a battery to the positive terminal, but most wiring diagrams assume "conventional theory" which assumes that electricity flows from positive to negative.

Source: James D. Halderman and Chase D. Mitchell, Jr., *Automotive Technology: Principles, Diagnosis, and Service,* Second Edition © 2003. Reprinted by permission of Pearson Education, Inc., Upper Saddle River, NJ.

The flow of electrons in a wire can be thought of as the flow of water through a pipe. The flow of water in a pipe can be measured in gallons per minute. That is, the amount of water that passes through a pipe in a given time can be measured. When considering the flow of electrons in a wire, this is called current. The **current** through a wire is the amount of electrical charge that passes through a conductor in a given time. Current is measured in amperes, or **amps.**

Of course, no water flows through a pipe if there is no pressure applied to the water at some point in the system. While water pressure is measured in pounds per square inch (psi), electrical pressure is called electromotive force or **potential,** and is measured in **volts.**

Because it takes some effort to pump water through a pipe, there must be some factor that resists the flow of the water. All common electrical components including light bulbs, motors, and even wires tend to resist the flow of electrons, as shown in Figure 14.2. This electrical tendency to restrict electron flow is called **resistance,** and is measured in **ohms** (the unit of ohms is given by the greek letter omega, Ω).

FIGURE 14.2 While current is a measure of electron flow, voltage is a measure of the force pushing the electrons against the resistance found in every circuit.

Source: James D. Halderman and Chase D. Mitchell, Jr., *Automotive Technology: Principles, Diagnosis, and Service,* Second Edition © 2003. Reprinted by permission of Pearson Education, Inc., Upper Saddle River, NJ.

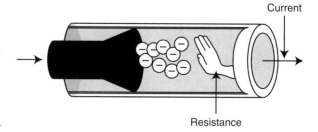

Electrical current, potential, and resistance are very closely associated. Think again about the flow of water through a pipe. If the size of the pipe is reduced (resistance increased), it will certainly reduce the flow of water (current) unless a higher pressure (voltage) is used. Similarly, if the pressure (voltage) is increased as water flows through a pipe, the flow of water (current) will increase. This relationship is summarized using **Ohm's law:**

$$\text{Potential (volts)} = \text{Current (amps)} \times \text{Resistance (ohms)}$$

This relationship can also be written as:

$$\text{Current (amps)} = \frac{\text{Potential (volts)}}{\text{Resistance (ohms)}}$$

$$\text{Resistance (ohms)} = \frac{\text{Potential (volts)}}{\text{Current (amps)}}$$

Ohm's law can be easily remembered in a way similar to that used in our study of force-pressure-area relationships. See Figure 14.3.

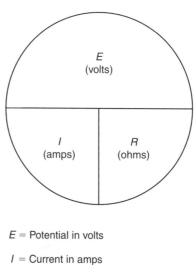

E = Potential in volts

I = Current in amps

R = Resistance in ohms

FIGURE 14.3 The relationships of volts, amps, and ohms are given in this handy diagram.

Here, E represents potential, in volts; I represents current, in amps, and R represents resistance, in ohms. Ohm's law can be used to find the any value of these three when the other two are known.

ELECTRICAL SYSTEMS

Example 14-1 Use the Ohm's law circle to determine the formula used to find current.

We want to find current (I). The other two variables occur in a way so that E is over R, so we can conclude

$$I = \frac{E}{R} \quad \text{or Current (amps)} = \frac{\text{Potential (volts)}}{\text{Resistance (ohms)}}$$

Example 14-2 A 12.6-volt battery is used to power a lightbulb that draws 500 milliamps of current. What is the resistance of the bulb?

Since we want to find resistance, cover the R on the Ohm's law circle. We then see that

$$R = \frac{E}{I} \quad \text{or Resistance (ohms)} = \frac{\text{Potential (volts)}}{\text{Current (amps)}}$$

Our current was given to be 500 milliamps (mA). We know that the prefix milli means 0.001 of our base units, so we have

$$500 \times 0.001 \text{ amps} = 0.500 \text{ or } 0.5 \text{ amps}$$

Then

$$R = \frac{12.6 \text{ volts}}{0.5 \text{ amps}} = 25.2 \text{ } \Omega$$

Remember that the greek letter Ω stands for the unit of ohms.

Example 14-3 An electric heater with a resistance of 0.9 Ω is plugged into a 12-volt power outlet. Will the heater work properly if the power outlet is protected by a 10-amp fuse?

Begin by finding the current through the heater. Using Ohm's law,

$$I = \frac{E}{R} \quad \text{or Current} = \frac{12 \text{ volts}}{0.9 \text{ } \Omega} = 13.33 \text{ amps}$$

Since the heater will draw 13.33 amps, the 10-amp fuse would blow, and the heater would not work.

ELECTRICAL POWER

You've probably noticed that many electrical components such as heaters, lightbulbs, stereos, and many power tools are rated according to the number of watts that are consumed. Voltage or current alone are not enough to determine how much electrical power some component consumes. While a spark plug requires very high voltage, it does not require much current. The starter motor on a vehicle requires very high current, but a low voltage. **Electrical power** is directly related to both the voltage and the current needed. Electrical power is measured in **watts**.

$$\text{Power (watts)} = \text{Current (amps)} \times \text{Potential (volts)}$$

This relationship leads to yet another relationship similar to the one described by Ohm's law. The relationship between power, current, and potential is given in Figure 14.4. This circle is used just as that given for Ohm's law.

Example 14-4 A typical household lightbulb is rated at 60 watts and operates at 120 volts. How much current is drawn by a typical lightbulb?

Using the following relationship, we see

$$I = \frac{P}{E} \quad \text{or Current (amps)} = \frac{\text{Power (watts)}}{\text{Potential (volts)}}$$

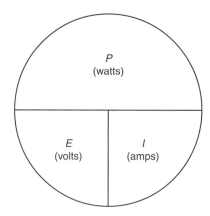

FIGURE 14.4 The relationships for power, current, and potential are given in this diagram.

Here, $P = 60$ watts and $V = 120$ volts.

$$I = \frac{60 \text{ watts}}{120 \text{ volts}} = 0.5 \text{ amps}$$

Example 14-5 A typical 14.4-volt alternator is rated at 85 amps. What is the power output of this alternator?

Since power is given by $P = I \times E$,

$$\text{Power} = 85 \text{ amps} \times 14.4 \text{ volts} = 1224 \text{ watts}$$

Example 14-6 A certain headlight is rated at 70 watts and operates on 12.6 volts. What is the resistance of this bulb?

We need to use Ohm's law to find resistance, but to do that we must know the current through the bulb. Begin by finding current.

$$I = \frac{P}{E}$$

or

$$I = \frac{70 \text{ watts}}{12.6 \text{ volts}} = 5.56 \text{ amps}$$

Now, we can find the resistance of the bulb.

$$R = \frac{12.6 \text{ volts}}{5.56 \text{ amps}} = 2.77 \text{ }\Omega$$

At this point it's easy to see that power, current, potential, and resistance are all related to one another. The two relationships we've seen in this section can be combined into a single chart as shown in Figure 14.5. If any one of the four values needs to be found, it can be done using any two of the others.

Example 14-7 Use the *P-I-E-R* chart in Figure 14.5 to find the resistance of a 70-watt bulb operating at 12.6 volts.

We want to find R when $P = 70$ watts and $E = 12.6$ volts. Look in the corner of the table labeled R for a formula using P and E. We see that

$$R = \frac{E^2}{P}$$

or

$$R = \frac{(12.6 \text{ volts})^2}{70 \text{ watts}} = \frac{158.76}{70} = 2.27 \text{ }\Omega$$

ELECTRICAL SYSTEMS

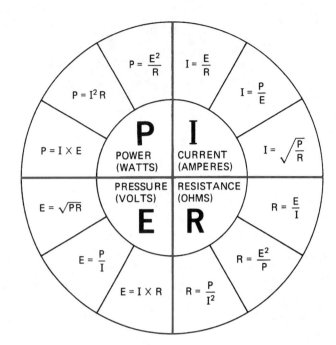

FIGURE 14.5 This *P-I-E-R* table can be used to solve for power, current, potential, or resistance when any two of the other values are known.

Source: James D. Halderman and Chase D. Mitchell, Jr., *Automotive Technology: Principles, Diagnosis, and Service*, Second Edition © 2003. Reprinted by permission of Pearson Education, Inc., Upper Saddle River, NJ.

Since electric motors and generators convert electrical power to mechanical and vice versa, it makes sense that we should be able to convert the units of mechanical power and electrical power.

$$1 \text{ horsepower} = 746 \text{ watts}$$

Example 14-8 An air compressor uses a motor rated at 2.5 horsepower. How many watts is this?

$$\frac{2.5 \text{ hp}}{1} \times \frac{746 \text{ watts}}{1 \text{ hp}} = \frac{1865 \text{ watts}}{1} = 1865 \text{ watts}$$

Example 14-9 A 12.6-volt alternator can produce up to 110 amps. How many horsepower are used in turning the alternator when producing maximum power?

Begin by finding the power output of the alternator, in watts.

$$\text{Power} = 110 \text{ amps} \times 12.6 \text{ volts} = 1386 \text{ watts}$$

Now, convert this to horsepower.

$$\frac{1386 \text{ watts}}{1} \times \frac{1 \text{ hp}}{746 \text{ watts}} = \frac{1386 \text{ hp}}{746} = 1.86 \text{ hp}$$

RESISTOR CIRCUITS

All electrical components such as lightbulbs, motors, and heating elements resist the flow of electricity, and are therefore resistors. Most electrical circuits contain more than one of these components. The headlights on a vehicle, for example, are part of the same circuit. Each of these bulbs has some resistance. In order to apply Ohm's law to circuits containing multiple resistors, we must understand the ways that these components can be connected.

The first type of circuit is a series circuit. In a **series circuit** electricity must pass through every resistor, one after another, while flowing though the circuit. In a series circuit if one resistor breaks the circuit, the entire circuit will be open. You might think of a series circuit as a single-lane road. Every vehicle passes through the same lane, and if that lane becomes blocked, all traffic stops. Figure 14.6 shows a simple example of a series circuit.

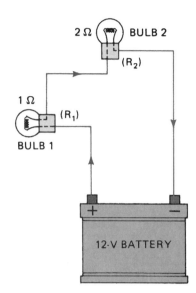

FIGURE 14.6 This series circuit contains two lightbulbs that act as resistors. As the current flows through the circuit, it must pass through each bulb before completing the loop.

Source: James D. Halderman and Chase D. Mitchell, Jr., *Automotive Technology: Principles, Diagnosis, and Service,* Second Edition © 2003. Reprinted by permission of Pearson Education, Inc., Upper Saddle River, NJ.

The total resistance in a series circuit is found by adding the total resistance of each of the resistors.

$$\text{Total resistance (resistors in series)} = R_1 + R_2 + R_3 + \ldots$$

Example 14-10 Find the current in the circuit shown below.

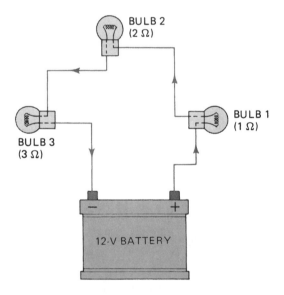

Source: James D. Halderman and Chase D. Mitchell, Jr., *Automotive Technology: Principles, Diagnosis, and Service,* Second Edition © 2003. Reprinted by permission of Pearson Education, Inc., Upper Saddle River, NJ.

We'll need to use Ohm's law to find the total current. To do this, we need to know the total resistance of the circuit. Since this is a series circuit we'll simply add the three resistances.

$$\text{Total resistance (series)} = 1\,\Omega + 2\,\Omega + 3\,\Omega = 6\,\Omega$$

Using Ohm's law,

$$I = \frac{E}{R} = \frac{12 \text{ volts}}{6\,\Omega} = 2 \text{ amps}$$

Series circuits are not commonly used because of one major disadvantage. If one resistor fails (such as a bulb burning out), none of the other components on the circuit will

ELECTRICAL SYSTEMS

work because the circuit has been broken. A more common circuit design is based on resistors in parallel. A **parallel circuit** consists of multiple resistors alongside each other, providing the current with several possible paths to follow, as shown in Figure 14.7. Parallel circuits are like multilane highways. If one lane becomes blocked, the traffic will still flow by using the other available lanes.

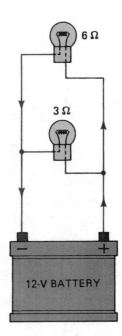

FIGURE 14.7 This parallel circuit provides two possible paths for current to follow. Some electrons will flow through the top bulb, while some will flow through the bottom bulb.

Source: James D. Halderman and Chase D. Mitchell, Jr., *Automotive Technology: Principles, Diagnosis, and Service*, Second Edition © 2003. Reprinted by permission of Pearson Education, Inc., Upper Saddle River, NJ.

In a parallel circuit, adding resistors actually reduces the total resistance in the same way that adding lanes to a road makes traffic flow more easily. Total resistance for resistors in parallel is given by the following:

$$\text{Total resistance (resistors in parallel)} = \frac{1}{\frac{1}{R_1} + \frac{1}{R_2} + \frac{1}{R_3} + \cdots}$$

Example 14-11 Find the total resistance of the parallel circuit shown in Figure 14.7.

Since the circuit contains one 3 Ω and one 6 Ω resistor in parallel, we can apply the formula for total resistance as follows:

$$\text{Total resistance (parallel)} = \frac{1}{\frac{1}{3} + \frac{1}{6}} = \frac{1}{\frac{2}{6} + \frac{1}{6}} = \frac{1}{\frac{3}{6}} = \frac{6}{3} = 2\,\Omega$$

This can be approximated by using decimals. Special care must be taken when rounding, however.

$$\text{Total resistance (parallel)} = \frac{1}{\frac{1}{3} + \frac{1}{6}} = \frac{1}{0.333 + 0.167} = \frac{1}{0.5} = 2\,\Omega$$

Either method is suitable for finding total resistance, keeping in mind that decimal values are often not exact.

Schematic drawings are often used to simplify the representation of circuits. Some common symbols are shown in Figure 14.8.

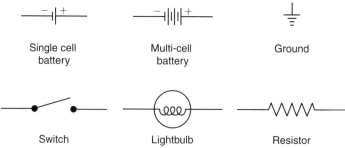

FIGURE 14.8 Common symbols found in schematic drawings of circuits.

Example 14-12 A circuit consists of a 6-volt power source and 3 Ω, 4 Ω, and 6 Ω resistors in parallel. Find the total current in this circuit.

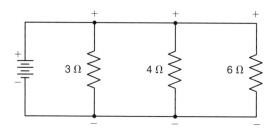

Begin by finding the total resistance of the circuit, so that we may then apply Ohm's law to find current.

$$\text{Total resistance (parallel)} = \frac{1}{\frac{1}{3} + \frac{1}{4} + \frac{1}{6}} = \frac{1}{\frac{4}{12} + \frac{3}{12} + \frac{2}{12}} = \frac{1}{\frac{9}{12}}$$
$$= \frac{12}{9}\,\Omega = 1.333\,\Omega$$

Using decimals, we have

$$\text{Total resistance (parallel)} = \frac{1}{\frac{1}{3} + \frac{1}{4} + \frac{1}{6}} = \frac{1}{0.333 + 0.25 + 0.167}$$
$$= \frac{1}{0.75} = 1.333\,\Omega$$

Now we can use this total resistance value in Ohm's law to find total current.

$$I = \frac{E}{R} = \frac{6\text{ volts}}{1.333\,\Omega} = 4.500\text{ amps}$$

Example 14-13 Find the total power consumed by the combination circuit shown below.

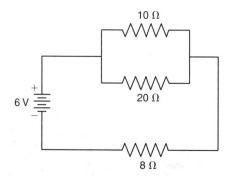

ELECTRICAL SYSTEMS

Begin by finding the total resistance of the circuit. Notice that the 10 Ω and 20 Ω resistors are in parallel, while the 8 Ω resistor is in series. Find the resistance of the parallel part of the circuit first.

$$\text{Total resistance (parallel)} = \frac{1}{\frac{1}{10} + \frac{1}{20}} = \frac{1}{0.1 + 0.05} = \frac{1}{0.15} = 6.667 \, \Omega$$

Since the 8 Ω resistor is in series, we can simply add this to the parallel leg.

$$\text{Total resistance} = 6.667 \, \Omega + 8 \, \Omega = 14.667 \, \Omega.$$

Use Ohm's law to find current.

$$I = \frac{E}{R} = \frac{6 \text{ volts}}{14.667 \, \Omega} = 0.409 \text{ amps}$$

Finally, the total power is given by $P = I \times E$, so

$$\text{Power} = 0.409 \text{ amps} \times 6 \text{ volts} = 2.5 \text{ watts}$$

As mentioned earlier in this chapter, wire itself acts as a resistor. As the length of a wire increases, the resistance also increases. This increase in resistance with length must be offset by the use of a thicker wire. Table 14.1 shows what gauge wire is needed to carry a given current (in the left column, in amps) through a given wire length (in the top row, in feet).

Note that the gauge number increases as the wire thickness decreases. Low gauge sizes represent thicker wires. If the length of a wire is between values listed on the table, round the length up to the next longer value on the table. There's no problem in using a thicker wire. Using a wire that is too thin, however, generates extra heat that can melt the wire's insulation. Table 14.1 applies to 12-volt circuits only.

Example 14-14 An additional 12-volt power outlet is being added to a vehicle. The 9-ft long wire must he able to carry 20 amps. What gauge wire should be used?

There is no column for 9-ft long wires on our table. The next longest length that is listed is 10 ft. If we look in the spot where the 10-ft column meets the 20-amp row, we see that we need 16 gauge wire to safely carry the load.

TABLE 14.1 Minimum Wire Gauge based on Length and Current Load

12 V	Wire Gauge (for Length in Feet)*						
Amps	3'	5'	7'	10'	15'	20'	25'
5	18	18	18	18	18	18	18
7	18	18	18	18	18	18	16
10	18	18	18	18	16	16	16
12	18	18	18	18	16	16	14
15	18	18	18	18	14	14	12
18	18	18	16	16	14	14	12
20	18	18	16	16	14	12	10
22	18	18	16	16	12	12	10
24	18	18	16	16	12	12	10
30	18	16	16	14	10	10	10
40	18	16	14	12	10	10	8
50	16	14	12	12	10	10	8
100	12	12	10	10	6	6	4
150	10	10	8	8	4	4	2
200	10	8	8	6	4	4	2

*When mechanical strength is a factor, use the next larger wire gauge.

CHAPTER 14 *Practice Problems*

Current, Voltage, Resistance, and Ohm's Law

A basic fuse-protected circuit for problems 1 and 2.

Source: James D. Halderman and Chase D. Mitchell, Jr., *Automotive Technology: Principles, Diagnosis, and Service*, Second Edition © 2003. Reprinted by permission of Pearson Education, Inc., Upper Saddle River, NJ.

A basic circuit is shown above. The load (a bulb) in this circuit is protected by a fuse.

1. Suppose the battery shown provides 12 volts of potential, and the resistance of the lightbulb is 30 Ω. What is the current passing through the bulb?

2. As a battery loses charge, its voltage drops. Suppose the battery shown produces only 8.6 volts, but the resistance of the bulb is still 30 Ω. What is the current in the bulb? Will the current increase or decrease as the battery loses charge?

3. In order to properly raise and lower a power window, a small electric motor must draw 4.0 amps at 12.6 volts. What is the resistance of this motor?

4. A starter motor draws 150 amps from the vehicle's 12.6-volt battery. What is the resistance of this starter motor?

5. A cabin light is rated at 100 milliamps (mA) and has a resistance of 144 Ω. What voltage is required to make this bulb glow brightly?

6. The resistance of a starter solenoid is measured and is found to be 8 Ω. The solenoid will operate if it draws at least 750 milliamps (mA) of current. How many volts are needed?

7. An electric motor is defective, and has a resistance value of only 0.72 Ω. When powered by the vehicle's 14.4-volt electrical system, will the motor cause a 15-amp fuse to blow? How much current is being drawn by the motor?

8. The headlamp circuit on a vehicle is designed to use a 15-amp fuse. If the headlights (and all other lights on that circuit) have a total resistance of 2.13 Ω in this 12-volt system, would the circuit still operate if a 5-amp fuse were accidentally inserted in the fuse holder?

ELECTRICAL SYSTEMS

••9. Use Ohm's law to complete the table below.

Current (Amps)	Potential (Volts)	Resistance (Ohms)
Increased	Increased	No change
No change		Increased
	Increased	No change
Decreased	No change	
	No change	Increased
No change	Decreased	

Electrical Power

A typical starter might draw 100 amperes or more in a 12.6-volt system.

Source: James D. Halderman and Chase D. Mitchell, Jr., *Automotive Technology: Principles, Diagnosis, and Service*, Second Edition © 2003. Reprinted by permission of Pearson Education, Inc., Upper Saddle River, NJ.

•1. A starter motor is part of a 12.6-volt system, and draws 100 amps. How many watts is the starter using?

•2. An air conditioning compressor clutch draws 6 amps in a 12.6-volt system. How many watts is this?

•3. Which uses more power: an electric saw using 3.0 amps at 120 volts, or a windshield wiper motor drawing 18 amps at 12 volts?

•4. Which lightbulb consumes more power: a floodlight using 0.7 amps at 120 volts, or a headlight using 4.3 amps at 12 volts?

•5. How much current is drawn by a 100-watt stereo operating on 12 volts? Round to the nearest hundredth amp.

•6. Is a 10-amp fuse in a 12.6-volt system adequate to power headlights drawing a total of 170 watts?

•7. A small air compressor is rated at 1.5 horsepower. How many watts is this? How many kilowatts is this?

•8. A starter motor can produce up to 5 horsepower. How many watts is this? If the motor operates on 12.6 volts, how many amps does the motor draw?

For the following problems, use the chart in Figure 14.5 to find the necessary equation to solve the problem.

●●9. What is the resistance of an 80-watt headlight operating at 12.6 volts?

●●10. How much current will be used by a 25-watt lightbulb if its resistance is 278 Ω?

●●11. What voltage is necessary to operate a 16-watt solenoid having a resistance of 4.3 Ω?

A typical automotive solenoid consumes several watts of electrical power.

Source: James D. Halderman and Chase D. Mitchell, Jr., *Automotive Technology: Principles, Diagnosis, and Service,* Second Edition © 2003. Reprinted by permission of Pearson Education, Inc., Upper Saddle River, NJ.

●●12. How many watts are needed to operate a power window motor having a resistance of 5.6 Ω when operating as part of a 12.6-volt system?

Resistor Circuits

For each of the circuits shown below, determine: (a) the total resistance of the circuit, (b) the total current in the circuit, and (c) the total power consumed by the circuit. Round all values to the nearest tenth.

●1.

ELECTRICAL SYSTEMS

●2.

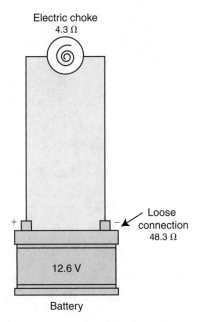

●3.

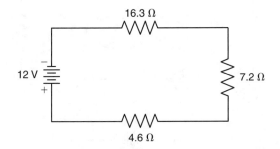

●4.

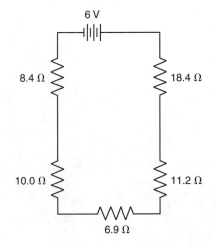

●●5.

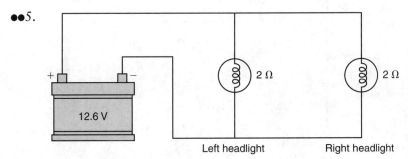

●●6.
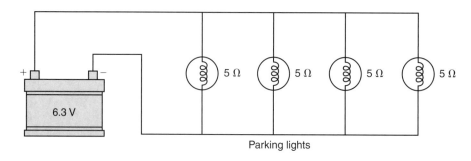

●●7.
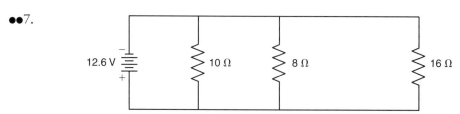

●●8.
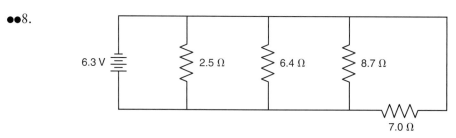

●●9.
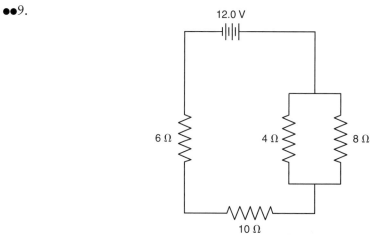

●●10.
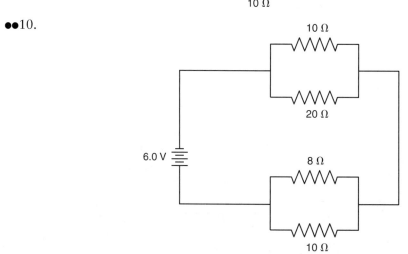

ELECTRICAL SYSTEMS

The current in a series circuit is the same in every part of the circuit. Then for any series circuit we can use the current and the resistance of each bulb to find the voltage drop across each bulb. Use this fact to answer problems 11 and 12.

●●●11. a. What is the voltage drop across each bulb in problem 1?
b. What is the total voltage drop across both bulbs?
c. How does this compare to the battery voltage?

●●●12. a. What is the voltage drop across each resistor in problem 3?
b. What is the total voltage drop across all resistors?
c. How does this compare to the battery voltage?

In a parallel circuit, the voltage across every resistor is equal to the source voltage. Using the resistance of each bulb and the source voltage, the current through each resistor can be found.

●●●13. a. What is the current through each bulb in problem 5?
b. What is the total current passing through both bulbs?
c. How does this compare with the total current in the circuit you found in problem 5?

●●●14. a. What is the total current passing through each bulb in problem 7?
b. What is the total current flowing through all resistors?
c. How does this compare to the total current found in problem 7?

As the length of a wire increases, the resistance also increases. This increase in resistance with length must be offset with the use of a thicker wire. Table 14.1 shows what gauge wire is needed to carry a given current (in the left column, in amps) through a given wire length (in the top row, in feet). Note that the thicker a wire is, the lower its gauge number will be. If a wire needs to be longer than a table value, use the gauge of the next longer table value wire. Use Table 14.1 to answer questions 15 through 18. Assume all circuits use 12 volts.

●15. What gauge wire is needed for a 10-ft long wire supplying power to a motor that draws 40 amps?

●16. A solenoid draws about 20 amps of power. A 5-ft long wire is used to activate the solenoid. What gauge should be used?

●17. A 12-ft wire needs to carry 50 amps from an alternator. What gauge should be used?

●18. What gauge is necessary if a 22-ft wire is used to provide power to an antenna motor drawing 20 amps?

CHAPTER 15

Motion

DISTANCE, SPEED, AND TIME RELATIONSHIPS

In Chapter 12 we discussed some of the factors that can create an incorrect speedometer reading. With a basic understanding of the physics of motion, a technician can better understand the information being reported on a speedometer or odometer.

Suppose on a road trip, a vehicle on a highway passes mile marker 12 at exactly 9:00 AM. At noon, the driver passes mile marker 168, as shown in Figure 15.1. Apparently, the driver has been driving for 3 hours, and has driven a total of $168 - 12 = 156$ miles. These two values can be written as a ratio of whole numbers:

$$\frac{156 \text{ mi}}{3 \text{ hour}}$$

or, as a unit rate, 52 miles/hour or 52 miles per hour (MPH).

This ratio of distance traveled to the time needed is called the **average speed**.

$$\text{Average speed} = \frac{\text{Total distance traveled}}{\text{Total time}}$$

In the metric system, speed usually has units of

$$\frac{\text{m}}{\text{second}} \text{ or } \frac{\text{km}}{\text{hour}}$$

In the customary system, speed often has units of

$$\frac{\text{ft}}{\text{second}} \text{ or } \frac{\text{mi}}{\text{hour}}$$

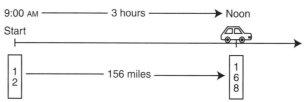

FIGURE 15.1 Average speed is calculated by dividing the total distance traveled by the total time spent.

It is very unlikely that the driver maintained exactly 52 miles per hour for the entire trip. At times she may have driven as fast as 60 miles per hour, and at times she may have come to a stop. The **instantaneous speed** is the speed at any given moment. The driver had an instantaneous speed of zero while stopped at a stop sign, but the average speed of 52 MPH indicates that for most of the drive she was making good progress. The speedometer on a vehicle, such as the one in Figure 15.2, reports the approximate instantaneous speed at any time. Some digital instrument panels will calculate and report the average speed over some time.

FIGURE 15.2 The speedometer on a vehicle reports the instantaneous speed of the vehicle at any moment.

Source: Jerry D. Wilson and Anthony J. Buffa, *College Physics*, Fourth Edition © 2000. Reprinted by permission of Pearson Education, Upper Saddle River, NJ.

Example 15-1 On a drive to work a commuter travels 23.7 miles. One morning this trip takes 45 minutes. What is the commuter's average speed in miles per hour?

To find the speed in miles per hour, we need the distance traveled to have units of miles (which it does) and the time used to have units of hours. Convert 45 minutes to hours.

$$\frac{45 \text{ minutes}}{1} \times \frac{1 \text{ hour}}{60 \text{ minutes}} = 0.75 \text{ hours}$$

Now, find the average speed.

$$\text{Average speed} = \frac{23.7 \text{ miles}}{0.75 \text{ hours}} = 31.6 \text{ MPH}$$

Example 15-2 After getting a speeding ticket, a customer complains that his speedometer is incorrect. To check this claim, a technician takes the vehicle on the interstate. After setting the cruise control at 60 MPH, he uses a stopwatch to measure the time needed to travel between mile marker 207 and 212. It takes 4 minutes and 27 seconds. Is the speedometer calibrated correctly?

Begin by finding the average velocity. The distance traveled is $212 - 207 = 5$ miles. The time needed is $4\frac{27}{60}$ or 4.45 minutes.

$$\frac{4.45 \text{ minutes}}{1} \times \frac{1 \text{ hour}}{60 \text{ minutes}} = 0.0742 \text{ hours}$$

$$\text{Average speed} = \frac{5 \text{ mi}}{0.0742 \text{ hours}} = 67.4 \text{ MPH}$$

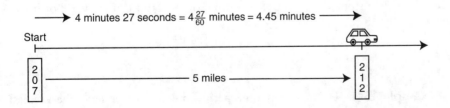

The customer is correct. The vehicle's average speed was 67.4 MPH, even though the cruise was set (and the speedometer reading must have been) 60 MPH. The vehicle has a higher speed than the speedometer indicates.

If we create a graph with time along the horizontal axis and location along the vertical axis, time is said to be the **independent variable** because the driver's location will depend on time, but the time does not depend on the driver's location. This means that the location is the **dependent variable**.

Example 15-3 While on a trip, a driver notes the nearest mile marker on the roadside every 30 minutes. Graph the driver's progress.

Time	Mile Marker
Noon	120
12:30	141
1:00	168
1:30	194
2:00	235
2:30	239
3:00	257
3:30	280
4:00	254

Using the information in the table above, we chart the independent variable, time, against the dependent variable, location.

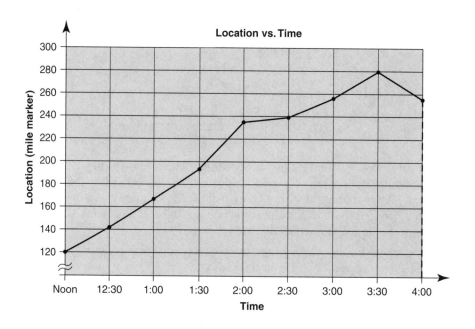

The chart of location vs. time in the example above shows us information about the driver's progress. Note that in some places, the graph is steeper than in others. The steepest portion of the graph means that the driver covered the greatest distance in a given time. The **slope** of some part of a line on a graph is a measure of how steep the line is.

$$\text{Slope} = \frac{\text{Rise}}{\text{Run}}$$

Example 15-4 During what time period did the driver have the greatest average speed?

The portion of the trip with the greatest average speed can be seen on the graph.
In this case the steepest part of the graph is between 1:30 and 2:00. To find the average speed during this time, we'll calculate the slope.

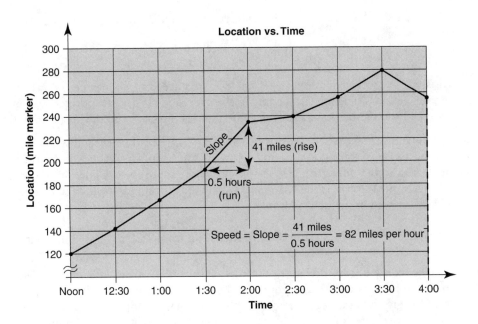

The graph rises from 194 to 235 during this time. The rise is $235 - 194 = 41$ miles. The run is the change in time in this case, $2:00 - 1:30 = 30$ minutes $= 0.5$ hours.

$$\text{Slope} = \frac{41 \text{ mi}}{0.5 \text{ hours}} = 82 \text{ MPH}$$

Then the average speed is 82 MPH.

Example 15-5 During which portion of the trip did the driver have the lowest average speed?

By looking at the graph we see that the graph is not very steep between 2:00 and 2:30. This means the driver did not travel very far in that time period. To find the speed, we'll calculate the slope.

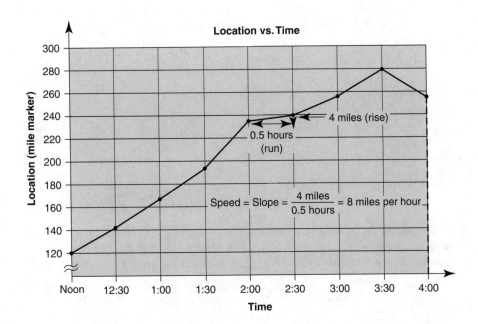

Here the rise is $239 - 235 = 4$ miles. The run is again 30 minutes, or 0.5 hours.

$$\text{Slope} = \frac{4 \text{ mi}}{0.5 \text{ hours}} = 8 \text{ MPH}$$

It is not likely that the driver only traveled at a speed of 8 miles per hour during that entire period. Remember that this is the average velocity. The driver was probably stopped for a good portion of those 30 minutes, probably getting a speeding ticket.

We have seen that when the slope is steep, the speed is higher, and when the slope is less steep, the speed during that period is lower. What happens when a driver travels back in the direction they came from? They may have the same speed as before, but in the opposite direction. **Velocity** is the term used when a direction is associated with some speed. We'll see this used in the next example.

Example 15-6 What is the slope of the graph from 3:30 to 4:00, and what does this mean?

We can see from the graph that the driver travels from mile marker 280 to 254 between 3:30 and 4:00. The graph does not rise at all, but actually falls. We say then that the graph has a negative rise, and thus a negative slope.

The rise here is $254 - 280 = -26$ miles. The run is again 0.5 hours.

$$\text{Slope} = \frac{-26 \text{ mi}}{0.5 \text{ hours}} = -52 \text{ MPH}$$

It doesn't make sense to say that the driver's speed was -52 MPH. Recall from our work with signed numbers that the negative sign really just means the driver changed directions, and is headed back to where he started. We say that the driver's speed is 52 MPH, but that his velocity is -52 MPH.

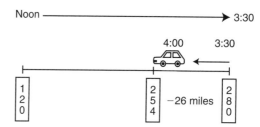

Example 15-7 What was the driver's average *speed* for the whole trip?

The total distance traveled was from mile marker 120 to 280, then back to 254 for a total of $160 + 26 = 186$ miles. The entire trip took 4 hours.

$$\text{Average speed} = \frac{186 \text{ mi}}{4.0 \text{ hours}} = 46.5 \text{ MPH}$$

The equation

$$\text{Average speed} = \frac{\text{Distance traveled}}{\text{Time}}$$

can be rewritten to give another well-known relationship. Multiply both sides of the equation by the variable "time," and we see that

$$\boxed{\text{Distance traveled} = \text{Average speed} \times \text{Time}}$$

Example 15-8 On the interstate, a driver may be able to average 54 MPH over the course of a day including gas stops. If a family can spend up to 9 hours per day in a car without getting angry at each other, how far can they travel in a day?

$$\text{Distance traveled} = \frac{54 \text{ mi}}{\text{hour}} \times 9 \text{ hours} = 486 \text{ mi}$$

MOTION

VELOCITY, ACCELERATION, AND TIME RELATIONSHIPS

Two cars were road tested by an automobile magazine. The first, a Subaru Imprezza, was powered by a turbocharged 4-cylinder engine. The second, a Ford Mustang, was powered by a V-8 engine. The results of the road tests are shown in Table 15.1.

TABLE 15.1 Subaru Impreza and Ford Mustang Road Test Results*

	Subaru Impreza WRX	Ford Mustang GT
0–60 MPH	4.9 seconds	5.0 seconds
Top speed	145 MPH	147 MPH

* Subaru Impreza performance data from Car and Driver Magazine, *February 2005*. Ford Mustang performance data taken from Car and Driver Magazine, *June 2005*.

The question is, which is faster? The Impreza is able to more quickly reach 60 MPH from rest, while the Mustang has a higher top speed. The answer shows that there are different ways to measure how fast a vehicle is. In the previous section we explored the idea of speed. If the question is simply which car can reach a higher speed, the answer is clearly the Mustang. The 0–60 times, though, represent the ability of the car to accelerate quickly. **Acceleration** is measure of how quickly a car can change its speed. Because speed can decrease or increase, the change in speed must be represented by a positive number (if there is an increase) or a negative number (if there's a decrease). At this point we'll use the term velocity instead of speed to account for this indicator of direction.

$$\text{Average acceleration} = \frac{\text{Change in velocitiy}}{\text{Time}}$$

The units of acceleration in the metric system are

$$\frac{\text{meters}}{\text{second}^2}$$

while in the customary system they are

$$\frac{\text{ft}}{\text{second}^2}$$

Example 15-9 Find the average acceleration from 0 to 60 MPH of the Mustang.

So that we arrive at units of ft/second², we need to convert the change in speed (velocity) units to ft/second. Time is already in seconds.

Change in velocity =

$$\frac{60 \text{ mi}}{1 \text{ hour}} \times \frac{5280 \text{ ft}}{1 \text{ mile}} \times \frac{1 \text{ hour}}{60 \text{ minute}} \times \frac{1 \text{ minute}}{60 \text{ second}} = \frac{316{,}800 \text{ ft}}{3600 \text{ second}} = 88 \text{ ft/second}$$

Then the acceleration of this vehicle is

$$\text{Average acceleration} = \frac{88 \text{ ft/second}}{5.0 \text{ second}} = 17.6 \text{ ft/sec}^2$$

We can rearrange the basic equation

$$\text{Average acceleration} = \frac{\text{Change in velocity}}{\text{Time}}$$

by multiplying both sides by "time" to see the following:

$$\text{Change in velocity} = \text{Average acceleration} \times \text{Time}$$

Example 15-10 A very fast car is capable of accelerating at an average rate of 21.5 ft/second² at speeds below 80 MPH. If, from a standing start, a driver accelerates at this rate for 5.0 seconds, what will be the final velocity in miles per hour?

$$\text{Change in velocity} = \frac{21.5 \text{ ft}}{\text{second}^2} \times 5.0 \text{ second} = 107.5 \text{ ft/second}$$

Now, convert the units of this velocity to MPH.

$$\frac{107.5 \text{ ft}}{1 \text{ second}} \times \frac{1 \text{ mi}}{5280 \text{ ft}} \times \frac{60 \text{ seconds}}{1 \text{ minute}} \times \frac{60 \text{ minutes}}{1 \text{ hour}} = \frac{38,700 \text{ mi}}{5280 \text{ hours}} = 73.3 \text{ MPH}$$

Example 15-11 On a road test, a small truck is accelerated from rest. Its velocity is recorded every two seconds. The results are shown below. Make a graph of velocity versus time. Again, time is on the horizontal axis, while velocity is on the vertical axis.

Time (seconds)	Velocity (MPH)	Velocity (ft per second)
0	0	0
2.0	19	27.9
4.0	31	45.5
6.0	40	58.7
8.0	46	67.5
10.0	35	51.3

Each velocity, in MPH, has been converted to feet per second for the purposes of graphing.

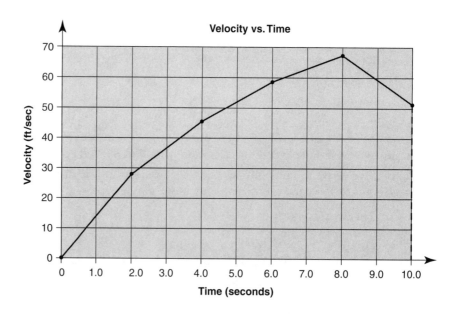

Example 15-12 During which two-second interval did the vehicle have the greatest average acceleration? What was the average acceleration during that interval?

We can see from the graph that our curve rises the most between 0 and 2.0 seconds. If we find the slope of the line segment from 0 to 2.0 seconds, we'll also be finding the average acceleration.

From 0 to 2.0 seconds, the velocity changes from 0 to 27.9 ft/second.

$$\text{Slope} = \frac{\text{Rise}}{\text{Run}} = \frac{\text{Change in velocity}}{\text{Time}} = \frac{27.9 \text{ ft/second}}{2.0 \text{ second}} = 13.95 \frac{\text{ft}}{\text{second}^2}$$

Example 15-13 Find the slope of the line segment between 8.0 and 10.0 seconds. What does this slope show about the truck's motion?

From the graph, we see that the curve falls between 8.0 and 10.0 seconds. The velocity changes from 67.5 ft/second to 51.3 ft/second. This means that the overall change in speed (and rise) was negative. This gives a change in velocity of −16.2 ft/second.

$$\text{Slope} = \frac{\text{Rise}}{\text{Run}} = \frac{-16.2 \text{ ft/second}}{2.0 \text{ second}} = -8.1 \frac{\text{ft}}{\text{second}^2}$$

The slope of a curve on a speed vs. time graph also tells us our acceleration, so the acceleration is -8.1 ft/second^2.

It is important to note in the previous example that the vehicle does NOT change direction. This negative acceleration simply means the vehicle is slowing down, not speeding up. We can see from the graph that the car is simply moving at a lower velocity at 10.0 seconds than it was at 8.0 seconds. This negative slope just means the driver has used his brakes.

Example 15-14 If the driver continues to brake, creating an acceleration of -8.1 ft/second^2 as shown in the previous example, how fast (in MPH) will the vehicle be moving when the time reaches 16.0 seconds?

The driver will continue to accelerate at -8.1 ft/second^2 from 10.0 to 16.0 seconds, a total time of 6.0 seconds.

Recall that

$$\text{Change in velocity} = \text{Average acceleration} \times \text{Time}$$

Then

$$\text{Change in velocity} = -8.1 \text{ ft/second}^2 \times 6.0 \text{ second} = -48.6 \text{ ft/second}$$

Converting this to MPH, we see that the driver's velocity will change by

$$\frac{-48.6 \text{ ft}}{1 \text{ second}} \times \frac{1 \text{ mi}}{5280 \text{ ft}} \times \frac{60 \text{ second}}{1 \text{ minute}} \times \frac{60 \text{ minute}}{1 \text{ hour}} = -33.1 \text{ MPH}$$

From the table of data, we see that at 10.0 seconds the driver had a velocity of 35.0 MPH. Then at 16.0 seconds his velocity will be

$$35.0 - 33.1 = 1.9 \text{ MPH}$$

The driver has almost stopped.

CHAPTER 15 *Practice Problems*

Distance, Speed, and Time Relationships

- 1. A motorcyclist passes mile marker 83 at 1:30 in the afternoon and marker 179 at 3:30 that afternoon. What is her average speed?

- 2. On the way home, the same rider crosses mile marker 172 at 7:15 in the evening, and exits at mile marker 64 at 10:45. What was her average speed on the return trip?

- 3. A technician uses a stopwatch to measure the time needed to travel between mile marker 37 and mile marker 47 on the interstate. It takes 11 minutes and 30 seconds. What is the average speed (in MPH), rounded to the nearest tenth?

- 4. Suspecting that putting new tires on his truck has altered his speedometer's accuracy, a driver notes that it takes 3 minutes and 40 seconds to travel four miles. What is the average speed?

•5. A driver with the cruise control set at 58 MPH (dangerously) uses his cell phone while driving. His call lasts 54 minutes. How far did he travel during that call?

•6. A traveler is cruising at 68 MPH when Led Zeppelin's "Stairway to Heaven" comes on the radio. The song lasts 7 minutes and 55 seconds. How far does the driver travel while this song is on? Round to the nearest tenth of a mile.

••7. When checking the calibration of a speedometer, a technician sets the cruise control at 53 MPH on a highway. It takes exactly 6 minutes and 24 seconds to travel five miles. Is the speedometer correct? If not, does it give a high or low readout, and by how much? Round to the nearest tenth.

••8. A customer complains that the speedometer in his car "just doesn't seem right." On the freeway, while traveling consistently at 58 MPH it takes 6 minutes and 12 seconds to travel from mile marker 97 to mile marker 91. Is the speedometer in error? If so, is it too high or too low, and by how much?

••9. At 8:30 one morning, a technician in Minnesota calls in sick and decides to drive to Daytona, Florida with his wife and daughter to watch the race that weekend. They leave at 8:45 AM. They travel 26 miles and drop off their dog at 9:30. At 10:00 they're back on the road, and travel 183 miles before stopping for lunch at 1:00 PM. At 1:45 they're back on the road, and cover 273 miles before stopping for the day at 7:30 PM. What was their average speed for the entire day? Assume the trip began when they left at 8:45. Round to the nearest tenth.

••10. A parts supplier leaves to deliver an alternator at 10:15 AM. He drives 8.3 miles to the dealership and arrives at 10:31. At 10:47 he leaves the dealership and returns to the warehouse, covering another 8.3 miles. He returns at 11:03. What is his average speed for the entire trip?

••11. On a road trip, you log the time you pass each of the following mile markers.

Time	Mile Marker
9:30 AM	36
10:00	56
10:30	79
11:00	107
11:30	108
12:00 PM	120
12:30	155
1:00	189
1:30	209
2:00	234
2:30	215
3:00	182

a. Make a graph of location versus time on the graph below.
b. During which time period was the average speed highest? What was the average speed during this time?
c. During which time period was the average speed lowest? What was the average speed during this time?

MOTION

d. At what time did the driver turn around and drive back toward the starting point?
e. What was the overall average speed for this trip?

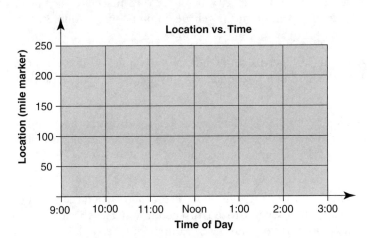

●●12. On the same trip a week later, you record the same information.

Time	Mile Marker
9:12 AM	27
9:50	51
10:20	73
10:36	86
11:12	101
11:50	148
12:25 PM	165
1:07	190
1:39	207
2:20	236
2:40	236
3:15	236

a. Make a graph of location versus time on the graph below.
b. During which time period was the average speed highest? What was the average speed during this time?
c. During which time period was the average speed lowest? What was the average speed during this time?
d. At what time did the driver quit driving?
e. What was the overall average speed for this trip?

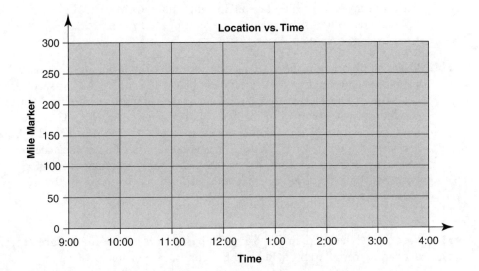

Velocity, Acceleration and Time Relationships

•1. A vehicle increases velocity from 0 to 42 ft/second in 4.8 seconds. What is the average acceleration of the vehicle, in units of ft/second^2?

•2. A sprint car increases its velocity very quickly. If a sprint car goes from 0 to 150 ft/second in 6.0 seconds, what is the average acceleration in ft/second^2?

•3. While braking, a car decreases in velocity from 88 ft/second to 16 ft/second. If it takes 4.8 seconds to do this, what is the average acceleration of the vehicle? Report your answer in units of ft/second^2.

•4. Over the course of 120 seconds, a truck climbs a large hill and decreases in velocity from 80 ft/second to 65 ft/second. What is the average acceleration of this vehicle? Report your answer with units of ft/second^2.

•5. If a vehicle at rest begins accelerating at 7.3 ft/second^2 steadily for 15.3 seconds, what will be the final speed, in ft/second?

•6. A vehicle is traveling at 50 ft/second then brakes at a rate of 10 ft/second^2 for 3 seconds. What is the final velocity?

••7. Suppose that for 2.0 seconds a vehicle brakes at a rate of −21.0 ft/second^2 from an initial speed of 60 MPH. What is the new velocity of the vehicle, in units of MPH?

••8. A lug nut is dropped from a tall bridge. Under the influence of gravity it accelerates at a rate of −32.2 ft/second^2. It falls for 4.5 seconds before hitting the water below. What was the final speed of the lug nut just as hits the water? Report the answer in MPH.

••9. A typical car can accelerate from 0 to 60 MPH in 8.0 seconds. Find the average acceleration, in ft/second^2.

••10. An economy car accelerates on the interstate from 50 MPH to 75 MPH in 8.7 seconds. What is the average acceleration, in ft/second^2?

••11. When an object is dropped, it accelerates under the influence of gravity. If air resistance is ignored, gravity accelerates objects at a rate of 32.2 ft/second^2 toward the ground. The acceleration rate of a vehicle or aircraft is often compared to the acceleration of gravity. An airplane that accelerates with 2 G's, for example, would accelerate at a rate of 2 × 32.2 ft/second^2 = 64.4 ft/second^2. A car that accelerates at a rate of 16.1 ft/sec^2 accelerates half as fast as the acceleration rate of gravity, thus it develops $\frac{16.1}{32.2}$ = 0.5 G's. Find the acceleration, in G's, for the Ford Mustang featured in Table 15.1. It can accelerate from 0 to 60 MPH in 5.0 seconds.

••12. Find the acceleration, in G's, for the Subaru Impreza featured in Table 15.1. It can accelerate from 0 to 60 MPH in 4.9 seconds. (See problem 11.)

MOTION

●●13. While accelerating on a freeway entrance ramp, the velocity of a vehicle is recorded every 5 seconds. The data is shown below.

Time (seconds)	Velocity (MPH)	Velocity (feet per second)
0	5	
5.0	47	
10.0	63	
15.0	71	
20.0	71	
25.0	68	
30.0	70	

a. Calculate the velocity, in ft/second, for each entry in the table.
b. Make a graph of Velocity vs. Time for the data above. Use the velocity values you calculated in ft/second.

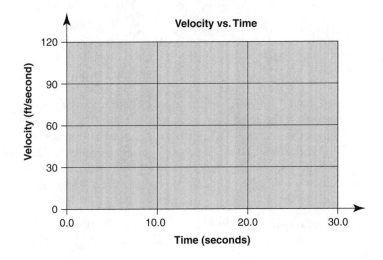

c. During what interval was the average acceleration the greatest? What is the average acceleration over this interval?
d. What is the slope of the line segment from 10.0 seconds to 15.0 seconds?
e. What is the average acceleration from 15.0 seconds to 20.0 seconds?
f. At some points on the curve the slope is 0. What does this indicate about the motion of the vehicle?

●●●14. If the vehicle in problem 13 were to begin accelerating from rest at a rate of 4.4 ft/second^2 when the time was 30.0 seconds, and kept accelerating until the time was 40.0 seconds, what would be the final speed, in MPH?

●●15. During a braking test, a vehicle is slowed, and the velocity is recorded every 0.5 seconds. The data is shown below.

Time (seconds)	Velocity (MPH)	Velocity (feet per second)
0	60	
0.5	58	
1.0	55	
1.5	49	
2.0	35	
2.5	20	
3.0	11	
3.5	6	
4.0	0	

a. Calculate the velocity, in ft/second, for each entry on the table.
b. Make a graph of Velocity vs. Time for the data above. Use the velocity values you calculated in ft/second.

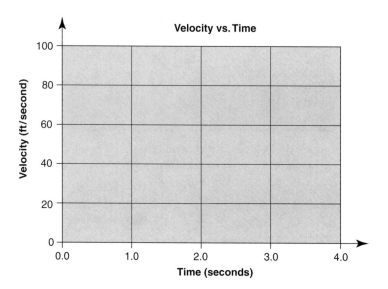

c. During what interval was the average deceleration (negative acceleration) the greatest? What is the average acceleration over this interval?
d. What is the slope of the line segment from 1.0 seconds to 1.5 seconds?
e. What is the average acceleration from 2.5 seconds to 3.0 seconds?
f. If the curve were to continue beyond 4.0 seconds, and begin to rise again, what could you conclude about the driver's action at that point?

●●●16. Just before the driver in problem 15 reached 60 miles per hour, she accelerated at a rate of 3.0 ft/second^2 for 4.0 seconds. What was her velocity, in MPH, before she began accelerating?

CHAPTER 16

Repair Orders

FILLING OUT A REPAIR ORDER

The process of billing a customer when a repair has been completed may be the most basic step of running a successful business. Not only must the labor involved with the repair be billed fairly, but sales tax and part costs must be accurately calculated. While this is now often done using a computer database, a basic understanding of the billing process is essential for any technician.

There are several companies that produce manuals that give recommended time allowances for repairs. One example is printed as an appendix, the motor parts and labor guide for the General Motors Saturn L-series. This appendix will be used for all examples and practice problems.

Example 16-1 What time is reasonable for a technician to remove and replace an EGR valve on a 2000 Saturn L-series?

Look on page 53-3 of the Appendix under Emission System.
The job number is 2-2 (Section 2, job 2): EGR VALVE, R&R.
(R&R means "remove and replace.")
Notice that the time allowances are given as a decimal portion of an hour. To replace this EGR valve, the manual allows approximately 0.4 hours. Notice here that two times are given. The Motor Time is the time recommended by the publisher of this manual for a vehicle that is worn and may require slightly more shop time than the Factory Time of 0.3 hours.

$$\frac{0.4 \text{ hours}}{1} \times \frac{60 \text{ minutes}}{1 \text{ hour}} = \frac{24 \text{ minutes}}{1} = 24 \text{ minutes}$$

Example 16-2 At $65.00 per hour, what is the labor charge for this job?

$$\frac{0.4 \text{ hours}}{1} \times \frac{\$65.00}{1 \text{ hour}} = \frac{\$26.00}{1} = \$26.00$$

Or, more simply, 0.4 hours × $65.00/hour = $26.00.

Example 16-3 What is the part number and price for the EGR valve?

Again, look on page 53-3 of the Parts & Labor Guide.
Look under section 2, PARTS.
For the EGR valve on this vehicle with the 3.0 L engine, the guide shows

Part Number: 09120453   Price $150.00

Example 16-4 If the sales tax rate is 6.5%, what is the sales tax on the EGR valve?

Sales Tax: $150.00 × 0.065 = $9.75

Example 16-5 Assuming labor is tax-free, what is the total amount due from the customer?

Labor:	$ 26.00
Parts:	$150.00
Tax:	$ 9.75
Total	$185.75

Example 16-6 A technician spends 3 hours and 40 minutes replacing a head gasket. How much time is this, to the nearest tenth of an hour?

We must convert 40 minutes to a decimal part of an hour.

$$\frac{40 \text{ minutes}}{1} \times \frac{1 \text{ hour}}{60 \text{ minutes}} = \frac{40 \text{ hour}}{60} = 0.667 \text{ hours} \approx 0.7 \text{ hours}$$

With the 3 full hours, it takes about 3.7 hours.

Example 16-7 Calculate the customer's total charges for this repair. A customer, Katie Daniels, brings her 2000 Saturn LS with the 3.0 L engine and automatic transaxle in for repairs. She agrees to have the following services performed:

Lube & filter service   Replace serpentine belt
Replace both coolant hoses   Replace thermostat
Replace both valve cover gaskets   Replace shifter control cable

Assume a labor rate of $65.00 per hour and a sales tax rate of 6%. Assume there is no sales tax on labor in this state.

This vehicle takes four quarts of motor oil at $2.79 per quart. Since the coolant will already be drained, it will not take as long to replace the thermostat as the manual recommends. Assume it takes 30 minutes.

Begin by filling in all information about the customer and the vehicle, and list all services to be performed on the vehicle. See Figure 16.1.

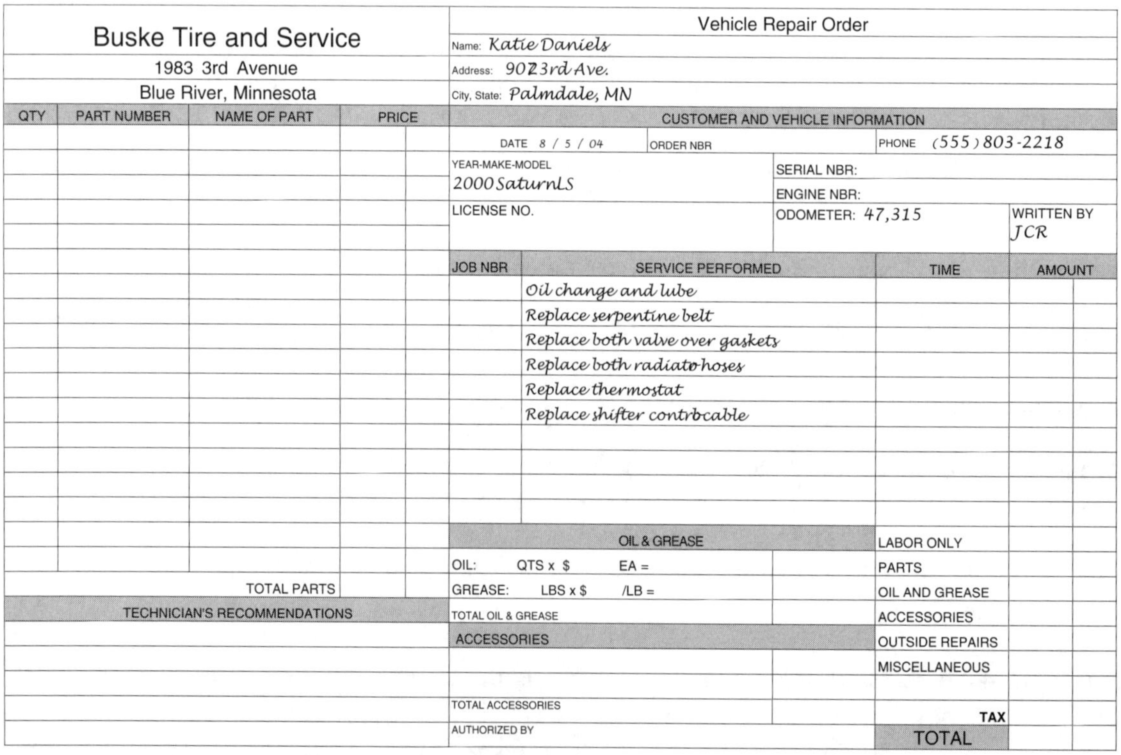

FIGURE 16.1 Step 1: List all services to be performed.

REPAIR ORDERS

Look up the job number and labor charges for each service, and list it on the repair order. See Figure 16.2.

\multicolumn{4}{c	}{Buske Tire and Service}	\multicolumn{5}{c	}{Vehicle Repair Order}
\multicolumn{4}{c	}{1983 3rd Avenue}	\multicolumn{5}{l	}{Name: Katie Daniels}
\multicolumn{4}{c	}{Blue River, Minnesota}	\multicolumn{5}{l	}{Address: 907 23rd Ave.}
				\multicolumn{5}{l	}{City, State: Palmdale, MN}
QTY	PART NUMBER	NAME OF PART	PRICE	\multicolumn{5}{c	}{CUSTOMER AND VEHICLE INFORMATION}
				DATE 8/5/04	ORDER NBR	\multicolumn{3}{l	}{PHONE (555) 803-2218}
				\multicolumn{2}{l	}{YEAR-MAKE-MODEL 2000 Saturn LS}	\multicolumn{3}{l	}{SERIAL NBR:}
						\multicolumn{3}{l	}{ENGINE NBR:}
				\multicolumn{2}{l	}{LICENSE NO.}	ODOMETER: 47,315	\multicolumn{2}{l	}{WRITTEN BY JCR}
				JOB NBR	SERVICE PERFORMED		TIME	AMOUNT
				1-10	Oil change and lube		0.9	
				1-3	Replace serpentine belt		1.2	
				10-11	Replace both valve cover gaskets		3.5	
				8-1	Replace both radiator hoses		1.1	
				*	Replace thermostat		0.5	
				14-1	Replace shifter control cable		2.7	
				\multicolumn{3}{c	}{OIL & GREASE}	\multicolumn{2}{l	}{LABOR ONLY}
				\multicolumn{3}{l	}{OIL: QTS x $ EA =}	\multicolumn{2}{l	}{PARTS}
		\multicolumn{2}{r	}{TOTAL PARTS}	\multicolumn{3}{l	}{GREASE: LBS x $ /LB =}	\multicolumn{2}{l	}{OIL AND GREASE}
\multicolumn{4}{c	}{TECHNICIAN'S RECOMMENDATIONS}	\multicolumn{3}{l	}{TOTAL OIL & GREASE}	\multicolumn{2}{l	}{ACCESSORIES}
				\multicolumn{3}{l	}{ACCESSORIES}	\multicolumn{2}{l	}{OUTSIDE REPAIRS}
							\multicolumn{2}{l	}{MISCELLANEOUS}
				\multicolumn{3}{l	}{TOTAL ACCESSORIES}	\multicolumn{2}{l	}{TAX}
				\multicolumn{3}{l	}{AUTHORIZED BY}	\multicolumn{2}{l	}{TOTAL}

FIGURE 16.2 Step 2: Add the job number and time for each service.

List all parts that are used in this service. Look up their part numbers and prices, and list them on the repair order. Include the 4 quarts of oil under the "OIL & GREASE" section. See Figure 16.3.

\multicolumn{4}{c	}{Buske Tire and Service}	\multicolumn{5}{c	}{Vehicle Repair Order}
\multicolumn{4}{c	}{1983 3rd Avenue}	\multicolumn{5}{l	}{Name: Katie Daniels}
\multicolumn{4}{c	}{Blue River, Minnesota}	\multicolumn{5}{l	}{Address: 907 23rd Ave.}
				\multicolumn{5}{l	}{City, State: Palmdale, MN}
QTY	PART NUMBER	NAME OF PART	PRICE	\multicolumn{5}{c	}{CUSTOMER AND VEHICLE INFORMATION}
1	21018826	Oil filter	4 99	DATE 8/5/04	ORDER NBR	\multicolumn{3}{l	}{PHONE (555) 803-2218}
1	24401180	Serpentine belt	29 99	\multicolumn{2}{l	}{YEAR-MAKE-MODEL 2000 Saturn LS}	\multicolumn{3}{l	}{SERIAL NBR:}
2	90502291	Valve cover gskt	36 00			\multicolumn{3}{l	}{ENGINE NBR:}
1	50571378	Lwr radiator hose	15 00	\multicolumn{2}{l	}{LICENSE NO.}	ODOMETER: 47,315	\multicolumn{2}{l	}{WRITTEN BY JCR}
1	90571380	Upper radiator hose	15 00					
1	09120455	Thermostat	22 49	JOB NBR	SERVICE PERFORMED		TIME	AMOUNT
1	90523858	Shifter control cable	75 00	1-10	Oil change and lube		0.9	
				1-3	Replace serpentine belt		1.2	
				10-11	Replace both valve cover gaskets		3.5	
				8-1	Replace both radiator hoses		1.1	
				*	Replace thermostat		0.5	
				14-1	Replace shifter control cable		2.7	
				\multicolumn{3}{c	}{OIL & GREASE}	\multicolumn{2}{l	}{LABOR ONLY}
				\multicolumn{2}{l	}{OIL: 4 QTS x $ 2.79 EA =}	11.16	\multicolumn{2}{l	}{PARTS}
		\multicolumn{2}{r	}{TOTAL PARTS}	\multicolumn{3}{l	}{GREASE: LBS x $ /LB =}	\multicolumn{2}{l	}{OIL AND GREASE}
\multicolumn{4}{c	}{TECHNICIAN'S RECOMMENDATIONS}	\multicolumn{2}{l	}{TOTAL OIL & GREASE}	11.16	\multicolumn{2}{l	}{ACCESSORIES}
				\multicolumn{3}{l	}{ACCESSORIES}	\multicolumn{2}{l	}{OUTSIDE REPAIRS}
							\multicolumn{2}{l	}{MISCELLANEOUS}
				\multicolumn{3}{l	}{TOTAL ACCESSORIES}	\multicolumn{2}{l	}{TAX}
				\multicolumn{3}{l	}{AUTHORIZED BY}	\multicolumn{2}{l	}{TOTAL}

FIGURE 16.3 Step 3: List all parts by quantity, number, name, and price.

Calculate the labor charge for each portion of the job, using the shop rate of $65.00 per hour. Also, find the total labor charges and the total parts charges, and list them in the subtotal section of the invoice. See Figure 16.4.

Buske Tire and Service 1983 3rd Avenue Blue River, Minnesota				Vehicle Repair Order Name: Katie Daniels Address: 907 23rd Ave. City, State: Palmdale, MN				
QTY	PART NUMBER	NAME OF PART	PRICE	CUSTOMER AND VEHICLE INFORMATION				
1	21018826	Oil filter	4 | 99	DATE 8/5/04 ORDER NBR			PHONE (555) 803-2218	
1	24401180	Serpentine belt	29 | 99	YEAR-MAKE-MODEL 2000 Saturn LS		SERIAL NBR:		
2	90502291	Valve cover gskt	36 | 00			ENGINE NBR:		
1	50571378	Lwr radiator hose	15 | 00	LICENSE NO.		ODOMETER: 47,315	WRITTEN BY JCR	
1	90571380	Upper radiator hose	15 | 00					
1	09120455	Thermostat	22 | 49	JOB NBR	SERVICE PERFORMED		TIME	AMOUNT
1	90523858	Shifter control cable	75 | 00	1-10	Oil change and lube		0.9	58 | 50
				1-3	Replace serpentine belt		1.2	78 | 00
				10-11	Replace both valve cover gaskets		3.5	227 | 50
				8-1	Replace both radiator hoses		1.1	71 | 50
				*	Replace thermostat		0.5	32 | 50
				14-1	Replace shifter control cable		2.7	175 | 50
				OIL & GREASE			LABOR ONLY	643 | 50
				OIL: 4 QTS x $2.79 EA =		11.16	PARTS	198 | 47
	TOTAL PARTS		198 | 47	GREASE: LBS x $ /LB =			OIL AND GREASE	11 | 16
	TECHNICIAN'S RECOMMENDATIONS			TOTAL OIL & GREASE		11.16	ACCESSORIES	
				ACCESSORIES			OUTSIDE REPAIRS	
							MISCELLANEOUS	
				TOTAL ACCESSORIES			TAX	
				AUTHORIZED BY			TOTAL	

FIGURE 16.4 Step 4: For each service calculate the labor amount by multiplying the time by the labor rate. At this stage, add up the total parts and total labor, and calculate the amount for oil and grease.

Buske Tire and Service 1983 3rd Avenue Blue River, Minnesota				Vehicle Repair Order Name: Katie Daniels Address: 907 23rd Ave. City, State: Palmdale, MN				
QTY	PART NUMBER	NAME OF PART	PRICE	CUSTOMER AND VEHICLE INFORMATION				
1	21018826	Oil filter	4 | 99	DATE 8/5/04 ORDER NBR			PHONE (555) 803-2218	
1	24401180	Serpentine belt	29 | 99	YEAR-MAKE-MODEL 2000 Saturn LS		SERIAL NBR:		
2	90502291	Valve cover gskt	36 | 00			ENGINE NBR:		
1	50571378	Lwr radiatior hose	15 | 00	LICENSE NO.		ODOMETER: 47,315	WRITTEN BY JCR	
1	90571380	Upper radiator hose	15 | 00					
1	09120455	Thermostat	22 | 49	JOB NBR	SERVICE PERFORMED		TIME	AMOUNT
1	90523858	Shifter control cable	75 | 00	1-10	Oil change and lube		0.9	58 | 50
				1-3	Replace serpentine belt		1.2	78 | 00
				10-11	Replace both valve cover gaske		3.5	227 | 50
				8-1	Replace both radiator hoses		1.1	71 | 50
				*	Replace thermostat		0.5	32 | 50
				14-1	Replace shifter control cable		2.7	175 | 50
				OIL & GREASE			LABOR ONLY	643 | 50
				OIL: 4 QTS x $2.79 EA =		11.16	PARTS	198 | 47
	TOTAL PARTS		198 | 47	GREASE: LBS x $ /LB =			OIL AND GREASE	11 | 16
	TECHNICIAN'S RECOMMENDATIONS			TOTAL OIL & GREASE		11.16	ACCESSORIES	
				ACCESSORIES			OUTSIDE REPAIRS	
							MISCELLANEOUS	
				TOTAL ACCESSORIES			TAX	12 | 58
				AUTHORIZED BY			TOTAL	865 | 71

FIGURE 16.5 Step 5: Finally, calculate the tax and add the final total for all parts and labor for the order.

Determine the correct amount of sales tax. See Figure 16.5. The parts and oil are taxable, but labor is not. Then the sales tax is found by finding the total taxable amount

$$\$198.47 + \$11.16 = \$209.63$$

REPAIR ORDERS

and multiplying by 6%

$$\$209.63 \times 0.06 = \$12.58$$

Finally, add to find the total charges.

CHAPTER 16 *Practice Problems*

Filling Out a Repair Order

Convert each of the decimal times below to hours and minutes.

- 1. 0.6 hours to replace a belt

- 2. 0.2 hours to grease the chassis

- 3. 1.3 hours to replace a starter

- 4. 12.4 hours to replace an engine

- 5. 8.8 hours to replace a cylinder head

- 6. 2.1 hours to install shock absorbers

- 7. 5.9 hours to repair a transmission

- 8. 6.0 hours to remove a broken head bolt

Round each of these times to the nearest tenth of an hour.

- 9. 30 minutes to change oil

- 10. 48 minutes to replace a tire

- 11. 16 minutes to replace an air filter

- 12. 1 hour 18 minutes to replace spark plugs

- 13. 5 hours 52 minutes to replace a transaxle

- 14. 2 hours 15 minutes to install a hitch

CHAPTER 16

Look up the price of each part below in the Appendix. Assume the vehicle is a 2000 Saturn LS, with the 3.0 L engine and automatic transaxle.

- ●15. Oxygen sensor (on exhaust manifold)

- ●16. Fuel rail

- ●17. Left engine mount

- ●18. Rear brake drum

- ●19. Front left brake caliper

- ●20. Power steering pressure hose

Calculate the labor for each job below by looking up the suggested time allowance in the Appendix. Assume the vehicle is a 2000 Saturn LS with automatic transmission, 3.0 L engine, front wheel drive, and ABS. The shop rate is $58.00 per hour.

- ●●21. Remove and replace (R&R) both front coil springs

- ●●22. Remove and replace the brake system master cylinder

- ●●23. Overhaul the transaxle

- ●●24. Remove and replace lower engine/transaxle mount

- ●●25. Remove and replace muffler

- ●●26. Remove and replace starter

- ●●27. Kevin Vang brings his 2000 Saturn LW with 96,312 miles in for the following repairs:

    Replace the air filter
    Replace the muffler
    Rotate tires (four wheels only)

    The vehicle has the 3.0 L engine with manual transmission and no ABS. Fill out a complete repair order and find the total amount due from the customer for this work assuming the shop charges $50.00 per hour, and the sales tax rate (on parts only) is 5.5%. Use one of the blank repair orders at the end of this chapter.

REPAIR ORDERS

••28. Tiffany Benson brings in her Saturn LS in for a drivability problem. The vehicle has the 3.0 L engine, automatic transaxle, and 73,090 miles. Fill out a repair order for the work shown:

Replace rear exhaust manifold gasket
Replace one fuel injector

Use a blank repair order at the end of this chapter. Assume a shop rate of $48.00 per hour and 6% tax on everything except labor.

•••29. Mr. Matt Hanson has a 2000 Saturn LW in need of major repairs. His vehicle has the 3.0 L engine with manual transmission, no ABS, and 103,681 miles. Fill out a complete repair order assuming the following work is done.

Oil, lube, and filter service (4 quarts of oil, $2.95 each)
Replace fuel filter
Replace front exhaust manifold (Replace the gasket also)
Replace clutch disc and flywheel
Replace power steering pump
Replace passenger-side window regulator for manual window (not power window)

The shop rate is $68.00 per hour, and sales tax of 5.5% is charged on everything except labor.

•••30. Mia Valenta has a 2000 Saturn LS with an automatic transaxle, 3.0 L engine, ABS, and 80,640 miles. Her vehicle needs the following repairs:

Replace complete left headlamp assembly, and align the headlights
Replace throttle body
Replace water pump
Replace rear main bearing seal

The shop rate is $64.50 per hour, and the sales tax rate is 6.5% on parts only.

Vehicle Repair Order

Buske Tire and Service
1983 3rd Avenue
Blue River, Minnesota

CUSTOMER AND VEHICLE INFORMATION

Name:
Address:
City, State:

ORDER NBR
PHONE ( )
DATE / /
YEAR-MAKE-MODEL
SERIAL NBR:
ENGINE NBR:
LICENSE NO.
ODOMETER:
WRITTEN BY

SERVICE PERFORMED

JOB NBR		TIME	AMOUNT

OIL & GREASE

OIL: QTS x $ EA =
GREASE: LBS x $ /LB =
TOTAL OIL & GREASE

ACCESSORIES

TOTAL ACCESSORIES

AUTHORIZED BY

QTY	PART NUMBER	NAME OF PART	PRICE

TOTAL PARTS

TECHNICIAN'S RECOMMENDATIONS

LABOR ONLY
PARTS
OIL AND GREASE
ACCESSORIES
OUTSIDE REPAIRS
MISCELLANEOUS
TAX
TOTAL

REPAIR ORDERS

Buske Tire and Service
1983 3rd Avenue
Blue River, Minnesota

Vehicle Repair Order

CUSTOMER AND VEHICLE INFORMATION

Name:
Address:
City, State:

DATE / /	ORDER NBR	PHONE ( )
YEAR-MAKE-MODEL		SERIAL NBR:
		ENGINE NBR:
LICENSE NO.		ODOMETER:

WRITTEN BY

JOB NBR	SERVICE PERFORMED	TIME	AMOUNT

QTY	PART NUMBER	NAME OF PART	PRICE
		TOTAL PARTS	

OIL & GREASE

OIL:	QTS x $	EA =
GREASE:	LBS x $	/LB =
TOTAL OIL & GREASE		

ACCESSORIES

TOTAL ACCESSORIES		

TECHNICIAN'S RECOMMENDATIONS

AUTHORIZED BY

LABOR ONLY	
PARTS	
OIL AND GREASE	
ACCESSORIES	
OUTSIDE REPAIRS	
MISCELLANEOUS	
TAX	
TOTAL	

Buske Tire and Service

1983 3rd Avenue
Blue River, Minnesota

Vehicle Repair Order

CUSTOMER AND VEHICLE INFORMATION

Name:
Address:
City, State:

DATE / /	ORDER NBR	PHONE ( )
YEAR-MAKE-MODEL		SERIAL NBR:
		ENGINE NBR:
LICENSE NO.		ODOMETER:

QTY	PART NUMBER	NAME OF PART	PRICE

TOTAL PARTS

TECHNICIAN'S RECOMMENDATIONS

JOB NBR	SERVICE PERFORMED	TIME	AMOUNT	WRITTEN BY

OIL & GREASE

OIL:	QTS x $	EA =
GREASE:	LBS x $	/LB =

TOTAL OIL & GREASE

ACCESSORIES

TOTAL ACCESSORIES

AUTHORIZED BY

LABOR ONLY	
PARTS	
OIL AND GREASE	
ACCESSORIES	
OUTSIDE REPAIRS	
MISCELLANEOUS	
TAX	
TOTAL	

APPENDIX

53 GENERAL MOTORS 53
SATURN L-SERIES

SECTION INDEX

1 - MAINTENANCE & LUBRICATION 53-3	3.0L 53-12
2 - EMISSION SYSTEM 53-3	11 - CLUTCH 53-14
3 - ELECTRICAL 53-4	12 - MANUAL TRANSAXLE 53-15
4 - ALTERNATOR 53-7	14 - AUTOMATIC TRANSAXLE 53-15
5 - STARTER 53-7	18 - BRAKES 53-15
6 - FUEL SYSTEM 53-7	19 - FRONT SUSPENSION 53-17
7 - EXHAUST SYSTEM 53-8	20 - FRONT DRIVE AXLE 53-18
8 - COOLING SYSTEM 53-9	21 - STEERING 53-18
9 - HVAC 53-9	22 - STEERING COLUMN 53-19
10 - ENGINE	24 - REAR SUSPENSION 53-20
2.2L 53-11	25 - BODY HARDWARE 53-21

ALPHABETICAL INDEX

-A-

AIR FILTER	53-3
ALL HOSES	53-9
ALTERNATOR	53-7
ANTENNA BASE	53-4
ANTENNA CABLE	53-4
ANTENNA MAST	53-4
AXLE ASSY	53-18

-B-

BACKING PLATE	53-15
BACK-UP SWITCH	53-4
BAFFLE PLATE	
2.2L	53-11
3.0L	53-12
BALANCE SHAFT	53-11
BATTERY	53-4
BATTERY TERMINALS ...	53-4
BEARING SET	53-12
BEARINGS	53-11
BLOWER MOTOR	53-9
BLOWER MOTOR SWITCH	53-9
BOOSTER CHECK VALVE	53-15
BRAKE BOOSTER	53-15
BRAKE SWITCH	53-7
BRAKES	53-15
BUSHINGS	53-20

-C-

CABLE	53-15
CALIPER ASSY	53-15
CAMSHAFT	53-11
CAMSHAFT GEAR	
2.2L	53-11
3.0L	53-12
CAMSHAFT SEAL	53-12
CATALYTIC CONVERTER	53-8
CHAIN	53-11
CHAIN GUIDE	53-11
CHARGING CIRCUIT	53-7
CHASSIS	53-3
CIRCUIT	53-7
CLOCKSPRING	53-4
CLUSTER BULB	53-4
CLUTCH	53-14
CLUTCH PLATE & HUB	
ASSY	53-15
COIL SPRING	
FRONT SUSPENSION	53-17
REAR SUSPENSION	53-20
COMBINATION SWITCH	53-4
COMPLETE EXHAUST	
SYSTEM	53-8
COMPOSITE ASSY	53-4
COMPRESSION	53-4
COMPRESSOR	53-9
CONDENSER	53-9

-C- Con't

CONNECTING ROD	
2.2L	53-11
3.0L	53-12
CONTROL MODULE	53-15
CONVERTER HEAT SHIELD	53-8
COOLANT TEMP SENSOR	53-4
COOLING SYSTEM	53-3
COOLING SYSTEM	
PRESSURE	53-9
CRANKSHAFT	
2.2L	53-11
3.0L	53-12
CRANKSHAFT GEAR	
2.2L	53-11
3.0L	53-12
CRANKSHAFT POSITION	
SENSOR	53-4
CYLINDER BLOCK	53-11
CYLINDER HEAD	
2.2L	53-11
3.0L	53-12

-D-

DASH CONTROL UNIT ...	53-9
DIAGNOSTIC CIRCUIT ...	53-4
DIAGNOSTIC MODULE ..	53-4
DISCHARGE HOSE	53-9
DOOR ACTUATOR	53-9
DOOR JAMB SWITCH ...	53-4
DRIER	53-9
DRUM	53-15

-E-

ECM	53-4
EGR VALVE	53-3
ENGAGEMENT SWITCH	53-7
ENGINE	
2.2L	53-11
3.0L	53-12
EVAPORATOR CORE	53-9
EXHAUST CAMSHAFT ...	53-12
EXHAUST MANIFOLD ...	53-8
EXHAUST VALVE	
2.2L	53-11
3.0L	53-12
EXPANSION PLUG	53-9
EXPANSION VALVE	53-9
EXTERIOR BULBS	53-4

-F-

FAN BLADE	53-9
FAN MOTOR	53-9
FAN RELAY	53-9
FLEX HOSE	53-15
FLUID LEVEL SWITCH ..	53-15
FOG LAMP ASSY	53-4

-F- Con't

FOG LAMP BULB	53-4
FOG LAMP SWITCH	53-4
FRONT COVER	
2.2L	53-11
3.0L	53-12
FRONT COVER GASKET	
2.2L	53-11
3.0L	53-12
FRONT CRANK SEAL	
2.2L	53-11
3.0L	53-12
FRONT SPEED SENSOR	53-15
FRONT SPEED SENSOR	
HARNESS	53-15
FRONT WASHER PUMP...	53-4
FRONT WHEEL BEARING	53-17
FUEL FILTER	53-3
FUEL GAUGE SENDING	
UNIT	53-4
FUEL PUMP	53-7
FUEL PUMP RELAY	53-7
FUEL RAIL	53-7
FUEL TANK	53-7

-G-

GEAR	53-11

-H-

HAZARD FLASHER	53-4
HAZARD SWITCH	53-4
HEAD GASKET	
2.2L	53-11
3.0L	53-12
HEADLAMP	53-4
HEADLAMP BULB	53-4
HEADLAMP DIMMER	
SWITCH	53-4
HEADLAMP SWITCH	53-4
HEATER CORE	53-9
HEATER HOSE	53-9
HORN	53-4
HORN RELAY	53-4
HORN SWITCH	53-4
HUB	53-17
HUB & BEARING ASSY ...	53-20

-I-

IDLE AIR CONTROL VALVE	53-7
IDLER PULLEY	53-12
IGNITION COIL	53-4
IGNITION LOCK CYLINDER	53-4
IGNITION MODULE	53-4
IGNITION SWITCH	53-4
INFLATOR MODULE	53-4
INJECTOR	53-7
INNER TIE ROD	53-18

-I- Con't

INNER TIE ROD BOOT	53-18
INSTRUMENT CLUSTER....	53-4
INSTRUMENT LIGHT	
RHEOSTAT	53-4
INTAKE CAMSHAFT	53-12
INTAKE PIPE	53-7
INTAKE TEMP SENSOR	53-4
INTAKE VALVE	
2.2L	53-11
3.0L	53-12
INTERMEDIATE HEAT	
SHIELD	53-8
INTERMEDIATE PIPE	53-8
INTERMEDIATE SHAFT	53-19

-K-

KNOCK SENSOR	53-4
KNUCKLE	53-17

-L-

LATCH	53-21
LIQUID LINE	53-9
LOCK ACTUATOR	53-4
LOCK CYLINDER	53-21
LOCK MODULE	53-19
LOCK RELAY	53-4
LOCK SWITCH	53-4
LONG BLOCK	53-11
LOWER CONTROL ARM	
FRONT SUSPENSION	53-17
REAR SUSPENSION	53-20
LOWER COVER	53-19
LOWER HOSE	53-9
LUBE & FILTER	53-3

-M-

MAINTENANCE SCHEDULE	53-3
MANIFOLD GASKET	
FUEL SYSTEM	53-7
EXHAUST SYSTEM	53-8
MAP SENSOR	53-4
MASTER CYLINDER	
CLUTCH	53-14
BRAKES	53-15
MIRROR SWITCH	53-4
MODULATOR	53-15
MODULE	53-7
MOTOR	53-4
MOUNT	
2.2L	53-11
3.0L	53-12
MUFFLER	53-8
MUFFLER HEAT SHIELD	53-8

-N-

NEGATIVE CABLE	53-4

53 GENERAL MOTORS
SATURN L-SERIES

-N- Con't
NEUTRAL SAFETY SWITCH	53-4

-O-
OIL FILTER	53-3
OIL PAN	
2.2L	53-11
3.0L	53-12
OIL PICK-UP	53-12
OIL PRESSURE SENDING UNIT	53-4
OIL PUMP	53-12
O-RING	53-9
OUTER TIE ROD	53-18
OVERHAUL GASKET SET	
2.2L	53-11
3.0L	53-12
OXYGEN SENSOR	53-4

-P-
PARK BRAKE WARNING SWITCH	53-4
PARKING BRAKE CONTROL	53-15
PISTON RINGS	
2.2L	53-11
3.0L	53-12
PISTONS	
2.2L	53-11
3.0L	53-12
PLENUM	53-7
POSITIVE CABLE	53-4
PRESSURE REGULATOR	53-7
PRESSURE RELIEF VALVE	53-9
PRESSURE SENSOR	53-3
P/S PRESSURE HOSE	53-18
P/S PUMP	53-18
P/S RETURN HOSE	53-18
PULLEY	
2.2L	53-11
3.0L	53-12
PURGE SOLENOID	53-3

-R-
RADIATOR	53-9
RADIO	53-4

-R- Con't
REAR MAIN SEAL	
2.2L	53-11
3.0L	53-12
REAR SENSOR HARNESS	53-15
REAR SPEED SENSOR	53-15
REAR TRAILING ARM	53-20
REAR WASHER PUMP	53-4
RELAY	
BRAKES	53-15
ELECTRICAL	53-4
HVAC	53-9
RELEASE CABLE	53-21
RESERVOIR	53-15
RESISTOR	53-9
ROCKER ARMS	53-11
ROTOR	53-15

-S-
SCHRADER VALVE CORE	53-9
SEAL KIT	53-18
SERPENTINE BELT	53-3
SERPENTINE IDLER PULLEY	53-3
SERPENTINE TENSIONER	53-3
SHAFT COVER	53-19
SHIFT CONTROL CABLE	53-15
SHORT BLOCK	
2.2L	53-11
3.0L	53-12
SLAVE CYLINDER	53-14
SPARK PLUGS	53-4
SPEAKER	53-4
SPROCKET	53-11
STABILIZER BAR	53-20
STARTER	53-7
STARTER DRAW	53-7
STEERING COLUMN	53-19
STEERING GEAR	53-18
STEERING WHEEL	53-19
STOPLAMP SWITCH	53-4
STRIKER	53-21
STRUT	
FRONT SUSPENSION	53-17
REAR SUSPENSION	53-20

-S- Con't
SUCTION HOSE	53-9
SUSPENSION	
FRONT SUSPENSION	53-17
REAR SUSPENSION	53-20
SUSPENSION CROSSMEMBER	53-20
SWITCH	53-4
SYSTEM	
BRAKES	53-15
EMISSION SYSTEM	53-3
ELECTRICAL	53-4
FUEL SYSTEM	53-7
HVAC	53-9

-T-
TENSIONER	53-11
THERMOSTAT	53-9
THERMOSTAT GASKET	53-9
THROTTLE BODY	53-7
THROTTLE CABLE	53-7
THROTTLE POSITION SENSOR	53-4
TIMING BELT	53-12
TIMING CHAIN	
2.2L	53-11
TIMING TENSIONER	53-12
TOE-IN	53-17
TRANS FLUID	53-3
TRANSAXLE	
MANUAL TRANSAXLE	53-15
AUTOMATIC TRANSAXLE	53-15
TURN SIGNAL FLASHER	53-4
TURN SIGNAL SWITCH	53-4

-U-
UPPER CONTROL ARM	53-20
UPPER COVER	53-19
UPPER HOSE	53-9

-V-
VALVE COVER	
2.2L	53-11
3.0L	53-12

-V- Con't
VALVE COVER GASKET	
2.2L	53-11
3.0L	53-12
VALVE GRIND GASKET KIT	
2.2L	53-11
3.0L	53-12
VALVE GUIDE	53-12
VALVE KEEPER	
2.2L	53-11
3.0L	53-12
VALVE LIFTERS	
2.2L	53-11
3.0L	53-12
VALVE SEALS	
2.2L	53-11
3.0L	53-12
VALVE SEATS	53-12
VALVE SPRING RETAINERS	
2.2L	53-11
3.0L	53-12
VALVE SPRING SEATS	53-12
VALVE SPRINGS	
2.2L	53-11
3.0L	53-12
VAPOR CANISTER	53-3
VENT CONTROL SOLENOID	53-3

-W-
WATER PUMP	53-9
WHEEL ALIGNMENT	53-20
WHEEL CYLINDER	53-15
WHEELS	
FRONT SUSPENSION	53-17
MAINTENANCE & LUBRICATION	53-3
WINDOW REGULATOR	
BODY HARDWARE	53-21
ELECTRICAL	53-4
WINDOW RELAY	53-4
WINDOW SWITCH	53-4
WIPER ARM	53-4
WIPER BLADE	53-4
WIPER MOTOR	53-4
WIPER SWITCH	53-4
WIPER TRANSMISSION	53-4

GENERAL MOTORS 53
SATURN L-SERIES

LABOR 1 — MAINTENANCE & LUBRICATION

OPERATION INDEX

Operation	Ref
AIR FILTER, R&R	6
CHASSIS, LUBRICATE	1
COOLING SYSTEM, SERVICE	9
FUEL FILTER, R&R	7
LUBE & FILTER, SERVICE	10
MAINTENANCE SCHEDULE, SERVICE	11
OIL FILTER, R&R	8
SERPENTINE BELT, R&R	3
SERPENTINE IDLER PULLEY, R&R	4
SERPENTINE TENSIONER, R&R	5
TRANS FLUID, CHANGE	12
WHEELS, ROTATE	2

CHASSIS

CHASSIS & WHEELS

	Factory Time	Motor Time
1-CHASSIS, LUBRICATE		
Saturn	00 C	0.6

NOTES FOR: CHASSIS, LUBRICATE
Includes: Lubricate All Fittings, Inspect & Correct All Fluid Levels And Tire Pressure.

	Factory Time	Motor Time
2-WHEELS, ROTATE		
Saturn		
4 Wheels	00 C	0.4
5 Wheels	00 C	0.5

ENGINE SERVICE

BELTS & PULLEYS

	Factory Time	Motor Time
3-SERPENTINE BELT, R&R		
Saturn		
2.2L	00 C (0.3)	0.4
3.0L	00 C (0.9)	1.2
4-SERPENTINE IDLER PULLEY, R&R		
Saturn		
2.2L	00 C (0.5)	0.6
3.0L	00 C (0.9)	1.3
5-SERPENTINE TENSIONER, R&R		
Saturn		
2.2L	00 C (0.5)	0.6
3.0L	00 C (0.9)	1.3

FILTERS

	Factory Time	Motor Time
6-AIR FILTER, R&R		
Saturn	00 C (0.2)	0.2
7-FUEL FILTER, R&R		
Saturn	00 C (0.5)	0.6
8-OIL FILTER, R&R		
Saturn	00 C (0.2)	0.3

NOTES FOR: OIL FILTER, R&R
Filter Only.
To Change Oil, Add 00 0.1

PERIODICAL MAINTENANCE

PERIODICAL MAINTENANCE

	Factory Time	Motor Time
9-COOLING SYSTEM, SERVICE		
Saturn	00 C (1.2)	1.6

NOTES FOR: COOLING SYSTEM, SERVICE
Includes: Pressure Test System For Leaks, Check Thermostat & Heater Operation, Inspect All Hoses & Belts, Drain & Flush System And Add Coolant.

	Factory Time	Motor Time
10-LUBE & FILTER, SERVICE		
All Models	00 C	0.9

NOTES FOR: LUBE & FILTER, SERVICE
Includes: Inspect And Correct All Fluid Levels And Tire Pressure.
To Install Grease Fittings, Add 00 0.1

	Factory Time	Motor Time
11-MAINTENANCE SCHEDULE, SERVICE		
6,000 MILES		
Saturn	00 C (0.9)	1.2
12,000 MILES		
Saturn	00 C (0.4)	0.5
18,000 MILES		
Saturn	00 C (0.9)	1.2
24,000 MILES		
Saturn	00 C (0.4)	0.5
30,000 MILES		
Saturn	00 C (1.0)	1.2
36,000 MILES		
Saturn	00 C (0.4)	0.5
42,000 MILES		
Saturn	00 C (0.9)	1.2
48,000 MILES		
Saturn	00 C (0.4)	0.5
54,000 MILES		
Saturn	00 C (0.9)	1.2
60,000 MILES		
Saturn	00 C (1.0)	1.2
66,000 MILES		
Saturn	00 C (0.9)	1.2
72,000 MILES		
Saturn	00 C (0.4)	0.5
78,000 MILES		
Saturn	00 C (0.9)	1.2
84,000 MILES		
Saturn	00 C (0.4)	0.5
90,000 MILES		
Saturn	00 C (1.0)	1.2
96,000 MILES		
Saturn	00 C (0.4)	0.5
100,000 MILES		
Saturn		
2.2L	00 C (2.0)	2.3
3.0L	00 C (6.3)	7.6

NOTES FOR: MAINTENANCE SCHEDULE, SERVICE
For Procedure Information On Service Maintenance Schedules Refer To Appropriate MOTOR Auto Repair/Tune-Up Manuals.

TRANSAXLE

AUTOMATIC TRANSAXLE

	Factory Time	Motor Time
12-TRANS FLUID, CHANGE		
Saturn	00 C (0.5)	0.6

NOTES FOR: TRANS FLUID, CHANGE
To Change Filter, Add 00 0.1

PARTS 1 — MAINTENANCE & LUBRICATION

PARTS INDEX

Part	Ref
AIR FILTER	3
FUEL FILTER	4
OIL FILTER	5
PAN GASKET	6
SERPENTINE BELT	1
SERPENTINE TENSIONER	2
TRANS FILTER KIT	7

ENGINE SERVICE

BELTS & PULLEYS

	Part No.	Price
1-SERPENTINE BELT		
2.2L	90537461	9.99
3.0L	24401180	29.99
2-SERPENTINE TENSIONER		
2.2L	90537585	

FILTERS

	Part No.	Price
3-AIR FILTER		
Saturn	90571362	18.25
4-FUEL FILTER		
Saturn	25319686	17.99
5-OIL FILTER		
Saturn		
2.2L	90537438	10.99
3.0L	21018826	4.99

TRANSAXLE

AUTOMATIC TRANSAXLE

	Part No.	Price
6-PAN GASKET		
Saturn	24203590	
7-TRANS FILTER KIT		
Saturn	24203770	10.49

LABOR 2 — EMISSION SYSTEM

OPERATION INDEX

Operation	Ref
EGR VALVE, R&R	2
PRESSURE SENSOR, R&R	3
PURGE SOLENOID, R&R	4
SYSTEM, DIAGNOSIS	1
VAPOR CANISTER, R&R	5
VENT CONTROL SOLENOID, R&R	6

EMISSION SYSTEM

EMISSION SYSTEM

	Factory Time	Motor Time
1-SYSTEM, DIAGNOSIS		
Saturn	00 B	0.8

NOTES FOR: SYSTEM, DIAGNOSIS
To Re-Test System, Add 00 0.2

EGR SYSTEM

	Factory Time	Motor Time
2-EGR VALVE, R&R		
Saturn	00 B (0.3)	0.4

VAPOR CANISTER

	Factory Time	Motor Time
3-PRESSURE SENSOR, R&R		
Saturn	00 B (1.6)	2.1
4-PURGE SOLENOID, R&R		
Saturn	00 B (0.2)	0.3
5-VAPOR CANISTER, R&R		
Saturn	00 B (1.6)	2.1
6-VENT CONTROL SOLENOID, R&R		
Saturn	00 B (0.6)	0.7

PARTS 2 — EMISSION SYSTEM

PARTS INDEX

Part	Ref
EGR VALVE	2
PURGE CONTROL VALVE	5
SEPARATOR CHECK VALVE	1
TUBE	3
VALVE GASKET	4
VAPOR CANISTER	6

EMISSION SYSTEM

EMISSION SYSTEM

	Part No.	Price
1-SEPARATOR CHECK VALVE		
Saturn		
3.0L	90502186	65.00

EGR SYSTEM

	Part No.	Price
2-EGR VALVE		
Saturn		
3.0L	09120453	150.00
3-TUBE		
Saturn		
3.0L	90571548	35.00
4-VALVE GASKET		
Saturn		
3.0L	90467547	3.25

VAPOR CANISTER

	Part No.	Price
5-PURGE CONTROL VALVE		
Saturn		
2.2L	09158137	29.95
3.0L	90572416	39.95
6-VAPOR CANISTER		
Saturn	21019497	79.95

53 GENERAL MOTORS
SATURN L-SERIES

LABOR — 3 ELECTRICAL 3 — LABOR

OPERATION INDEX

ANTENNA BASE, R&R	68
ANTENNA CABLE, R&R	69
ANTENNA MAST, R&R	70
BACK-UP SWITCH, R&R	38
BATTERY TERMINALS, CLEAN	21
BATTERY, CHARGE/TEST	19
BATTERY, R&R	20
CLOCKSPRING, R&R	64
CLUSTER BULB, R&R	24
COMBINATION SWITCH, R&R	39
COMPOSITE ASSY, R&R	32
COMPRESSION, TEST	1
COOLANT TEMP SENSOR, R&R	10
CRANKSHAFT POSITION SENSOR, R&R	11
DIAGNOSTIC CIRCUIT, INSPECT	12
DIAGNOSTIC MODULE, R&R	65
DOOR JAMB SWITCH, R&R	40
ECM, R&R	13
EXTERIOR BULBS, R&R	25
FOG LAMP ASSY, ALIGN	34
FOG LAMP ASSY, R&R	35
FOG LAMP BULB, R&R	26
FOG LAMP SWITCH, R&R	41
FRONT WASHER PUMP, R&R	52
FUEL GAUGE SENDING UNIT, R&R	36
HAZARD FLASHER, R&R	30
HAZARD SWITCH, R&R	42
HEADLAMP BULB, R&R	27
HEADLAMP DIMMER SWITCH, R&R	43
HEADLAMP SWITCH, R&R	44
HEADLAMP, ALIGN	33
HORN RELAY, R&R	29
HORN SWITCH, R&R	45
HORN, R&R	28
IGNITION COIL, R&R	2
IGNITION LOCK CYLINDER, R&R	8
IGNITION MODULE, R&R	3
IGNITION SWITCH, R&R	9
INFLATOR MODULE, R&R	66
INSTRUMENT CLUSTER, R&R	63
INSTRUMENT LIGHT RHEOSTAT, R&R	46
INTAKE TEMP SENSOR, R&R	14
KNOCK SENSOR, R&R	15
LOCK ACTUATOR, R&R	74
LOCK ACTUATOR, R&R	80
LOCK RELAY, R&R	75
LOCK SWITCH, R&R	76
MAP SENSOR, R&R	16
MIRROR SWITCH, R&R	84
MOTOR, R&R	82
NEGATIVE CABLE, R&R	22
NEUTRAL SAFETY SWITCH, R&R	47
OIL PRESSURE SENDING UNIT, R&R	37
OXYGEN SENSOR, R&R	17
PARK BRAKE WARNING SWITCH, R&R	48
POSITIVE CABLE, R&R	23
RADIO, R&I	71
RADIO, R&R	72
REAR WASHER PUMP, R&R	58
RELAY, R&R	85
SPARK PLUGS, R&I	4
SPARK PLUGS, R&R	5
SPEAKER, R&R	73
STOPLAMP SWITCH, R&R	49
SWITCH, R&R	83
SWITCH, R&R	86
SYSTEM, DIAGNOSIS	67
SYSTEM, DIAGNOSIS	6
SYSTEM, TUNE-UP	7
THROTTLE POSITION SENSOR, R&R	18
TURN SIGNAL FLASHER, R&R	31
TURN SIGNAL SWITCH, R&R	50
WINDOW REGULATOR, R&R	77
WINDOW REGULATOR, R&R	81
WINDOW RELAY, R&R	78
WINDOW SWITCH, R&R	79
WIPER ARM, R&R	53
WIPER ARM, R&R	59
WIPER BLADE, R&R	54
WIPER BLADE, R&R	60
WIPER MOTOR, R&I	55
WIPER MOTOR, R&I	61
WIPER MOTOR, R&R	56
WIPER MOTOR, R&R	62
WIPER SWITCH, R&R	51
WIPER TRANSMISSION, R&R	57

IGNITION SYSTEM

IGNITION SYSTEM

		Factory Time	Motor Time
1-COMPRESSION, TEST			
Saturn			
2.2L	00 B		0.8
3.0	00 B		3.0
NOTES FOR: COMPRESSION, TEST			
Includes: R&I Spark Plugs.			
2-IGNITION COIL, R&R			
Saturn			
2.2L			
One	00 B	(0.2)	0.3
Each Additional	00 B	(0.2)	0.3
3.0L			
Front Bank	00 B	(0.7)	0.9
Rear Bank	00 B	(1.4)	1.8
Both Banks	00	(1.9)	2.7
3-IGNITION MODULE, R&R			
Saturn	00 B	(0.2)	0.3
4-SPARK PLUGS, R&I			
Saturn			
2.2L			
Each	00 C	(0.3)	0.4
All	00 C	(0.4)	0.5
3.0L			
Front Bank	00 B	(0.8)	1.1
Rear Bank	00 B	(1.4)	1.9
Both Banks	00	(2.0)	2.8
5-SPARK PLUGS, R&R			
Saturn			
2.2L			
Each	00 C	(0.3)	0.4
All	00 C	(0.4)	0.5
3.0L			
Front Bank	00 B	(0.8)	1.1
Rear Bank	00 B	(1.4)	1.9
Both Banks	00	(2.0)	2.8
6-SYSTEM, DIAGNOSIS			
Saturn	00 B		0.6
7-SYSTEM, TUNE-UP			
Saturn			
See Service Maintenance Schedule	00		

IGNITION LOCK

8-IGNITION LOCK CYLINDER, R&R			
Saturn	00 B	(0.4)	0.5
NOTES FOR: IGNITION LOCK CYLINDER, R&R			
To Code Cylinder, Add	00	(0.2)	0.2
9-IGNITION SWITCH, R&R			
Saturn	00 B	(1.2)	1.8

POWERTRAIN CONTROL

POWERTRAIN CONTROL

10-COOLANT TEMP SENSOR, R&R			
Saturn			
2.2L	00 B	(0.4)	0.5
3.0L	00 B	(0.2)	0.3
11-CRANKSHAFT POSITION SENSOR, R&R			
Saturn			
2.2L	00 B	(0.7)	0.9
3.0L	00 B	(0.3)	0.4
12-DIAGNOSTIC CIRCUIT, INSPECT			
Saturn	00 B		1.3
NOTES FOR: DIAGNOSTIC CIRCUIT, INSPECT			
Includes: Time To Hook-Up And Disconnect Test Equipment And Perform Test. Does Not Include: Pinpoint Test Or Re-Test Upon Completion Of Repair.			
13-ECM, R&R			
Saturn			
2.2L	00 B	(0.6)	0.7
3.0L	00 B	(0.2)	0.3
14-INTAKE TEMP SENSOR, R&R			
Saturn	00 B	(0.2)	0.3
15-KNOCK SENSOR, R&R			
Saturn			
2.2L	00 B	(0.2)	0.3
3.0L			
Front Bank	00 B	(0.5)	0.6
Rear Bank	00 B	(1.2)	1.8
Both Banks	00	(1.5)	2.1
16-MAP SENSOR, R&R			
Saturn	00 B	(0.2)	0.3
17-OXYGEN SENSOR, R&R			
Saturn			
2.2L			
Manifold	00 B	(0.4)	0.5
Pipe	00 B	(0.5)	0.6
3.0L			
Manifold			
Each	00 B	(0.4)	0.5
Pipe			
Front	00 B	(0.5)	0.6
Rear	00 B	(0.6)	0.7
18-THROTTLE POSITION SENSOR, R&R			
Saturn	00 B	(0.3)	0.4

CHASSIS ELECTRICAL

BATTERY

19-BATTERY, CHARGE/TEST			
Saturn	00 B		0.4
NOTES FOR: BATTERY, CHARGE/TEST			
Includes: Test Battery And Charge If Necessary. Does Not Include: Battery, Renew.			
20-BATTERY, R&R			
Saturn	00 C		0.3
NOTES FOR: BATTERY, R&R			
Does Not Include: Any Test, Check Or Charge.			
21-BATTERY TERMINALS, CLEAN			
All Models	00 C		0.3
22-NEGATIVE CABLE, R&R			
Saturn	00 B	(0.4)	0.5
23-POSITIVE CABLE, R&R			
Saturn	00 B	(0.5)	0.6

BULBS

24-CLUSTER BULB, R&R			
Saturn			
One Or All	00 B	(0.5)	0.6
25-EXTERIOR BULBS, R&R			
Saturn			
One	00 C	(0.2)	0.3
26-FOG LAMP BULB, R&R			
Saturn			
One Side	00 B	(0.2)	0.3
27-HEADLAMP BULB, R&R			
Saturn			
One Side	00 C	(0.2)	0.3

HORN

28-HORN, R&R			
Saturn	00 C	(0.3)	0.4
29-HORN RELAY, R&R			
Saturn	00 C	(0.2)	0.3

FLASHERS

30-HAZARD FLASHER, R&R			
Saturn			
One	00 C	(0.2)	0.3
31-TURN SIGNAL FLASHER, R&R			
Saturn			
One	00 C	(0.2)	0.3

HEADLAMPS

32-COMPOSITE ASSY, R&R			
Saturn			
One Side	00 C	(0.4)	0.5
33-HEADLAMP, ALIGN			
Saturn	00 B	(0.4)	0.5

FOG LAMPS

34-FOG LAMP ASSY, ALIGN			
Saturn	00 B	(0.3)	0.4
35-FOG LAMP ASSY, R&R			
Saturn			
One Side	00 B	(0.4)	0.5

SENDERS

36-FUEL GAUGE SENDING UNIT, R&R			
Saturn	00 B	(1.6)	2.0
37-OIL PRESSURE SENDING UNIT, R&R			
Saturn	00 C	(0.4)	0.5

SWITCHES

38-BACK-UP SWITCH, R&R			
MANUAL TRANS			
Saturn	00 B	(0.2)	0.3
39-COMBINATION SWITCH, R&R			
Saturn	00 B	(0.2)	1.3

(CONTINUED)

GENERAL MOTORS 53
SATURN L-SERIES

ELECTRICAL - Time Cont'd

	Factory Time	Motor Time
40-DOOR JAMB SWITCH, R&R		
Saturn One	00 C (0.2)	0.3
41-FOG LAMP SWITCH, R&R		
Saturn	00 B (0.2)	0.3
42-HAZARD SWITCH, R&R		
Saturn	00 B (0.2)	0.3
43-HEADLAMP DIMMER SWITCH, R&R		
Saturn	00 B	1.3
44-HEADLAMP SWITCH, R&R		
Saturn	00 B	1.3
45-HORN SWITCH, R&R		
Saturn	00 C (0.9)	1.3
46-INSTRUMENT LIGHT RHEOSTAT, R&R		
Saturn	00 B (0.3)	0.4
47-NEUTRAL SAFETY SWITCH, R&R		
Saturn	00 B (0.4)	0.5
48-PARK BRAKE WARNING SWITCH, R&R		
Saturn	00 B (0.8)	1.0
49-STOPLAMP SWITCH, R&R		
Saturn	00 B (0.4)	0.5
50-TURN SIGNAL SWITCH, R&R		
Saturn	00 B	1.3
51-WIPER SWITCH, R&R		
Saturn	00 B	1.3

FRONT WIPERS

	Factory Time	Motor Time
52-FRONT WASHER PUMP, R&R		
Saturn	00 C (0.5)	0.6
53-WIPER ARM, R&R		
Saturn Each	00 C	0.2
54-WIPER BLADE, R&R		
Saturn Each	00 C	0.1
55-WIPER MOTOR, R&I		
Saturn	00 B (0.7)	0.9
56-WIPER MOTOR, R&R		
Saturn	00 B (0.7)	0.9
57-WIPER TRANSMISSION, R&R		
Saturn	00 B (0.5)	0.6
NOTES FOR: WIPER TRANSMISSION, R&R		
To R&R Wiper Motor If Nec., Add	00 (0.2)	0.3

REAR WIPERS

	Factory Time	Motor Time
58-REAR WASHER PUMP, R&R		
Saturn	00 C (0.5)	0.6
59-WIPER ARM, R&R		
Saturn Each	00 C (0.2)	0.2
60-WIPER BLADE, R&R		
Saturn Each	00 C	0.1
61-WIPER MOTOR, R&I		
Saturn	00 B (0.5)	0.6
62-WIPER MOTOR, R&R		
Saturn	00 B (0.5)	0.6

INSTRUMENTS & GAUGES

INSTRUMENTS & GAUGES

	Factory Time	Motor Time
63-INSTRUMENT CLUSTER, R&R		
Saturn	00 B (1.2)	1.6

RESTRAINT SYSTEMS

SUPPLEMENTAL RESTRAINT SYSTEM

	Factory Time	Motor Time
64-CLOCKSPRING, R&R		
Saturn	00 B (0.9)	1.3

NOTES FOR: CLOCKSPRING, R&R
WARNING: Before Repairing Any Air Restraint System, The Battery Cables And Any Back-up Power Supplies To The System Must Be Disconnected In Order To Prevent Accidental Deployment. Includes: Disable/Enable SIR System.

	Factory Time	Motor Time
65-DIAGNOSTIC MODULE, R&R		
Saturn	00 B (0.8)	1.0

NOTES FOR: DIAGNOSTIC MODULE, R&R
WARNING: Before Repairing Any Air Restraint System, The Battery Cables And Any Back-up Power Supplies To The System Must Be Disconnected In Order To Prevent Accidental Deployment.

	Factory Time	Motor Time
66-INFLATOR MODULE, R&R		
Saturn		
Driver Side	00 B (0.5)	0.6
Passenger Side	00 B (0.5)	0.6

NOTES FOR: INFLATOR MODULE, R&R
WARNING: Before Repairing Any Air Restraint System, The Battery Cables And Any Back-up Power Supplies To The System Must Be Disconnected In Order To Prevent Accidental Deployment. Includes: Disable/Enable SIR System.

	Factory Time	Motor Time
67-SYSTEM, DIAGNOSIS		
Saturn	00 B	0.5

NOTES FOR: SYSTEM, DIAGNOSIS
WARNING: Before Repairing Any Air Restraint System, The Battery Cables And Any Back-up Power Supplies To The System Must Be Disconnected In Order To Prevent Accidental Deployment.

BODY ELECTRICAL

ANTENNA & RADIO

	Factory Time	Motor Time
68-ANTENNA BASE, R&R		
Saturn	00 B (0.4)	0.5
69-ANTENNA CABLE, R&R		
Saturn		
Sedan	00 B (0.8)	1.0
Wagon	00 B (1.1)	1.2
70-ANTENNA MAST, R&R		
Saturn	00 B (0.2)	0.2
71-RADIO, R&I		
Saturn	00 B (0.3)	0.4
72-RADIO, R&R		
Saturn	00 B (0.3)	0.4
73-SPEAKER, R&R		
Saturn		
Front		
Woofer Each	00 B (0.5)	0.6
Tweeter Each	00 B (0.2)	0.2
Rear		
Woofer Each	00 B (0.4)	0.6
Tweeter Each	00 B (0.2)	0.2
Subwoofer		
Sedan	00 B (0.5)	0.6
Wagon	00 B (0.3)	0.4

FRONT DOOR

	Factory Time	Motor Time
74-LOCK ACTUATOR, R&R		
Saturn One	00 B (0.8)	1.0
75-LOCK RELAY, R&R		
Saturn	00 B (0.2)	0.3
76-LOCK SWITCH, R&R		
Saturn One	00 B (0.2)	0.3
77-WINDOW REGULATOR, R&R		
Saturn One Side	00 B (1.1)	1.3
78-WINDOW RELAY, R&R		
Saturn	00 B (0.2)	0.3
79-WINDOW SWITCH, R&R		
Saturn One	00 B (0.2)	0.3

REAR DOOR

	Factory Time	Motor Time
80-LOCK ACTUATOR, R&R		
Saturn One Side	00 B (0.8)	1.0
81-WINDOW REGULATOR, R&R		
Saturn One Side	00 B (0.8)	1.0

SUNROOF

	Factory Time	Motor Time
82-MOTOR, R&R		
Saturn	00 B (1.3)	1.6
83-SWITCH, R&R		
Saturn	00 B (0.2)	0.2

MIRRORS

	Factory Time	Motor Time
84-MIRROR SWITCH, R&R		
Saturn	00 B (0.2)	0.2

WINDOW DEFROSTER

	Factory Time	Motor Time
85-RELAY, R&R		
Saturn	00 B (0.2)	0.2
86-SWITCH, R&R		
Saturn	00 B (0.2)	0.2

PARTS — 3 ELECTRICAL 3 — PARTS

PARTS INDEX

Part	No.
ANTENNA BASE	45
ANTENNA MAST	46
BACK-UP SWITCH	26
CLOCKSPRING	41
COMPOSITE ASSY	22
CONTROL MODULE	47
COOLANT SENSOR	5
CRANKSHAFT POSITION SENSOR	6
DIAGNOSTIC MODULE	42
DOOR JAMB SWITCH	27
DRIVER INFLATOR MODULE	43
ECM	7
ELEMENT	56
FOG LAMP BULB	17
FOG LAMP SWITCH	28
FRONT MOTOR	36
FUEL GAUGE SENDING UNIT	23
HAZARD SWITCH	29
HEADLAMP BULB	18
HEADLAMP SWITCH	30
HORN	20
IGNITION COIL	1
IGNITION LOCK CYLINDER	3
IGNITION MODULE	2
IGNITION SWITCH	4
INSTRUMENT CLUSTER	19
INSTRUMENT CLUSTER	40
INSTRUMENT LIGHT RHEOSTAT	31
INTAKE TEMP SENSOR	8
KNOCK SENSOR	9
LOCK ACTUATOR	65
LOCK SWITCH	48
MAP SENSOR	10
MASS AIR FLOW SENSOR	11
MIRROR MOTOR	58
MIRROR SWITCH	59
MODULE	62
MOTOR	54
MOTOR	60
NEGATIVE CABLE	15
NEUTRAL SAFETY SWITCH	32
OIL PRESSURE SENDING UNIT	24
OXYGEN SENSOR	12
PARK BRAKE WARNING SWITCH	33
PASSENGER INFLATOR MODULE	44
POSITIVE CABLE	16
REAR MOTOR	37
STOPLAMP SWITCH	34
SWITCH	55
SWITCH	57
SWITCH	61
SWITCH	64
TEMP SENDING UNIT	25
THROTTLE POSITION SENSOR	13
TRANSMITTER	63
TURN SIGNAL FLASHER	21
VEHICLE SPEED SENSOR	14
WASHER PUMP	38
WINDOW MOTOR	50
WINDOW MOTOR	52
WINDOW REGULATOR	51
WINDOW REGULATOR	53
WINDOW SWITCH	49
WIPER SWITCH	35
WIPER TRANSMISSION	39

IGNITION SYSTEM

IGNITION SYSTEM

	Part No.	Price
1-IGNITION COIL		
Saturn		
3.0L		
Front	00 09118114	95.00
Rear	00 09118115	95.00
2-IGNITION MODULE		
Saturn		
2.2L	00 01104070	175.00

NOTES FOR: IGNITION MODULE
Includes Ignition Coil.

(CONTINUED)

53 GENERAL MOTORS
SATURN L-SERIES

ELECTRICAL - Parts Cont'd

IGNITION LOCK

3-IGNITION LOCK CYLINDER
		Part No.	Price
Saturn	00	21019052	34.95

4-IGNITION SWITCH
		Part No.	Price
Saturn	00	26062239	

POWERTRAIN CONTROL

POWERTRAIN CONTROL

5-COOLANT SENSOR
Saturn			
2.2L	00	NOT LISTED	
3.0L	00	90541937	14.99

6-CRANKSHAFT POSITION SENSOR
Saturn			
2.2L	00	24576398	15.00
3.0L	00	09134756	60.00

7-ECM
Saturn			
2.2L	00	09377750	725.00
3.0L	00	09153242	375.00

NOTES FOR: ECM
Order By Core ID# On Original Module, Parts & Prices For Estimating Purposes Only.

8-INTAKE TEMP SENSOR
Saturn			
2.2L	00	12160244	9.99

9-KNOCK SENSOR
Saturn			
2.2L	00	24577244	
3.0L	00	90541521	115.00

10-MAP SENSOR
Saturn			
2.2L	00	16212460	39.99
3.0L	00	90541409	74.99

11-MASS AIR FLOW SENSOR
Saturn			
3.0L	00	90528435	75.00

12-OXYGEN SENSOR
Saturn			
2.2L			
Front	00	24577552	49.99
Rear	00	25315218	65.00
3.0L			
At Exhaust Pipe			
Front	00	24403860	60.00
Rear	00	09182000	60.00
At Exhaust Manifold	00	09129871	195.00

13-THROTTLE POSITION SENSOR
Saturn			
2.2L	00	25319901	20.00

14-VEHICLE SPEED SENSOR
Saturn			
Manual Trans	00	NOT LISTED	
Auto Trans	00	24207507	

CHASSIS ELECTRICAL

BATTERY

15-NEGATIVE CABLE
Saturn			
2.2L			
Manual Trans	00	15357862	20.00
Auto Trans	00	15357861	20.00
3.0L	00	15357859	20.00

16-POSITIVE CABLE
Saturn			
2.2L	00	13537860	
3.0L	00	15357858	55.00

BULBS

17-FOG LAMP BULB
Saturn	00	09442876	9.99

18-HEADLAMP BULB
Saturn			
Low Beam	00	9441733	8.27
High Beam	00	9441732	8.27

19-INSTRUMENT CLUSTER
Saturn			
White	00	21019273	3.00
Yellow	00	21019274	1.48

HORN

20-HORN
Saturn	00	09229753	27.50

FLASHERS

21-TURN SIGNAL FLASHER
Saturn	00	NOT LISTED	

		Part No.	Price
Saturn			
Right	00	90584065	
Left	00	90584064	

HEADLAMPS

22-COMPOSITE ASSY
Saturn			
Right	00	90583595	150.00
Left	00	90583594	150.00

SENDERS

23-FUEL GAUGE SENDING UNIT
Saturn	00	90584032	105.00

24-OIL PRESSURE SENDING UNIT
Saturn			
2.2L	00	NO LISTED	
3.0L	00	90507539	6.24

25-TEMP SENDING UNIT
Saturn			
2.2L	00	NOT LISTED	
3.0L	00	90541937	14.99

SWITCHES

26-BACK-UP SWITCH
Saturn			
Manual Trans	00	21019137	15.00
Auto Trans			
See Neutral Safety Switch			

27-DOOR JAMB SWITCH
Saturn	00	90504151	8.00

28-FOG LAMP SWITCH
Saturn			
w/o Heated Seats	00	09135228	26.00
w/Heated Seats	00	09135224	26.00

29-HAZARD SWITCH
Saturn	00	90584072	

30-HEADLAMP SWITCH
Saturn	00	90584096	

31-INSTRUMENT LIGHT RHEOSTAT
Saturn			
w/o Remote Release	00	09178538	25.00
w/Remote Release	00	09104891	24.99

NOTES FOR: INSTRUMENT LIGHT RHEOSTAT
Order By Color.

32-NEUTRAL SAFETY SWITCH
Saturn			
Manual Trans	00	14094368	5.30
Auto Trans	00	12450161	

33-PARK BRAKE WARNING SWITCH
Saturn	00	90584275	0.25

34-STOPLAMP SWITCH
Saturn	00	90504499	4.00

35-WIPER SWITCH
Saturn			
w/o Rear Wiper	00	90584097	
w/Rear Wiper	00	90585753	

WIPERS

36-FRONT MOTOR
Saturn	00	21019036	

37-REAR MOTOR
Saturn	00	90584925	125.00

38-WASHER PUMP
Saturn			
Sedan	00	90586631	28.00
Wagon	00	90586632	28.00

39-WIPER TRANSMISSION
Saturn	00	21019035	90.00

INSTRUMENTS & GAUGES

INSTRUMENTS & GAUGES

40-INSTRUMENT CLUSTER
Saturn	00	21024241	

RESTRAINT SYSTEMS

SUPPLEMENTAL RESTRAINT SYSTEM

41-CLOCKSPRING
Saturn	00	90585754	65.00

42-DIAGNOSTIC MODULE
Saturn	00	21018733	

43-DRIVER INFLATOR MODULE
Saturn	00	24404317	

NOTES FOR: DRIVER INFLATOR MODULE
Order By Color.

		Part No.	Price

44-PASSENGER INFLATOR MODULE
Saturn	00	90584514	

NOTES FOR: PASSENGER INFLATOR MODULE
Order By Color.

BODY ELECTRICAL

ANTENNA & RADIO

45-ANTENNA BASE
Saturn			
Wagon	00	90585253	23.75

46-ANTENNA MAST
Saturn			
Sedan	00	10280480	
Wagon	00	24409086	19.00

ANTI-THEFT COMPONENTS

47-CONTROL MODULE
Saturn			
2.2L			
w/o Keyless Entry	00	90584372	
w/Keyless Entry	00	24412527	189.99
3.0L	00	24401694	189.99

DOOR

48-LOCK SWITCH
Saturn	00	90363758	9.00

NOTES FOR: LOCK SWITCH
Order By Color.

49-WINDOW SWITCH
Saturn			
Right	00	90363749	20.00
Left	00	90363746	26.00

NOTES FOR: WINDOW SWITCH
Order By Color.

FRONT DOOR

50-WINDOW MOTOR
Saturn			
Part Of Window Regulator	00		

51-WINDOW REGULATOR
Saturn			
Right	00	24405464	
Left	00	24405465	

REAR DOOR

52-WINDOW MOTOR
Saturn			
Part Of Window Regulator	00		

53-WINDOW REGULATOR
Saturn			
Right	00	24404447	124.95
Left	00	24404447	124.95

SUNROOF

54-MOTOR
Saturn	00	90503754	130.00

55-SWITCH
Saturn	00	09135541	14.99

NOTES FOR: SWITCH
Order By Color.

HEATED SEATS

56-ELEMENT
Saturn			
Upper	00	90585222	95.00
Lower	00	90585223	

57-SWITCH
Saturn			
w/o Fog Lamps	00	09135227	26.00
w/Fog Lamps	00	09135224	26.00

MIRRORS

58-MIRROR MOTOR
Saturn			
Right	00	21019066	25.00
Left	00	21019067	25.00

59-MIRROR SWITCH
Saturn	00	90363755	9.00

POWER SEATS

60-MOTOR
Saturn			
Adjuster Motor	00	09135285	125.00
Recliner Motor	00	90585054	85.00

61-SWITCH
Saturn	00	90595235	64.95

NOTES FOR: SWITCH
Order By Color.

(CONTINUED)

GENERAL MOTORS 53
SATURN L-SERIES

ELECTRICAL - Parts Cont'd

KEYLESS ENTRY COMPONENTS

62-MODULE
Saturn
		Part No.	Price
2.2L	00	24412527	189.99
3.0L	00	24401694	189.99

63-TRANSMITTER
		Part No.	Price
Saturn	00	24401698	31.69

WINDOW DEFROSTER

64-SWITCH
Saturn
Part Of Heater/AC Control.

FUEL DOOR

65-LOCK ACTUATOR
		Part No.	Price
Saturn	00	09180534	24.99

LABOR — 4 ALTERNATOR 4 — LABOR

OPERATION INDEX
ALTERNATOR, R&I	1
ALTERNATOR, R&R	2
CHARGING CIRCUIT, INSPECT	3

ALTERNATOR
ALTERNATOR

1-ALTERNATOR, R&I
Saturn
		Factory Time	Motor Time
2.2L	00 B	(0.8)	0.9
3.0L	00 B	(1.1)	1.2

NOTES FOR: ALTERNATOR, R&I
To Replace Fusible Link,
| Add | 00 | (0.1) | 0.2 |

2-ALTERNATOR, R&R
Saturn
		Factory Time	Motor Time
2.2L	00 B	(0.8)	0.9
3.0L	00 B	(1.1)	1.2

NOTES FOR: ALTERNATOR, R&R
To Replace Fusible Link,
| Add | 00 | (0.1) | 0.2 |

3-CHARGING CIRCUIT, INSPECT
| Saturn | 00 B | | 0.5 |

NOTES FOR: CHARGING CIRCUIT, INSPECT
Includes: Check Alternator Output, Regulator & Belt Tension.

PARTS — 4 ALTERNATOR 4 — PARTS

PARTS INDEX
ALTERNATOR	1

ALTERNATOR
ALTERNATOR

1-ALTERNATOR
Saturn
		Part No.	Price
2.2L	00	90585955	190.00
3.0L	00	09227893	190.00

LABOR — 5 STARTER 5 — LABOR

OPERATION INDEX
CIRCUIT, INSPECT	1
STARTER DRAW, TEST	5
STARTER, R&I	2
STARTER, R&I	4
STARTER, R&R	3

STARTER
STARTER

1-CIRCUIT, INSPECT
		Factory Time	Motor Time
Saturn	00 B		0.5

2-STARTER, R&I
Saturn
		Factory Time	Motor Time
2.2L	00 B	(0.6)	0.7
3.0L	00 B	(1.2)	1.5

3-STARTER, R&R
Saturn
		Factory Time	Motor Time
2.2L	00 B	(0.6)	0.7
3.0L	00 B	(1.2)	1.5

4-STARTER, R&I
Saturn
		Factory Time	Motor Time
2.2L	00 B	(0.6)	0.7
3.0L	00 B	(1.2)	1.5

5-STARTER DRAW, TEST
| All Models | 00 B | | 0.3 |

PARTS — 5 STARTER 5 — PARTS

PARTS INDEX
STARTER	1

STARTER
STARTER

1-STARTER
Saturn
		Part No.	Price
2.2L	00	09000835	
3.0L	00	09136093	195.00

LABOR — 6 FUEL SYSTEM 6 — LABOR

OPERATION INDEX
BRAKE SWITCH, R&R	14
ENGAGEMENT SWITCH, R&R	15
FUEL PUMP RELAY, R&R	2
FUEL PUMP, R&R	1
FUEL RAIL, R&R	6
FUEL TANK, R&R	3
IDLE AIR CONTROL VALVE, R&R	7
INJECTOR, R&R	8
INTAKE PIPE, R&I	11
MANIFOLD GASKET, R&R	12
MODULE, R&R	16
PLENUM, R&I	13
PRESSURE REGULATOR, R&R	9
SYSTEM, SERVICE	4
THROTTLE BODY, R&R	10
THROTTLE CABLE, R&R	5

FUEL SUPPLY
FUEL SUPPLY

1-FUEL PUMP, R&R
		Factory Time	Motor Time
Saturn	00 B	(1.6)	2.2

NOTES FOR: FUEL PUMP, R&R
Does Not Include: Drain & Refill Fuel Tank.

2-FUEL PUMP RELAY, R&R
		Factory Time	Motor Time
Saturn	00 B	(0.2)	0.3

3-FUEL TANK, R&R
		Factory Time	Motor Time
Saturn	00 B	(1.6)	2.2

NOTES FOR: FUEL TANK, R&R
Does Not Include: Drain & Refill Fuel Tank.

4-SYSTEM, SERVICE
		Factory Time	Motor Time
Saturn	00 B		3.0

NOTES FOR: SYSTEM, SERVICE
Includes: R&I Fuel Tank, Blow Out Lines & Renew Fuel Filter.

5-THROTTLE CABLE, R&R
		Factory Time	Motor Time
Saturn	00 B	(0.6)	0.8

FUEL INDUCTION
FUEL INJECTION

6-FUEL RAIL, R&R
Saturn
		Factory Time	Motor Time
2.2L	00 B	(0.7)	0.9
3.0L	00 B	(1.0)	1.4

7-IDLE AIR CONTROL VALVE, R&R
		Factory Time	Motor Time
Saturn	00 B	(0.6)	0.8

8-INJECTOR, R&R
Saturn
2.2L
		Factory Time	Motor Time
One	00 B	(0.7)	0.9
All	00 B	(0.8)	1.1

3.0L
		Factory Time	Motor Time
One	00 B	(1.1)	1.4
All	00 B	(1.2)	1.8

9-PRESSURE REGULATOR, R&R
		Factory Time	Motor Time
Saturn	00 B	(0.4)	0.5

THROTTLE BODY

10-THROTTLE BODY, R&R
Saturn
		Factory Time	Motor Time
2.2L	00 B	(0.4)	0.5
3.0L	00 B	(0.5)	0.6

INTAKE

11-INTAKE PIPE, R&I
Saturn
		Factory Time	Motor Time
Front	00 B	(0.3)	0.4

(CONTINUED)

53 GENERAL MOTORS
SATURN L-SERIES

FUEL SYSTEM - Time Cont'd

		Factory Time	Motor Time
Rear	00 B	(0.9)	1.4

NOTES FOR: INTAKE PIPE, R&I
To Renew Intake Pipe, Add
| Rear | 00 | (0.1) | 0.1 |

12-MANIFOLD GASKET, R&R
Saturn
| 2.2L | 00 B | (0.8) | 1.1 |
| 3.0L | 00 B | (1.1) | 1.4 |

NOTES FOR: MANIFOLD GASKET, R&R
To Replace Intake Manifold, Add
| 2.2L | 00 | (0.1) | 0.1 |

		Factory Time	Motor Time
3.0L	00	(0.1)	0.1

13-PLENUM, R&I
Saturn
| 3.0L | 00 B | (0.8) | 1.1 |

NOTES FOR: PLENUM, R&I
To Renew Plenum, Add
| | 00 | (0.3) | 0.4 |

CRUISE CONTROL
CRUISE CONTROL

14-BRAKE SWITCH, R&R
| Saturn | 00 B | (0.4) | 0.5 |

		Factory Time	Motor Time
15-ENGAGEMENT SWITCH, R&R			
Saturn	00 B	(0.7)	0.8
16-MODULE, R&R			
Saturn	00 B	(0.5)	0.6

PARTS — 6 FUEL SYSTEM 6 — PARTS

PARTS INDEX
ACCELERATION SWITCH ASSY	1
BRAKE SWITCH	14
CLUTCH SWITCH	15
FUEL PUMP	2
FUEL RAIL	5
FUEL TANK	3
GASKET	10
IDLE AIR CONTROL VALVE	6
INJECTOR	7
INJECTOR O-RING KIT	8
INTAKE MANIFOLD	12
MANIFOLD GASKET	13
MODULE	16
ON/OFF SWITCH	17
PRESSURE REGULATOR	9
RELEASE SWITCH	18
SET/RESUME SWITCH	19
THROTTLE BODY	11
THROTTLE CABLE	4

FUEL SUPPLY
FUEL SUPPLY

		Part No.	Price
1-ACCELERATION SWITCH ASSY			
Saturn 3.0L	00	90571550	
2-FUEL PUMP			
Saturn	00	21018807	
3-FUEL TANK			
Saturn	00	90570788	250.00
4-THROTTLE CABLE			
Saturn 2.2L			
Manual Trans	00	90571551	12.00
Auto Trans	00	90571552	12.00

FUEL INDUCTION
FUEL INJECTION

		Part No.	Price
5-FUEL RAIL			
Saturn			
2.2L	00	12563434	125.00
3.0L	00	90570596	40.00
6-IDLE AIR CONTROL VALVE			
Saturn	00	25177797	36.34
7-INJECTOR			
Saturn			
2.2L	00	12563428	
3.0L	00	09120371	37.00
8-INJECTOR O-RING KIT			
Saturn			
2.2L	00	12563891	8.95
3.0L	00	09120374	5.00
9-PRESSURE REGULATOR			
Saturn			
2.2L	00	12563431	55.00
3.0L	00	09118850	37.00

THROTTLE BODY

		Part No.	Price
10-GASKET			
Saturn			
2.2L	00	24573948	
3.0L	00	90571137	1.00
11-THROTTLE BODY			
Saturn			
2.2L			
Manual Trans	00	17099120	150.00
Auto Trans	00	17099121	150.00
3.0L	00	90570604	475.00

INTAKE

		Part No.	Price
12-INTAKE MANIFOLD			
Saturn			
2.2L	00	90537742	
3.0L	00	90570595	310.00
13-MANIFOLD GASKET			
Saturn			
2.2L	00	90537260	
3.0L	00	09120290	7.00

CRUISE CONTROL
CRUISE CONTROL

		Part No.	Price
14-BRAKE SWITCH			
Saturn	00	90504499	4.00
15-CLUTCH SWITCH			
Saturn	00	14094368	5.30
16-MODULE			
Saturn	00	25315904	195.00

NOTES FOR: MODULE
Includes Servo.

		Part No.	Price
17-ON/OFF SWITCH			
Saturn	00	21060915	24.00

NOTES FOR: ON/OFF SWITCH
Order By Color.

18-RELEASE SWITCH			
Saturn			
2.2L			
Manual Trans	00	90492444	5.00
Auto Trans	00	90508804	5.00
3.0L	00	90508804	5.00
19-SET/RESUME SWITCH			
Saturn	00	12450190	25.00

NOTES FOR: SET/RESUME SWITCH
Order By Color.

LABOR — 7 EXHAUST SYSTEM 7 — LABOR

OPERATION INDEX
CATALYTIC CONVERTER, R&R	3
COMPLETE EXHAUST SYSTEM, R&R	4
CONVERTER HEAT SHIELD, R&R	5
EXHAUST MANIFOLD, R&R	1
INTERMEDIATE HEAT SHIELD, R&R	6
INTERMEDIATE PIPE, R&R	7
MANIFOLD GASKET, R&R	2
MUFFLER HEAT SHIELD, R&R	9
MUFFLER, R&R	8

EXHAUST MANIFOLD
EXHAUST MANIFOLD

		Factory Time	Motor Time
1-EXHAUST MANIFOLD, R&R			
Saturn			
2.2L	00 B	(1.1)	1.3
3.0L			
Front	00 B	(3.7)	4.6
Rear	00 B	(1.9)	2.4
2-MANIFOLD GASKET, R&R			
Saturn			
2.2L	00 B	(1.1)	1.3
3.0L			
Front	00 B	(3.7)	4.6
Rear	00 B	(1.9)	2.4

EXHAUST SYSTEM
EXHAUST SYSTEM

		Factory Time	Motor Time
3-CATALYTIC CONVERTER, R&R			
Saturn	00 C	(0.8)	1.0
4-COMPLETE EXHAUST SYSTEM, R&R			
Saturn	00 C	(0.9)	1.3

		Factory Time	Motor Time
5-CONVERTER HEAT SHIELD, R&R			
Saturn	00 C	(0.8)	1.0
6-INTERMEDIATE HEAT SHIELD, R&R			
Saturn	00 C	(0.8)	1.0
7-INTERMEDIATE PIPE, R&R			
Saturn	00 C	(0.7)	0.9
8-MUFFLER, R&R			
Saturn	00 C	(0.4)	0.5
9-MUFFLER HEAT SHIELD, R&R			
Saturn	00 C	(1.2)	1.6

PARTS — 7 EXHAUST SYSTEM 7 — PARTS

PARTS INDEX
CATALYTIC CONVERTER	3
CONVERTER HEAT SHIELD	4
EXHAUST MANIFOLD	1
FRONT PIPE	5
MANIFOLD GASKET	2
MUFFLER	6
REAR PIPE	7

EXHAUST MANIFOLD
EXHAUST MANIFOLD

		Part No.	Price
1-EXHAUST MANIFOLD			
Saturn			
2.2L			
Manual Trans	00	90537677	135.00
Auto Trans	00	90537679	135.00
3.0L			
Front	00	90570542	260.00
Rear	00	90570541	110.00
2-MANIFOLD GASKET			
Saturn			
2.2L	00	90537696	9.25
3.0L	00	90490379	9.25

EXHAUST SYSTEM
EXHAUST SYSTEM

		Part No.	Price
3-CATALYTIC CONVERTER			
Saturn			

Part Of Front Pipe.

4-CONVERTER HEAT SHIELD			
Saturn	00	21013143	

(CONTINUED)

GENERAL MOTORS 53
SATURN L-SERIES

PARTS

EXHAUST SYSTEM - Parts Cont'd

5-FRONT PIPE

	Part No.	Price
Saturn		
2.2L		
Manual Trans		
Federal	00 09228454	495.00
California	00 09228455	720.00
Auto Trans		
Federal	00 90570528	
California	00 09129202	
3.0L		
Front	00 09228006	920.00
Rear	00 90570529	395.00

6-MUFFLER

	Part No.	Price
Saturn		
2.2L	00 09128682	
3.0L	00 09128684	115.00

NOTES FOR: MUFFLER
Includes Tailpipe.

7-REAR PIPE

	Part No.	Price
Saturn		
2.2L	00 09128694	175.00
3.0L	00 09128695	175.00

NOTES FOR: REAR PIPE
Includes Resonator.

LABOR

8 COOLING SYSTEM 8

OPERATION INDEX

ALL HOSES, R&R	1
COOLING SYSTEM PRESSURE, TEST	6
EXPANSION PLUG, R&R	7
FAN BLADE, R&R	11
FAN MOTOR, R&R	12
FAN RELAY, R&R	13
LOWER HOSE, R&R	2
RADIATOR, R&I	3
RADIATOR, R&R	4
THERMOSTAT GASKET, R&R	9
THERMOSTAT, R&R	8
UPPER HOSE, R&R	5
WATER PUMP, R&R	10

COOLING SYSTEM
RADIATOR

		Factory Time	Motor Time
1-ALL HOSES, R&R			
Saturn			
2.2L	00 C	(0.7)	0.9
3.0L	00 C	(0.8)	1.1
2-LOWER HOSE, R&R			
Saturn			
2.2L	00 C	(0.4)	0.5
3.0L	00 C	(0.6)	0.8
3-RADIATOR, R&I			
Saturn			
2.2L	00 C	(1.5)	2.2
3.0L	00 C	(2.2)	3.0
NOTES FOR: RADIATOR, R&I			
w/AC, Add	00	(0.1)	0.1
w/Auto Trans, Add	00	(0.2)	0.2
4-RADIATOR, R&R			
Saturn			
2.2L	00 C	(1.5)	2.2
3.0L	00 C	(2.2)	3.0
NOTES FOR: RADIATOR, R&R			
w/AC, Add	00	(0.1)	0.1
w/Auto Trans, Add	00	(0.2)	0.2
5-UPPER HOSE, R&R			
Saturn			
2.2L	00 C	(0.4)	0.5
3.0L	00 C	(0.5)	0.6

COOLING SYSTEM

		Factory Time	Motor Time
6-COOLING SYSTEM PRESSURE, TEST			
Saturn	00 C		0.3
7-EXPANSION PLUG, R&R			
One	00 B		0.5
Each Additional	00 B		0.4
NOTES FOR: EXPANSION PLUG, R&R			
Does Not Include: Time To Gain Access.			

		Factory Time	Motor Time
8-THERMOSTAT, R&R			
Saturn	00 C	(0.6)	0.9
9-THERMOSTAT GASKET, R&R			
Saturn			
2.2L	00 C	(0.5)	0.6
3.0L	00 C	(2.7)	3.8

WATER PUMP

		Factory Time	Motor Time
10-WATER PUMP, R&R			
Saturn			
2.2L	00 C	(1.9)	2.6
3.0L	00 C	(2.2)	3.0

COOLING FAN

		Factory Time	Motor Time
11-FAN BLADE, R&R			
Saturn			
Pusher	00 C	(0.4)	0.5
Puller			
2.2L	00 C	(0.4)	0.5
3.0L	00 C	(1.1)	1.4
12-FAN MOTOR, R&R			
Saturn			
Pusher	00 C	(0.4)	0.5
Puller			
2.2L	00 C	(0.4)	0.5
3.0L	00 C	(1.1)	1.4
13-FAN RELAY, R&R			
Saturn	00 B	(0.2)	0.3

PARTS

8 COOLING SYSTEM 8

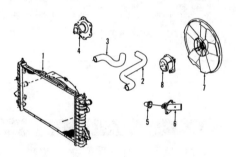

COOLING SYSTEM
RADIATOR

	Part No.	Price
1-RADIATOR		
Saturn		
2.2L		
Manual Trans	00	NOT LISTED
Auto Trans	00 09128553	244.95
3.0L	00 09128552	244.95
2-UPPER HOSE		
Saturn		
2.2L		
Manual Trans	00 90572512	15.00
Auto Trans	00 90571381	15.00
3.0L	00 90571380	15.00

	Part No.	Price
3-LOWER HOSE		
Saturn		
2.2L		
Manual Trans	00 09128713	25.00
Auto Trans	00 09128712	25.00
3.0L	00 90571378	15.00

COOLING SYSTEM

	Part No.	Price
5-THERMOSTAT		
Saturn		
2.2L	00 21018811	
3.0L	00 09120455	22.49
6-THERMOSTAT HOUSING		
Saturn		
2.2L	00 90537819	60.00
3.0L	00 90502201	

WATER PUMP

	Part No.	Price
4-WATER PUMP		
Saturn		
2.2L	00 90537805	84.99
3.0L	00 90543277	84.99

COOLING FAN

	Part No.	Price
7-FAN BLADE		
Saturn		
2.2L	00 09128558	140.00
3.0L	00 09128556	140.00
8-FAN MOTOR		
Saturn	00	NOT LISTED

LABOR

9 HVAC 9

OPERATION INDEX

BLOWER MOTOR SWITCH, R&R	1
BLOWER MOTOR, R&R	3
CLUTCH PLATE & HUB ASSY, R&R	20
COMPRESSOR, R&I	21
COMPRESSOR, R&R	22
CONDENSER, R&R	23
DASH CONTROL UNIT, R&R	2
DISCHARGE HOSE, R&R	9
DOOR ACTUATOR, R&R	6
DRIER, R&R	10
EVAPORATOR CORE, R&R	11
EXPANSION VALVE, R&R	12
HEATER CORE, R&R	7
HEATER HOSE, R&R	8
LIQUID LINE, R&R	13
O-RING, R&R	14
PRESSURE RELIEF VALVE, R&R	15
RELAY, R&R	16
RELAY, R&R	4
RESISTOR, R&R	5
SCHRADER VALVE CORE, R&R	17
SUCTION HOSE, R&R	18
SYSTEM, SERVICE	19

CONTROLS
CONTROLS

		Factory Time	Motor Time
1-BLOWER MOTOR SWITCH, R&R			
Saturn	00 B	(0.5)	0.6
2-DASH CONTROL UNIT, R&R			
Saturn	00 B	(0.5)	0.6

(CONTINUED)

53-9

53 GENERAL MOTORS
SATURN L-SERIES

HVAC - Time Cont'd

	Factory Time	Motor Time
BLOWER MOTOR		
3-BLOWER MOTOR, R&R		
Saturn	00 B (0.9)	1.4
4-RELAY, R&R		
Saturn	00 C (0.2)	0.3
5-RESISTOR, R&R		
Saturn	00 B (0.6)	0.7
MODE ACTUATORS		
6-DOOR ACTUATOR, R&R		
All Models		
Air Inlet (Recirculation)	00 B (0.5)	0.6
Mode Valve	00 B (0.5)	0.6
HEATER		
HEATER		
7-HEATER CORE, R&R		
Saturn	00 B (0.9)	1.4
8-HEATER HOSE, R&R		
Saturn		
Inlet		
2.2L	00 C (0.4)	0.5
3.0L	00 C (0.5)	0.6
Outlet		
2.2L	00 C (0.4)	0.5
3.0L	00 C (0.5)	0.6
AIR CONDITIONER		
AIR CONDITIONER		
9-DISCHARGE HOSE, R&R		
Saturn	00 B (1.2)	1.8
NOTES FOR: DISCHARGE HOSE, R&R		
Does Not Include: Refrigerant Recovery Or Evacuate & Recharge AC System.		
10-DRIER, R&R		
Saturn	00 B (1.4)	1.8
NOTES FOR: DRIER, R&R		
Does Not Include: Refrigerant Recovery Or Evacuate & Recharge AC System.		
11-EVAPORATOR CORE, R&R		
Saturn	00 B (3.0)	4.6
NOTES FOR: EVAPORATOR CORE, R&R		
Does Not Include: Refrigerant Recovery Or Evacuate &		

Recharge AC System. Includes: Disable/Enable SIR System On 1993-94 Models Only.

	Factory Time	Motor Time
12-EXPANSION VALVE, R&R		
Saturn	00 B (1.8)	2.7
NOTES FOR: EXPANSION VALVE, R&R		
Does Not Include: Refrigerant Recovery Or Evacuate & Recharge AC System.		
13-LIQUID LINE, R&R		
Saturn	00 B (1.8)	2.7
NOTES FOR: LIQUID LINE, R&R		
Does Not Include: Refrigerant Recovery Or Evacuate & Recharge AC System.		
14-O-RING, R&R		
COMPRESSOR		
Saturn		
Outlet	00 B (1.0)	1.4
Inlet	00 B (1.0)	1.4
CONDENSER		
Saturn		
Outlet Pipe	00 B (1.1)	1.4
Inlet Line	00 B (1.1)	1.4
RECEIVER-DRIER		
Saturn	00 B (1.1)	1.4
EXPANSION VALVE		
Saturn		
Outlet	00 B (1.1)	1.4
Inlet Line	00 B (1.2)	1.5
NOTES FOR: O-RING, R&R		
Does Not Include: Refrigerant Recovery Or Evacuate & Recharge AC System.		
15-PRESSURE RELIEF VALVE, R&R		
Saturn	00 B (1.0)	1.4
NOTES FOR: PRESSURE RELIEF VALVE, R&R		
Does Not Include: Refrigerant Recovery Or Evacuate & Recharge AC System.		
16-RELAY, R&R		
Saturn	00 C (0.2)	0.3
17-SCHRADER VALVE CORE, R&R		
Saturn	00 B (0.8)	1.1
NOTES FOR: SCHRADER VALVE CORE, R&R		
Does Not Include: Refrigerant Recovery Or Evacuate & Recharge AC System.		
18-SUCTION HOSE, R&R		
Saturn	00 B (1.1)	1.4
NOTES FOR: SUCTION HOSE, R&R		
Does Not Include: Refrigerant Recovery Or Evacuate & Recharge AC System.		
19-SYSTEM, SERVICE		
EVACUATE & RECHARGE		
Saturn	00 B	1.4

	Factory Time	Motor Time
REFRIGERANT, RECOVER (B)		
Saturn	00 B	0.4
PARTIAL REFRIGERANT CHARGE		
All Models	00 B	0.6
FLUSH (COMPLETE)		
All Models	00 B	0.3
NOTES FOR: SYSTEM, SERVICE		
Many Vehicles Are Now Using R134 Refrigerant In The AC System. Extra Care Must Be Observed When Servicing This Type Of System. R12 Refrigerant MUST NOT Be Used. See Manufacturers Service Manual For Specific Repair Procedures. With Any Operation Requiring A Refrigerant Line Disconnect, Add A/C Service, I.E. Evacuate, Recharge And Test For Leaks. Add For Refrigerant Cost. To Be Used In Conjunction With Component Replacement Which Could Contaminate System. Does Not Include: Evacuate & Recharge System.		
COMPRESSOR		
20-CLUTCH PLATE & HUB ASSY, R&R		
Saturn		
2.2L	00 B (1.7)	2.1
3.0L	00 B (2.6)	3.5
NOTES FOR: CLUTCH PLATE & HUB ASSY, R&R		
Does Not Include: Refrigerant Recovery Or Evacuate & Recharge AC System.		
21-COMPRESSOR, R&I		
Saturn		
2.2L	00 B (1.4)	1.8
3.0L	00 B (2.2)	3.0
NOTES FOR: COMPRESSOR, R&I		
Does Not Include: Refrigerant Recovery Or Evacuate & Recharge AC System.		
22-COMPRESSOR, R&R		
Saturn		
2.2L	00 B (1.4)	1.8
3.0L	00 B (2.2)	3.0
NOTES FOR: COMPRESSOR, R&R		
Does Not Include: Refrigerant Recovery Or Evacuate & Recharge AC System.		
CONDENSER		
23-CONDENSER, R&R		
Saturn	00 B (1.3)	1.8
NOTES FOR: CONDENSER, R&R		
Does Not Include: Refrigerant Recovery Or Evacuate & Recharge AC System.		

PARTS — 9 HVAC 9 — PARTS

PARTS INDEX

BLOWER MOTOR	3
BLOWER MOTOR SWITCH	1
COMPRESSOR	15
COMPRESSOR CLUTCH	16
CONDENSER	17
DASH CONTROL UNIT	2
DISCHARGE HOSE	6
DRIER	7
EVAPORATOR CORE	8
EXPANSION VALVE	9
HEATER CORE	5
LIQUID LINE	10
PRESSURE RELIEF VALVE	11
PRESSURE SENSOR	14
RESISTOR	4
SUCTION & DISCHARGE HOSE ASSY	12
SUCTION HOSE	13

CONTROLS

	Part No.	Price
1-BLOWER MOTOR SWITCH		
Saturn	90586835	18.00
2-DASH CONTROL UNIT		
Saturn	90583298	125.00

BLOWER MOTOR

	Part No.	Price
3-BLOWER MOTOR		
Saturn	90586833	
4-RESISTOR		
Saturn	90583344	35.00

HEATER

	Part No.	Price
5-HEATER CORE		
Saturn	90585608	125.00

AIR CONDITIONER

	Part No.	Price
6-DISCHARGE HOSE		
Saturn		
2.2L		
Manual Trans	90585632	44.00
Auto Trans	90585633	44.00
3.0L	90585627	44.00
7-DRIER		
Saturn	90585620	55.00
8-EVAPORATOR CORE		
Saturn	90585609	260.00
9-EXPANSION VALVE		
Saturn	90585610	69.95
10-LIQUID LINE		
Saturn	90583328	34.00
11-PRESSURE RELIEF VALVE		
Saturn	21031128	
12-SUCTION & DISCHARGE HOSE ASSY		
Saturn		
2.2L	NOT LISTED	
3.0L	90583325	44.00
13-SUCTION HOSE		
Saturn	21031289	

SWITCHES & SENSORS

	Part No.	Price
14-PRESSURE SENSOR		
Saturn	22601618	24.00

COMPRESSOR

	Part No.	Price
15-COMPRESSOR		
Saturn		
2.2L	21019269	375.00
3.0L	21019270	375.00
16-COMPRESSOR CLUTCH		
Saturn		
2.2L	21018760	129.99
3.0L	21018761	129.99

CONDENSER

	Part No.	Price
17-CONDENSER		
Saturn	90585619	

GENERAL MOTORS 53
SATURN L-SERIES

LABOR — 10 ENGINE – 2.2L 10 — LABOR

OPERATION INDEX

Operation	No.
CYLINDER BLOCK, R&R	1
ENGINE, OVERHAUL	4
ENGINE, R&I	2
ENGINE, R&R	3
FRONT COVER GASKET, R&R	11
FRONT COVER, R&R	10
LONG BLOCK, R&R	5
MOUNT, R&R	7
OIL PAN GASKET, R&R	14
OIL PAN, R&R	13
SHORT BLOCK, R&R	6
TRANS MOUNT, R&R	8
VALVE COVER, R&R	9
VIBRATION DAMPER, R&R	12

ENGINE

	Part No.	Factory Time	Motor Time
1-CYLINDER BLOCK, R&R			
Saturn			
Manual Trans	00 B		19.2
Auto Trans	00 B		19.2
NOTES FOR: CYLINDER BLOCK, R&R			
Includes: R&I Engine And Transfer All Necessary Parts.			
2-ENGINE, R&I			
Saturn			
Manual Trans	00 B	(6.5)	8.8
Auto Trans	00 B	(6.2)	8.8
NOTES FOR: ENGINE, R&I			
Does Not Include: Transfer Of Any Part Of Engine Or Replacement Of Optional Equipment.			
3-ENGINE, R&R			
Saturn			
Manual Trans	00 B		10.6
Auto Trans	00 B		10.6

NOTES FOR: ENGINE, R&R
Includes: Transfer All Fuel & Electrical Units. Does Not Include: Transfer Of Optional Equipment.

4-ENGINE, OVERHAUL
Saturn
Manual Trans .... 00 B 24.3
Auto Trans .... 00 B 24.3
NOTES FOR: ENGINE, OVERHAUL
Includes: Measure Cylinder Bores, Crankshaft & Pistons For Proper Size & Hone Cylinders. Renew Pistons, Rings, Pins, Main & Rod Bearings, Grind Valves & Tune-Up.

5-LONG BLOCK, R&R
Saturn
Manual Trans .... 00 B 13.6
Auto Trans .... 00 B 13.6
NOTES FOR: LONG BLOCK, R&R
Includes: R&I Engine And Transfer All Necessary Components Not Supplied With Long Block.

6-SHORT BLOCK, R&R
Saturn
Manual Trans .... 00 B 15.6
Auto Trans .... 00 B 15.6
NOTES FOR: SHORT BLOCK, R&R
Includes: R&I Engine And Transfer All Necessary Parts.

MOUNTS

7-MOUNT, R&R
Saturn .... 00 B (0.8) 1.1
8-TRANS MOUNT, R&R
Saturn
Manual
Front .... 00 B (0.9) 1.4
Rear .... 00 B (0.9) 1.4
Auto
Front .... 00 B (1.0) 1.4
Rear .... 00 B (1.1) 1.4
Right .... 00 B (1.1) 1.4
Lower .... 00 B (0.5) 0.6

CYLINDER HEAD & VALVES

9-VALVE COVER, R&R
Saturn .... 00 C (0.7) 0.9

CAMSHAFT & TIMING

10-FRONT COVER, R&R
Saturn .... 00 B (2.7) 4.0
11-FRONT COVER GASKET, R&R
Saturn .... 00 B (2.7) 4.0

CRANKSHAFT & BEARINGS

12-VIBRATION DAMPER, R&R
Saturn .... 00 B (1.1) 1.4

LUBRICATION

OIL PAN

13-OIL PAN, R&R
Saturn .... 00 C (1.5) 2.1
14-OIL PAN GASKET, R&R
Saturn .... 00 C (1.5) 2.1

PARTS — 10 ENGINE – 2.2L 10 — PARTS

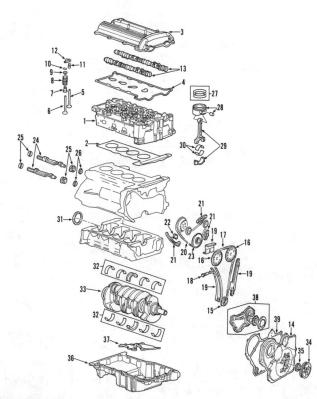

ENGINE

	Part No.	Price
CYLINDER BLOCK		
Saturn	00	NOT LISTED

NOTES FOR: CYLINDER BLOCK
Does Not Include Head, Pistons & Rods, Crankshaft, Bearings & Seal.

	Part No.	Price
ENGINE		
Saturn	00	NOT LISTED

NOTES FOR: ENGINE
Includes: Block, Cylinder Head, Rods & Pistons, Crankshaft, Bearings & Rear Seal, And Fuel System.

	Part No.	Price
LONG BLOCK		
Saturn	00	NOT LISTED

(CONTINUED)

53 GENERAL MOTORS
SATURN L-SERIES

ENGINE – 2.2L - Parts Cont'd

	Part No.	Price
NOTES FOR: LONG BLOCK		
Includes: Block, Cylinder Head, Rods & Pistons, Crankshaft, Bearings & Rear Seal.		
OVERHAUL GASKET SET		
Saturn	00	NOT LISTED
SHORT BLOCK		
Saturn	00	NOT LISTED
NOTES FOR: SHORT BLOCK		
Includes: Block, Pistons & Rods, Crankshaft, Bearings & Rear Seal.		

MOUNTS

MOUNTS

MOUNT
Saturn
Right		
Manual Trans	00	90575636	48.00
Auto Trans	00	90575638	48.00
Left		
Manual Trans	00	90576090	49.95
Auto Trans	00	90576091	49.95

CYLINDER HEAD & VALVES

CYLINDER HEAD & VALVES

VALVE GRIND GASKET KIT
Saturn .... 00 NOT LISTED
1-CYLINDER HEAD
Saturn .... 00 NOT LISTED
2-HEAD GASKET
Saturn .... 00 NOT LISTED
3-VALVE COVER
Saturn 00 90537249 135.00
4-VALVE COVER GASKET
Saturn 00 90537319 14.95
5-INTAKE VALVE
Saturn .... 00 NOT LISTED
6-EXHAUST VALVE
Saturn .... 00 NOT LISTED
7-VALVE SEALS
Saturn .... 00 NOT LISTED
8-VALVE SPRINGS
Saturn .... 00 NOT LISTED
9-VALVE SPRING RETAINERS
Saturn .... 00 NOT LISTED
10-VALVE KEEPER
Saturn .... 00 NOT LISTED
NOTES FOR: VALVE KEEPER
Price Is For Half Of One Key.

11-VALVE LIFTERS
Saturn .... 00 NOT LISTED
12-ROCKER ARMS
Saturn .... 00 NOT LISTED

CAMSHAFT & TIMING

CAMSHAFT & TIMING

13-CAMSHAFT
Saturn
Intake	00	NOT LISTED
Exhaust	00	NOT LISTED

14-FRONT COVER
Saturn 00 90537589 21.95
NOTES FOR: FRONT COVER
Includes Oil Pump.
15-CRANKSHAFT GEAR
Saturn .... 00 NOT LISTED
16-CAMSHAFT GEAR
Saturn .... 00 NOT LISTED
17-TIMING CHAIN
Saturn .... 00 NOT LISTED
18-TENSIONER
Saturn .... 00 NOT LISTED
19-CHAIN GUIDE
Saturn
Lower		
Tensioner Guide	00	NOT LISTED
Fixed Guide	00	NOT LISTED
Upper	00	NOT LISTED

39-FRONT COVER GASKET
Saturn 00 90537589 21.95

BALANCE SHAFTS

20-CHAIN
Saturn .... 00 NOT LISTED
21-CHAIN GUIDE
Saturn
Lower		
Tensioner Guide	00	NOT LISTED
Fixed Guide	00	NOT LISTED
Upper	00	NOT LISTED

22-TENSIONER
Saturn .... 00 NOT LISTED
23-GEAR
Saturn .... 00 NOT LISTED
24-BALANCE SHAFT
Saturn
Intake	00	NOT LISTED
Exhaust	00	NOT LISTED

25-BEARINGS
Saturn
Front	00	NOT LISTED
Rear	00	NOT LISTED

26-SPROCKET
Saturn .... 00 NOT LISTED

PISTONS, RINGS & BEARINGS

PISTONS, RINGS & BEARINGS

27-PISTON RINGS
Saturn .... 00 NOT LISTED
NOTES FOR: PISTON RINGS
Includes: One Standard Ring Set For One Piston.
28-PISTONS
Saturn .... 00 NOT LISTED
29-CONNECTING ROD
Saturn .... 00 NOT LISTED
30-BEARINGS
Saturn .... 00 NOT LISTED
NOTES FOR: BEARINGS
Includes: One Standard Upper & Lower Bearing For One Rod.

CRANKSHAFT & BEARINGS

CRANKSHAFT & BEARINGS

31-REAR MAIN SEAL
Saturn .... 00 NOT LISTED
32-BEARINGS
Saturn .... 00 NOT LISTED
NOTES FOR: BEARINGS
Includes: One Standard Upper & Lower Bearing.
33-CRANKSHAFT
Saturn .... 00 NOT LISTED
34-PULLEY
Saturn 00 90537704 34.65
35-FRONT CRANK SEAL
Saturn 00 90571925 9.95

LUBRICATION

OIL PAN

36-OIL PAN
Saturn 00 90537782 70.00
37-BAFFLE PLATE
Saturn .... 00 NOT LISTED
Saturn
Part Of Front Cover.

LABOR — ENGINE – 3.0L — LABOR

OPERATION INDEX

CAMSHAFT GEAR, R&R	18
CAMSHAFT, R&I	16
CAMSHAFT, R&R	17
CONNECTING ROD BEARING, R&R	24
CRANKSHAFT GEAR, R&R	19
CRANKSHAFT, R&R	26
CYLINDER BLOCK, R&R	1
CYLINDER HEAD, R&R	9
ENGINE, OVERHAUL	4
ENGINE, R&I	2
ENGINE, R&R	3
FRONT COVER SEAL, R&R	21
FRONT COVER, R&R	20
HEAD GASKET, R&R	10
LONG BLOCK, R&R	5
MAIN BEARINGS, R&R	27
MOUNT, R&R	7
OIL PAN GASKET, R&R	31
OIL PAN, R&R	30
OIL PICK-UP, R&R	32
OIL PUMP, R&R	33
PISTON RINGS, R&R	25
REAR MAIN SEAL, R&R	28
SHORT BLOCK, R&R	6
TIMING BELT, R&R	22
TIMING TENSIONER, R&R	23
TRANS MOUNT, R&R	8
VALVE COVER, R&R	11
VALVE LIFTERS, R&R	12
VALVE SEALS, R&R	13
VALVE SPRINGS, R&R	14
VALVES, GRIND	15
VIBRATION DAMPER, R&R	29

ENGINE

ENGINE

		(Factory Time)	Motor Time
1-CYLINDER BLOCK, R&R			
Saturn	00 A		23.8

NOTES FOR: CYLINDER BLOCK, R&R
Includes: R&I Engine And Transfer All Necessary Parts. Does Not Include: Transfer Of Any Part Of Engine, Replacement Of Optional Equipment.

2-ENGINE, R&I			
Saturn	00		8.4
3-ENGINE, R&R			
Saturn	00 B	(7.7)	10.8

NOTES FOR: ENGINE, R&R
Includes: Transfer All Fuel & Electrical Units. Does Not Include: Transfer Of Optional Equipment.

4-ENGINE, OVERHAUL			
Saturn	00 A		27.8

NOTES FOR: ENGINE, OVERHAUL
Includes: Measure Cylinder Bores, Crankshaft & Pistons For Proper Size & Hone Cylinders. Renew Pistons, Rings, Pins, Main & Rod Bearings, Grind Valves & Tune-Up.

5-LONG BLOCK, R&R			
Saturn	00 B		15.2

NOTES FOR: LONG BLOCK, R&R
Includes: R&I Engine And Transfer All Necessary Components Not Supplied With Long Block.

6-SHORT BLOCK, R&R			
Saturn	00 A		18.2

NOTES FOR: SHORT BLOCK, R&R
Includes: R&I Engine And Transfer All Necessary Parts.

MOUNTS

MOUNTS

		(Factory Time)	Motor Time
7-MOUNT, R&R			
Saturn	00 B	(0.9)	1.4
8-TRANS MOUNT, R&R			
Saturn			
Manual			
Front	00 B	(0.9)	1.4
Rear	00 B	(0.9)	1.4
Auto			
Front	00 B	(1.0)	1.4
Rear	00 B	(1.1)	1.4
Right	00 B	(1.1)	1.4
Lower	00 B	(0.5)	0.6

CYLINDER HEAD & VALVES

CYLINDER HEAD & VALVES

		(Factory Time)	Motor Time
9-CYLINDER HEAD, R&R			
Saturn			
Front Bank	00 B	(9.7)	13.7
Rear Bank	00 B	(10.2)	14.7
Both Banks	00 B	(13.7)	18.8

NOTES FOR: CYLINDER HEAD, R&R
Includes: R&I Cylinder Head, Grind Or Renew Valves, Transfer All Necessary Components & Make All Adjustments.

(CONTINUED)

GENERAL MOTORS 53
SATURN L-SERIES

ENGINE – 3.0L - Time Cont'd

	Factory Time	Motor Time
10-HEAD GASKET, R&R		
Saturn		
Front Bank	00 B (7.5)	10.8
Rear Bank	00 B (8.0)	10.9
Both Banks	00 B (9.6)	13.7

NOTES FOR: HEAD GASKET, R&R
Includes: Remove Carbon And Make Necessary Adjustments.

	Factory Time	Motor Time
11-VALVE COVER, R&R		
Saturn		
Front Bank	00 B (1.4)	1.8
Rear Bank	00 B (1.9)	2.7
Both Banks	00 B (2.5)	3.5
12-VALVE LIFTERS, R&R		
Saturn		
One Cylinder		
Front Bank	00 B (5.6)	8.0
Rear Bank	00 B (6.1)	8.8
Both Banks	00 B (6.9)	9.8
All Cylinders		
Front Bank	00 B (5.7)	8.1
Rear Bank	00 B (6.2)	8.9
Both Banks	00 B (7.1)	10.0
13-VALVE SEALS, R&R		
Saturn		
One Cylinder		
Front Bank	00 B (8.0)	10.8
Rear Bank	00 B (8.5)	11.8
Both Banks	00 B (10.5)	14.7
All Cylinders		
Front Bank	00 B (8.4)	11.8
Rear Bank	00 B (8.9)	12.7
Both Banks	00 B (11.3)	15.7
14-VALVE SPRINGS, R&R		
Saturn		
One Cylinder		
Front Bank	00 B (8.0)	10.8
Rear Bank	00 B (8.5)	11.8
Both Banks	00 B (10.5)	14.7
All Cylinders		
Front Bank	00 B (8.4)	11.8
Rear Bank	00 B (8.9)	12.7
Both Banks	00 B (11.3)	15.7
15-VALVES, GRIND		
Saturn		
Front Bank	00 B (9.6)	13.7
Rear Bank	00 B (10.1)	13.8
Both Banks	00 B (14.1)	20.0

NOTES FOR: VALVES, GRIND
Includes: R&I Cylinder Head, Grind All Valves & Seats And Make All Necessary Adjustments.

	Factory Time	Motor Time
To Knurl Valve Guide, Add		
One	00	0.1
To Renew Valve Guide, Add		
One	00	0.2
To Renew Camshaft, Add		
One	00	0.2
Each Additional	00	0.1
To Renew Timing Belt, Add	00	1.2

CAMSHAFT & TIMING

	Factory Time	Motor Time
16-CAMSHAFT, R&I		
Saturn		
Front Bank		
One	00 B (5.6)	8.0
Rear Bank		
One	00 B (6.1)	8.8
17-CAMSHAFT, R&R		
Saturn		
Front Bank		
One	00 B (5.6)	8.0
Rear Bank		
One	00 B (6.1)	8.8
18-CAMSHAFT GEAR, R&R		
Saturn		
Front Bank		
One	00 B (3.2)	4.6
Both	00 B (3.3)	4.7
Rear Bank		
One	00 B (3.2)	4.6
Both	00 B (3.3)	4.7
19-CRANKSHAFT GEAR, R&R		
Saturn	00 B (3.2)	4.6
20-FRONT COVER, R&R		
Saturn		
Front	00 B (2.1)	3.0
Rear	00 B (3.2)	4.7
21-FRONT COVER SEAL, R&R		
Saturn	00 B (3.3)	4.6
22-TIMING BELT, R&R		
Saturn	00 B (3.0)	4.6
23-TIMING TENSIONER, R&R		
Saturn	00 B (3.0)	4.6

PISTONS, RINGS & BEARINGS

	Factory Time	Motor Time
24-CONNECTING ROD BEARING, R&R		
Saturn		
One	00 A (11.9)	16.7
Each Additional	00 A (0.1)	0.2
All	00 A (12.2)	17.1

NOTES FOR: CONNECTING ROD BEARING, R&R
Includes: R&I Oil Pan & Plastigage Bearings.

	Factory Time	Motor Time
25-PISTON RINGS, R&R		
Saturn		
One Cylinder		
Front Bank	00 B (14.9)	21.3
Rear Bank	00 B (15.0)	21.3
Both Banks	00 B (16.4)	22.5
All Cylinders		
Front Bank	00 B (15.5)	21.9
Rear Bank	00 B (15.6)	21.9
Both Banks	00 B (17.5)	25.1
Front Bank	00	0.4
Rear Bank	00	0.7

NOTES FOR: PISTON RINGS, R&R
Includes: R&I Cylinder Head & Oil Pan, Hone Cylinder Walls And Plastigage Bearings.

	Factory Time	Motor Time
To Renew Piston, Pin Or Connecting Rod, Add		
One	00	0.2
Each Additional	00	0.1
All	00	0.5
To Renew Connecting Rod Bearings, Add		
Each	00	0.1
To Renew Timing Belt, Add	00	1.2
To Grind Valves, Add		
Front Bank	00	3.8
Rear Bank	00	4.0
Both Banks	00	5.6
To Renew Spark Plugs, Add		
Both Banks	00	1.0

CRANKSHAFT & BEARINGS

	Factory Time	Motor Time
26-CRANKSHAFT, R&R		
Saturn	00 A (11.9)	16.7

NOTES FOR: CRANKSHAFT, R&R
Includes: R&I Engine & Oil Pan, Renew All Bearings And Seals And Make All Necessary Adjustments.

	Factory Time	Motor Time
27-MAIN BEARINGS, R&R		
Saturn		
One	00 A (11.9)	16.7
Each Additional	00 A (0.1)	0.2
All	00 A (12.2)	17.1

NOTES FOR: MAIN BEARINGS, R&R
Includes: R&I Oil Pan & Plastigage Bearings.

	Factory Time	Motor Time
28-REAR MAIN SEAL, R&R		
Saturn	00 B (7.9)	10.8
29-VIBRATION DAMPER, R&R		
Saturn	00 B (1.3)	1.8

LUBRICATION

OIL PAN

	Factory Time	Motor Time
30-OIL PAN, R&R		
Saturn	00 C (1.0)	1.4
31-OIL PAN GASKET, R&R		
Saturn	00 C (1.0)	1.4

OIL PUMP

	Factory Time	Motor Time
32-OIL PICK-UP, R&R		
Saturn	00 B (1.1)	1.5

NOTES FOR: OIL PICK-UP, R&R
Includes: R&I Oil Pan.

	Factory Time	Motor Time
33-OIL PUMP, R&R		
Saturn	00 B (3.3)	4.6

PARTS — ENGINE – 3.0L — PARTS

ENGINE

	Part No.	Price
ENGINE		
Saturn	00 09192208	6000.00

NOTES FOR: ENGINE
Includes: Block, Cylinder Head, Rods & Pistons, Crankshaft, Bearings & Rear Seal, And Fuel System.

OVERHAUL GASKET SET
Saturn .... 00 NOT LISTED

SHORT BLOCK
Saturn .... 00 NOT LISTED

NOTES FOR: SHORT BLOCK
Includes: Block, Pistons & Rods, Crankshaft, Bearings & Rear Seal.

MOUNTS

	Part No.	Price
MOUNT		
Saturn		
Right	00 90575640	48.00
Left	00 90575643	49.95

CYLINDER HEAD & VALVES

	Part No.	Price
VALVE GRIND GASKET KIT		
Saturn	00	NOT LISTED
1-CYLINDER HEAD		
Saturn		
Front	00 09192235	800.00
Rear	00 09192234	800.00
2-HEAD GASKET		
Saturn	00 09118839	22.50
3-VALVE COVER		
Saturn		
Front	00 90572292	60.00
Rear	00 90572291	60.00
4-VALVE COVER GASKET		
Saturn	00 90502291	18.00
5-INTAKE VALVE		
Saturn		
Intake Valve (Standard)	00 90412277	10.00
6-EXHAUST VALVE		
Saturn		
Exhaust Valve (Standard)	00 09128626	35.00
7-VALVE SEALS		
Saturn	00 90410741	2.00
8-VALVE SPRINGS		
Saturn	00 90448050	
9-VALVE SPRING RETAINERS		
Saturn	00 90354649	2.00
10-VALVE KEEPER		
Saturn	00 90354648	1.00

NOTES FOR: VALVE KEEPER
Price Is For Half Of One Key.

	Part No.	Price
11-VALVE GUIDE		
Saturn		
Intake (Standard)	00 90502079	2.00
Exhaust (Standard)	00 90502080	2.00
12-VALVE SPRING SEATS		
Saturn	00 90410740	0.39
13-VALVE SEATS		
Saturn		
Intake (Standard)	00 90502081	2.00
Exhaust (Standard)	00 90502082	2.00
14-VALVE LIFTERS		
Saturn	00 09117904	19.99

CAMSHAFT & TIMING

	Part No.	Price
15-INTAKE CAMSHAFT		
Saturn	00 90540002	375.00

(CONTINUED)

53-13

53 GENERAL MOTORS
SATURN L-SERIES

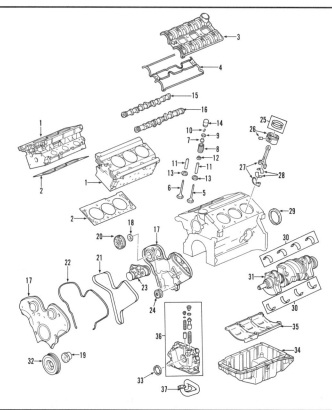

ENGINE – 3.0L - Parts Cont'd	Part No.	Price
16-EXHAUST CAMSHAFT		
Saturn 00	90512906	375.00
17-FRONT COVER		
Saturn		
Inner 00	90572785	45.00
Outer 00	90572779	69.00
18-CAMSHAFT SEAL		
Saturn 00	90285291	
19-CRANKSHAFT GEAR		
Saturn 00	90409548	18.00
20-CAMSHAFT GEAR		
Saturn		
Front Cyl Head 00	90466551	
Rear Cyl Head 00	90573286	39.99
21-TIMING BELT		
Saturn 00	09128500	
22-FRONT COVER GASKET		
Saturn 00	09128500	
23-TIMING TENSIONER		
Saturn 00	09202578	100.00
24-IDLER PULLEY		
Saturn 00	09120308	50.00

PISTONS, RINGS & BEARINGS	Part No.	Price
····· PISTONS, RINGS & BEARINGS ·····		
25-PISTON RINGS		
Saturn 00	09118853	149.99
NOTES FOR: PISTON RINGS		
Includes: One Standard Ring Set For One Piston.		
26-PISTONS		
Saturn 00	90571284	120.00
27-CONNECTING ROD		
Saturn 00	90541632	400.00
28-BEARING SET		
Saturn 00	90487297	48.00
NOTES FOR: BEARING SET		
Includes: Six Standard Upper & Lower Bearings For Six Rods.		
CRANKSHAFT & BEARINGS		
····· CRANKSHAFT & BEARINGS ·····		
29-REAR MAIN SEAL		
Saturn 00	90325572	10.00

	Part No.	Price
30-BEARING SET		
Saturn 00	09192965	28.00
NOTES FOR: BEARING SET		
Includes: Six Standard Upper & Lower Bearing. Order By Color.		
31-CRANKSHAFT		
Saturn 00	90509018	675.00
32-PULLEY		
Saturn 00	90469392	
33-FRONT CRANK SEAL		
Saturn 00	90322669	10.50
LUBRICATION		
············ OIL PAN ············		
34-OIL PAN		
Saturn 00	09157816	149.99
35-BAFFLE PLATE		
Saturn 00	90573197	
············ OIL PUMP ············		
36-OIL PUMP		
Saturn 00	09193203	200.00
37-OIL PICK-UP		
Saturn 00	90573257	16.49

LABOR 11 CLUTCH 11 LABOR

OPERATION INDEX	
CLUTCH, R&R	1
MASTER CYLINDER, R&R	2
SLAVE CYLINDER, R&R	3

CLUTCH & FLYWHEEL
········ CLUTCH & FLYWHEEL ········

		(Factory Time)	Motor Time
1-CLUTCH, R&R			
Saturn .... 00 B		(7.0)	9.8
NOTES FOR: CLUTCH, R&R			
Includes: R&I Transaxle.			
To Replace Flywheel, Add 00		(0.2)	0.2

HYDRAULIC SYSTEM
········ HYDRAULIC SYSTEM ········

		(Factory Time)	Motor Time
2-MASTER CYLINDER, R&R			
Saturn .... 00 B		(0.9)	1.4
3-SLAVE CYLINDER, R&R			
Saturn .... 00 B		(6.6)	8.8

GENERAL MOTORS 53
SATURN L-SERIES

PARTS

PARTS INDEX
DISC	1
FLYWHEEL	2
MASTER CYLINDER	4
PRESSURE PLATE	3
SLAVE CYLINDER	5

CLUTCH & FLYWHEEL

11 CLUTCH 11

	Part No.	Price
1-DISC		
Saturn 00	90578944	105.00
2-FLYWHEEL		
Saturn 00	90537283	149.99
3-PRESSURE PLATE		
Saturn 00	90578570	

PARTS

	Part No.	Price
HYDRAULIC SYSTEM		
4-MASTER CYLINDER		
Saturn 00	09181238	49.99
5-SLAVE CYLINDER		
Saturn 00	90578179	155.00

LABOR

OPERATION INDEX
TRANSAXLE, OVERHAUL	3
TRANSAXLE, R&I	1
TRANSAXLE, R&R	2

MANUAL TRANSAXLE

12 MANUAL TRANSAXLE 12

	(Factory Time)	Motor Time
1-TRANSAXLE, R&I		
Saturn .... 00 B		7.2
2-TRANSAXLE, R&R		
Saturn .... 00 B	(7.2)	9.8
3-TRANSAXLE, OVERHAUL		
Saturn .... 00 A	(8.9)	12.7

LABOR

PARTS

PARTS INDEX
FLYWHEEL	1
TRANSAXLE	2

MANUAL TRANSAXLE

12 MANUAL TRANSAXLE 12

	Part No.	Price
1-FLYWHEEL		
Saturn 00	90537283	149.99
2-TRANSAXLE		
Saturn 00	09126242	2100.00

PARTS

LABOR

OPERATION INDEX
SHIFT CONTROL CABLE, R&R	1
TRANSAXLE, OVERHAUL	4
TRANSAXLE, R&I	2
TRANSAXLE, R&R	3

AUTOMATIC TRANSAXLE

14 AUTOMATIC TRANSAXLE 14

	(Factory Time)	Motor Time
1-SHIFT CONTROL CABLE, R&R		
Saturn .... 00 B	(1.8)	2.7
2-TRANSAXLE, R&I		
Saturn		
2.2L .... 00 B		6.4
3.0L .... 00 B		7.6
NOTES FOR: TRANSAXLE, R&I		
To Replace Driveplate, Add 00		0.2

LABOR

	(Factory Time)	Motor Time
3-TRANSAXLE, R&R		
Saturn		
2.2L .... 00 B	(6.4)	8.8
3.0L .... 00 B	(7.6)	10.8
NOTES FOR: TRANSAXLE, R&R		
To Replace Driveplate, Add 00		0.2
4-TRANSAXLE, OVERHAUL		
Saturn		
2.2L .... 00 B	(15.2)	21.3
3.0L .... 00 B	(16.4)	22.5

PARTS

PARTS INDEX
DRIVE PLATE	1
SHIFT CONTROL CABLE	2
TRANSAXLE	3

AUTOMATIC TRANSAXLE

14 AUTOMATIC TRANSAXLE 14

	Part No.	Price
1-DRIVE PLATE		
Saturn		
2.2L 00	09126397	40.00
3.0L 00	09126346	45.00
2-SHIFT CONTROL CABLE		
Saturn 00	90523858	75.00
3-TRANSAXLE		
Saturn		
2.2L 00	24215280	3200.00

PARTS

	Part No.	Price
3.0L 00	24215281	3200.00

LABOR

OPERATION INDEX
BACKING PLATE, R&R	5
BOOSTER CHECK VALVE, R&R	12
BRAKE BOOSTER, R&I	13
BRAKE BOOSTER, R&R	14
BRAKES, ADJUST	40
BRAKES, R&R	1
CABLE, R&R	41
CALIPER ASSY, OVERHAUL	23
CALIPER ASSY, OVERHAUL	27
CALIPER ASSY, R&I	21
CALIPER ASSY, R&I	25
CALIPER ASSY, R&R	22
CALIPER ASSY, R&R	26
CONTROL MODULE, R&R	31
DRUM, R&I	6
DRUM, R&R	7
DRUM, REFACE	8
FLEX HOSE, R&R	24
FLEX HOSE, R&R	28
FLUID LEVEL SWITCH, R&R	15
FRONT SPEED SENSOR HARNESS, R&R	33
FRONT SPEED SENSOR, R&R	32
MASTER CYLINDER, OVERHAUL	18
MASTER CYLINDER, R&I	16
MASTER CYLINDER, R&R	17
MODULATOR, R&R	34

18 BRAKES 18

PARKING BRAKE CONTROL, R&R	42
REAR SENSOR HARNESS, R&R	35
REAR SPEED SENSOR, R&R	36
RELAY, R&R	37
RESERVOIR, R&R	38
ROTOR, R&I	2
ROTOR, R&I	9
ROTOR, R&R	10
ROTOR, R&R	3
ROTOR, REFACE	11
ROTOR, REFACE	4
SYSTEM, BLEED	19
SYSTEM, OVERHAUL	20
SYSTEM, TEST	39
WHEEL CYLINDER, R&I	29
WHEEL CYLINDER, R&R	30

BRAKE COMPONENTS

	(Factory Time)	Motor Time
1-BRAKES, R&R		
FRONT AXLE		
Saturn .... 00 B	(0.6)	0.8

LABOR

	(Factory Time)	Motor Time
REAR AXLE		
Pads .... 00 B	(0.6)	0.8
Shoes .... 00 B	(1.0)	1.3
NOTES FOR: BRAKES, R&R		
To Renew Caliper, Add		
Each 00		0.2
To Overhaul Caliper, Add		
Each 00		0.3
To Renew Wheel Cylinder, Add		
Each 00		0.1
To Overhaul Wheel Cylinder, Add		
Each 00		0.2
To Renew Rotor, Add		
Each 00		0.2
To Reface Rotor, Add		
Each 00		0.4
To Renew Drum, Add		
Each 00		0.1
To Reface Drum, Add		
Each 00		0.3
To Renew Brake Hose, Add		
Each 00		0.1

(CONTINUED)

53 GENERAL MOTORS
SATURN L-SERIES

BRAKES - Time Cont'd

	Factory Time	Motor Time
To Renew Parking Brake Cables, Add		
Rear		
Disc Brakes		
Each	00	0.3
Drum Brakes		
Each	00	0.5

FRONT BRAKES

	Factory Time	Motor Time
2-ROTOR, R&I		
Saturn		
One Side	00 B (0.4)	0.5
Both Sides	00 B (0.7)	0.9
3-ROTOR, R&R		
Saturn		
One Side	00 B (0.4)	0.5
Both Sides	00 B (0.7)	0.9
4-ROTOR, REFACE		
Saturn		
One Side	00 B	0.4
NOTES FOR: ROTOR, REFACE		
Does Not Include: R&I Rotor.		

REAR BRAKES

	Factory Time	Motor Time
5-BACKING PLATE, R&R		
Saturn		
One Side	00 B (1.1)	1.3
Both Sides	00 B (2.4)	2.6
6-DRUM, R&I		
Saturn		
One Side	00 B (0.4)	0.5
Both Sides	00 B (0.5)	0.6
7-DRUM, R&R		
Saturn		
One Side	00 B (0.4)	0.5
Both Sides	00 B (0.5)	0.6
8-DRUM, REFACE		
Saturn		
One Side	00 B	0.3
NOTES FOR: DRUM, REFACE		
Does Not Include: R&I Drum.		
9-ROTOR, R&I		
Saturn		
One Side	00 B (0.4)	0.5
Both Sides	00 B (0.7)	0.9
10-ROTOR, R&R		
Saturn		
One Side	00 B (0.4)	0.5
Both Sides	00 B (0.7)	0.9
11-ROTOR, REFACE		
Saturn		
One Side	00 B	0.4
NOTES FOR: ROTOR, REFACE		
Does Not Include: R&I Rotor.		

HYDRAULIC SYSTEM

HYDRAULIC SYSTEM

	Factory Time	Motor Time
12-BOOSTER CHECK VALVE, R&R		
Saturn	00 B (0.2)	0.3
13-BRAKE BOOSTER, R&I		
Saturn	00 B (1.9)	2.4
NOTES FOR: BRAKE BOOSTER, R&I		
To Adjust Stop Lamp Switch		
If Necessary, Add	00	(0.2) 0.3
14-BRAKE BOOSTER, R&R		
Saturn	00 B (1.9)	2.4
NOTES FOR: BRAKE BOOSTER, R&R		
To Adjust Stop Lamp Switch		
If Necessary, Add	00	(0.2) 0.3
15-FLUID LEVEL SWITCH, R&R		
Saturn	00 B (0.2)	0.3
16-MASTER CYLINDER, R&I		
Saturn		
w/o ABS	00 B (0.9)	1.3
w/ABS	00 B (1.6)	2.0
17-MASTER CYLINDER, R&R		
Saturn		
w/o ABS	00 B (0.9)	1.3
w/ABS	00 B (1.0)	2.0
18-MASTER CYLINDER, OVERHAUL		
Saturn		
w/o ABS	00 B (1.4)	1.6
w/ABS	00 B (2.0)	2.4
19-SYSTEM, BLEED		
Saturn		
w/o ABS	00 B (0.9)	1.3
w/ABS	00 B (1.0)	1.4
20-SYSTEM, OVERHAUL		
Saturn		
w/o Rear Disc	00 B	5.0
w/Rear Disc	00 B	5.5
NOTES FOR: SYSTEM, OVERHAUL		
Includes: Overhaul Calipers, Wheel Cylinders & Master Cylinder Drain, Flush, Refill & Bleed System.		

FRONT BRAKES

	Factory Time	Motor Time
21-CALIPER ASSY, R&I		
Saturn		
One Side	00 B (0.7)	0.9
Both Sides	00 B (1.0)	1.3
22-CALIPER ASSY, R&R		
Saturn		
One Side	00 B (0.7)	0.9
Both Sides	00 B (1.0)	1.3
23-CALIPER ASSY, OVERHAUL		
Saturn		
One Side	00 B (1.1)	1.3
Both Sides	00 B (1.8)	2.4
24-FLEX HOSE, R&R		
Saturn		
One Side	00 B (0.7)	0.9
Both Sides	00 B (0.9)	1.3

REAR BRAKES

	Factory Time	Motor Time
25-CALIPER ASSY, R&I		
Saturn		
One Side	00 B (0.7)	0.9
Both Sides	00 B (1.1)	1.3
26-CALIPER ASSY, R&R		
Saturn		
One Side	00 B (0.7)	0.9
Both Sides	00 B (1.1)	1.3
27-CALIPER ASSY, OVERHAUL		
Saturn		
One Side	00 B (1.1)	1.3
Both Sides	00 B (1.9)	2.4
28-FLEX HOSE, R&R		
Saturn		
One Side	00 B (0.7)	0.9
Both Sides	00 B (0.9)	1.3
29-WHEEL CYLINDER, R&I		
Saturn		
One Side	00 B (1.4)	1.6
Both Sides	00 B (2.1)	2.8
30-WHEEL CYLINDER, R&R		
Saturn		
One Side	00 B (1.4)	1.6
Both Sides	00 B (2.1)	2.8

ANTI-LOCK BRAKES

ANTI-LOCK BRAKES

	Factory Time	Motor Time
31-CONTROL MODULE, R&R		
Saturn	00 B (1.9)	2.4
32-FRONT SPEED SENSOR, R&R		
Saturn		
One Side	00 B (0.4)	0.5
Both Sides	00 B (0.6)	0.8
33-FRONT SPEED SENSOR HARNESS, R&R		
Saturn		
One Side	00 B (0.4)	0.5
34-MODULATOR, R&R		
Saturn	00 B (1.9)	2.7
35-REAR SENSOR HARNESS, R&R		
Saturn		
One Side	00 B (0.3)	0.4
36-REAR SPEED SENSOR, R&R		
Saturn		
One Side		
Disc Brakes	00 B (1.5)	2.0
Drum Brakes	00 B (1.0)	1.3
Both Sides		
Disc Brakes	00 B (2.6)	3.1
Drum Brakes	00 B (1.4)	1.6
37-RELAY, R&R		
Saturn	00 B (0.2)	0.3
38-RESERVOIR, R&R		
Saturn	00 B (1.1)	1.3
39-SYSTEM, TEST		
All Models	00 B	1.0

PARKING BRAKE

PARKING BRAKE

	Factory Time	Motor Time
40-BRAKES, ADJUST		
Saturn	00 B (0.3)	0.4
41-CABLE, R&R		
Saturn		
Front	00 B (1.1)	1.3
Intermediate	00 B (1.0)	1.3
Rear		
One Side		
Disc Brakes	00 B (0.5)	0.6
Drum Brakes	00 B (1.0)	1.3
Both Sides		
Disc Brakes	00 B (0.6)	0.8
Drum Brakes	00 B (1.7)	2.0
NOTES FOR: CABLE, R&R		
Includes: Adjust Parking Brake.		
42-PARKING BRAKE CONTROL, R&R		
Saturn	00 B (0.7)	0.9
NOTES FOR: PARKING BRAKE CONTROL, R&R		
Includes: Adjust Parking Brake.		

PARTS 18 BRAKES 18 PARTS

PARTS INDEX

BACKING PLATE	5
BRAKE BOOSTER	8
CALIPER ASSY	11
CALIPER ASSY	14
CALIPER OVERHAUL KIT	15
CONTROL MODULE	18
DRUM	6
FLEX HOSE	12
FLEX HOSE	16
FLUID LEVEL SWITCH	9
FRONT CABLE	22
FRONT PADS	1
FRONT SPEED SENSOR	19
INTERMEDIATE CABLE	23
MASTER CYLINDER	10
MODULATOR ASSY	20
OVERHAUL KIT	13
PARKING BRAKE CONTROL	24
REAR CABLE	25
REAR PADS	2
REAR SPEED SENSOR	21
ROTOR	4
ROTOR	7
SHOES	3
WHEEL CYLINDER	17

BRAKE COMPONENTS

BRAKE COMPONENTS

		Part No.	Price
1-FRONT PADS			
Saturn	00	21018993	60.00
2-REAR PADS			
Saturn	00	21019005	60.00
3-SHOES			
Saturn	00	21019008	37.00

FRONT BRAKES

		Part No.	Price
4-ROTOR			
Saturn	00	21019227	65.00

REAR BRAKES

		Part No.	Price
5-BACKING PLATE			
Saturn			
Rear Disc Brakes			
Right	00	21019208	35.00
Left	00	21019209	35.00
Rear Drum Brakes			
Right	00	21019206	35.00
Left	00	21019207	9.99
6-DRUM			
Saturn	00	21019260	74.99
7-ROTOR			
Saturn	00	21019258	60.00

HYDRAULIC SYSTEM

HYDRAULIC SYSTEM

		Part No.	Price
8-BRAKE BOOSTER			
Saturn	00	21018890	135.00

(CONTINUED)

GENERAL MOTORS 53
SATURN L-SERIES

BRAKES - Parts Cont'd

	Part No.	Price
9-FLUID LEVEL SWITCH		
Saturn 00	21010590	3.30
10-MASTER CYLINDER		
Saturn 00	21018891	135.00

FRONT BRAKES

	Part No.	Price
11-CALIPER ASSY		
Saturn		
Right 00	21018992	149.99
Left 00	21018991	149.99
12-FLEX HOSE		
Saturn 00	09127444	15.99
13-OVERHAUL KIT		
Saturn 00	21018995	25.00

REAR BRAKES

	Part No.	Price
14-CALIPER ASSY		
Saturn		
Right 00	21019004	149.99
Left 00	21019003	149.99
15-CALIPER OVERHAUL KIT		
Saturn 00	21019006	35.00
16-FLEX HOSE		
Saturn 00	21019248	23.00
17-WHEEL CYLINDER		
Saturn 00	21019205	28.00

ANTI-LOCK BRAKES

	Part No.	Price
18-CONTROL MODULE		
Saturn 00	21019062	475.00
19-FRONT SPEED SENSOR		
Saturn 00	90464775	95.00
20-MODULATOR ASSY		
Saturn 00	21019063	1000.00
21-REAR SPEED SENSOR		
Saturn Part Of Hub Assy.		

PARKING BRAKE

	Part No.	Price
22-FRONT CABLE		
Saturn 00	21019505	14.99
23-INTERMEDIATE CABLE		
Saturn 00	24401279	23.00
24-PARKING BRAKE CONTROL		
Saturn		
w/o Leather 00	21019501	39.95
w/Leather 00	21019502	69.95
NOTES FOR: PARKING BRAKE CONTROL		
Order By Color.		
25-REAR CABLE		
Saturn		
w/Drum Brakes 00	21019211	14.00
w/Disc Brakes 00	21019234	14.00

LABOR 19 FRONT SUSPENSION 19 LABOR

OPERATION INDEX

COIL SPRING, R&R	4
FRONT WHEEL BEARING, R&R	5
HUB, R&R	6
KNUCKLE, R&R	7
LOWER CONTROL ARM, R&R	9
STRUT, R&R	8
SUSPENSION, OVERHAUL	1
TOE-IN, ADJUST	2
WHEELS, ALIGN	3

FRONT SUSPENSION

SUSPENSION SERVICE

		(Factory Time)	Motor Time
1-SUSPENSION, OVERHAUL			
Saturn			
One Side	00 B		3.5
Both Sides	00 B		6.8
2-TOE-IN, ADJUST			
Saturn	00 B		0.7
3-WHEELS, ALIGN			
Saturn			
Front Wheel Alignment.	00 B		1.3

		(Factory Time)	Motor Time
Four Wheel Alignment.	00 B		2.3

SUSPENSION COMPONENTS

		(Factory Time)	Motor Time
4-COIL SPRING, R&R			
Saturn			
One Side	00 B	(1.0)	1.2
Both Sides	00 B	(1.7)	1.9
NOTES FOR: COIL SPRING, R&R			
Does Not Include: Wheel Alignment.			
5-FRONT WHEEL BEARING, R&R			
Saturn			
One Side	00 B	(1.4)	1.6
Both Sides	00 B	(2.9)	3.4
NOTES FOR: FRONT WHEEL BEARING, R&R			
Does Not Include: Wheel Alignment.			
w/ABS, Add	00	(0.1)	0.1
To Remove Inner Bearing Race From Hub, Add			
One Side	00	(0.2)	0.2
6-HUB, R&R			
Saturn			
One Side	00 B	(1.2)	1.6
Both Sides	00 B	(2.2)	2.6
NOTES FOR: HUB, R&R			
Does Not Include: Wheel Alignment.			
w/ABS, Add	00	(0.1)	0.1

		(Factory Time)	Motor Time
7-KNUCKLE, R&R			
Saturn			
One Side	00 B	(1.1)	1.2
Both Sides	00 B	(1.8)	2.3
NOTES FOR: KNUCKLE, R&R			
Does Not Include: Wheel Alignment.			
w/ABS, Add	00	(0.1)	0.1
To Remove Inner Bearing Race From Hub, Add			
One Side	00	(0.2)	0.2
8-STRUT, R&R			
Saturn			
One Side	00 B	(1.0)	1.2
Both Sides	00 B	(1.7)	1.9
NOTES FOR: STRUT, R&R			
Does Not Include: Wheel Alignment.			

LOWER CONTROL ARM

		(Factory Time)	Motor Time
9-LOWER CONTROL ARM, R&R			
Saturn			
One Side	00 B	(0.6)	0.7
Both Sides	00 B	(1.1)	1.2
NOTES FOR: LOWER CONTROL ARM, R&R			
Does Not Include: Wheel Alignment.			

PARTS 19 FRONT SUSPENSION 19 PARTS

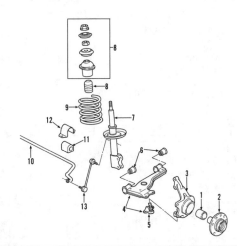

FRONT SUSPENSION

SUSPENSION COMPONENTS

	Part No.	Price
1-FRONT WHEEL BEARING		
Saturn 00	21018786	44.95
2-FRONT HUB		
Saturn 00	21019229	49.95
3-KNUCKLE		
Saturn		
Right 00	21018784	150.00
Left 00	21018785	150.00
7-STRUT		
Saturn		
2.2L		
Right 00	21019028	84.99
Left 00	21019029	84.99
3.0L		
Right 00	21019030	84.99
Left 00	21019031	84.99
8-STRUT MOUNT		
Saturn 00	21018787	39.95

	Part No.	Price
9-COIL SPRING		
Saturn		
2.2L		
Manual Trans 00	21018788	60.00
Auto Trans 00	21018789	60.00
3.0L 00	21018790	60.00

LOWER CONTROL ARM

	Part No.	Price
4-LOWER CONTROL ARM		
Saturn		
Right 00	09127244	
Left 00	09127243	
(CONTINUED)		

53-17

53 GENERAL MOTORS
SATURN L-SERIES

FRONT SUSPENSION - Parts Cont'd

NOTES FOR: LOWER CONTROL ARM
Includes Lower Ball Joint And Bushings.

		Part No.	Price
5-LOWER BALL JOINT			
Saturn	00	21018783	
6-BUSHINGS			
Saturn			
Front	00	90468636	8.95
Rear	00	90468638	8.95

STABILIZER BAR

		Part No.	Price
10-STABILIZER BAR			
Saturn	00	21018791	39.99
11-BUSHINGS			
Saturn	00	21018793	
12-BRACKET			
Saturn	00	90576835	3.00
13-LINK			
Saturn	00	21018793	

LABOR — 20 FRONT DRIVE AXLE 20 — LABOR

OPERATION INDEX

AXLE ASSY, R&I	1
AXLE ASSY, R&R	2

DRIVE AXLES

1-AXLE ASSY, R&I

		Factory Time	Motor Time
Saturn			
Right Side	00 B	(1.0)	1.4
Left Side	00 B	(1.1)	1.4
Both Sides	00 B	(1.9)	2.6

NOTES FOR: AXLE ASSY, R&I

		Factory Time	Motor Time
To Replace Shaft, Add One	00	(0.5)	0.6
To Replace Inner And/Or Outer Seal, Add One Side	00	(0.4)	0.5
To Replace Outer CV Joint, Add One	00	(0.3)	0.4
To Replace Inner Joint, Add One	00	(0.3)	0.4

2-AXLE ASSY, R&R

		Factory Time	Motor Time
Saturn			
Right Side	00 B	(1.1)	1.4
Left Side	00 B	(1.1)	1.4
Both Sides	00 B	(1.9)	2.6

PARTS — 20 FRONT DRIVE AXLE 20 — PARTS

DRIVE AXLES

1-AXLE ASSY

		Part No.	Price
Saturn			
Manual Trans			
Right	00	24404111	350.00
Left			
w/o ABS	00	24404108	350.00
w/ABS	00	24404107	350.00
Auto Trans			
w/o ABS			
Right	00	24404110	350.00
Left	00	24404109	350.00
w/ABS	00	24404110	350.00

AXLE SHAFTS & JOINTS

2-INNER BOOT
		Part No.	Price
Saturn			
Manual Trans	00	21018878	
Auto Trans	00	21018879	

3-OUTER BOOT
		Part No.	Price
Saturn	00	21018877	

4-INNER CV JOINT
		Part No.	Price
Saturn			
Manual Trans	00	21018882	130.00
Auto Trans	00	21018883	130.00

5-OUTER CV JOINT
		Part No.	Price
Saturn			
w/o ABS	00	21018881	160.00
w/ABS	00	21018880	160.00

6-SHAFT
		Part No.	Price
Saturn			
Manual Trans			
Right	00	21019012	35.00
Left	00	21018885	35.00
Auto Trans	00	21018884	35.00

LABOR — 21 STEERING – POWER 21 — LABOR

OPERATION INDEX

INNER TIE ROD BOOT, R&R	10
INNER TIE ROD, R&R	9
OUTER TIE ROD, R&R	11
P/S PRESSURE HOSE, R&R	1
P/S PUMP, OVERHAUL	4
P/S PUMP, R&I	2
P/S PUMP, R&R	3
P/S RETURN HOSE, R&R	5
RESERVOIR, R&R	6
SHAFT SEALS, R&R	7
STEERING GEAR, ADJUST	12
STEERING GEAR, R&I	13
STEERING GEAR, R&R	14
SYSTEM, DIAGNOSIS	8

P/S PUMP & HOSES

1-P/S PRESSURE HOSE, R&R

		Factory Time	Motor Time
Saturn			
2.2L	00 B	(0.8)	1.0
3.0L	00 B	(1.0)	1.3

2-P/S PUMP, R&I
		Factory Time	Motor Time
Saturn			
2.2L	00 B	(0.7)	0.9
3.0L	00 B	(1.1)	1.3

3-P/S PUMP, R&R
		Factory Time	Motor Time
Saturn			
2.2L	00 B	(0.7)	0.9
3.0L	00 B	(1.1)	1.3

4-P/S PUMP, OVERHAUL
		Factory Time	Motor Time
Saturn			
2.2L	00 B	(1.3)	1.6
3.0L	00 B	(1.7)	2.0

5-P/S RETURN HOSE, R&R
		Factory Time	Motor Time
Saturn			
2.2L	00 B	(0.8)	1.0
3.0L	00 B	(1.0)	1.3

6-RESERVOIR, R&R
		Factory Time	Motor Time
Saturn			
2.2L	00 B	(0.8)	1.0
3.0L	00 B	(0.6)	0.8

7-SHAFT SEALS, R&R
		Factory Time	Motor Time
Saturn			
2.2L	00 B	(0.8)	1.0
3.0L	00 B	(1.2)	1.6

8-SYSTEM, DIAGNOSIS
		Factory Time	Motor Time
All Models	00 B		0.5

NOTES FOR: SYSTEM, DIAGNOSIS
Includes: Test Pump And System Pressure. Check Pounds Pull On Steering Wheel And Check For Leaks.

STEERING GEAR & LINKAGE

9-INNER TIE ROD, R&R

		Factory Time	Motor Time
Saturn			
One Side			
2.2L	00 B	(4.1)	5.3
3.0L	00 B	(4.2)	5.3
Both Sides			
2.2L	00 B	(4.5)	5.9

(CONTINUED)

GENERAL MOTORS 53
SATURN L-SERIES

STEERING – POWER - Time Cont'd

		Factory Time	Motor Time
3.0L	00 B	(4.6)	5.9

NOTES FOR: INNER TIE ROD, R&R
Does Not Include: Wheel Alignment.

10-INNER TIE ROD BOOT, R&R

Saturn		Factory Time	Motor Time
One Side	00 B	(0.9)	1.3
Both Sides	00 B	(1.2)	1.6

NOTES FOR: INNER TIE ROD BOOT, R&R
Does Not Include: Wheel Alignment.

11-OUTER TIE ROD, R&R

Saturn		Factory Time	Motor Time
One Side	00 B	(0.4)	0.5
Both Sides	00 B	(0.6)	0.8

NOTES FOR: OUTER TIE ROD, R&R
Does Not Include: Wheel Alignment.

12-STEERING GEAR, ADJUST

Saturn	00 B	(0.4)	0.5

13-STEERING GEAR, R&I

Saturn		Factory Time	Motor Time
2.2L	00 B	(4.5)	5.9
3.0L	00 B	(4.6)	5.9

NOTES FOR: STEERING GEAR, R&I
Does Not Include: Wheel Alignment.

14-STEERING GEAR, R&R

Saturn		Factory Time	Motor Time
2.2L	00 B	(4.5)	5.9
3.0L	00 B	(4.6)	5.9

NOTES FOR: STEERING GEAR, R&R
Does Not Include: Wheel Alignment.

PARTS — 21 STEERING – POWER 21 — PARTS

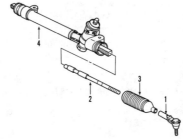

P/S PUMP & HOSES

P/S PRESSURE HOSE

		Part No.	Price
Saturn			
2.2L			
Manual Trans	00	90539615	
Auto Trans	00	90539617	
3.0L	00	24409007	44.99

P/S PUMP

Saturn			
2.2L	00	21019181	205.00
3.0L	00	21019173	205.00

P/S RETURN HOSE

		Part No.	Price
Saturn			
2.2L			
Manual Trans	00	09191633	12.99
Auto Trans	00	09191634	12.99
3.0L			
To Gear	00	09191939	29.99
To Pump	00	90539619	25.00

SEAL KIT

Saturn			
2.2L	00	26038091	19.00
3.0L	00	26053277	5.50

STEERING GEAR & LINKAGE

1-OUTER TIE ROD

		Part No.	Price
Saturn	00	21019184	47.50

2-INNER TIE ROD

Saturn	00	21019183	47.50

3-INNER TIE ROD BOOT

Saturn	00	21019185	22.50

4-STEERING GEAR

Saturn	00	26077036	570.00

LABOR — 22 STEERING COLUMN 22 — LABOR

OPERATION INDEX

INTERMEDIATE SHAFT, R&R	2
LOCK MODULE, R&R	3
LOWER COVER, R&R	4
SHAFT COVER, R&R	5
STEERING COLUMN, R&R	6
STEERING WHEEL, R&R	1
UPPER COVER, R&R	7

STEERING COLUMN

STEERING WHEEL

1-STEERING WHEEL, R&R

		Factory Time	Motor Time
Saturn	00 B	(0.5)	0.6

NOTES FOR: STEERING WHEEL, R&R
WARNING: Before Repairing Any Air Restraint System, The Battery Cables And Any Back-up Power Supplies To The System Must Be Disconnected In Order To Prevent Accidental Deployment.

STEERING COLUMN

2-INTERMEDIATE SHAFT, R&R

Saturn	00 B	(0.5)	0.6

3-LOCK MODULE, R&R

		Factory Time	Motor Time
Saturn	00 B	(1.1)	1.3

4-LOWER COVER, R&R

Saturn	0 B	(0.3)	0.4

5-SHAFT COVER, R&R

Saturn	00 B	(0.6)	0.8

6-STEERING COLUMN, R&R

Saturn	00 B	(1.6)	2.0

7-UPPER COVER, R&R

Saturn	00 B	(0.2)	0.3

PARTS — 22 STEERING COLUMN 22 — PARTS

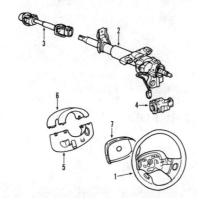

STEERING COLUMN

STEERING WHEEL

1-STEERING WHEEL

		Part No.	Price
Saturn			
w/o Leather			
w/o Cruise Control	00	21018731	
w/Cruise Control	00	21018726	
w/Leather	00	21018722	

NOTES FOR: STEERING WHEEL
Order By Color.

(CONTINUED)

53-19

53 GENERAL MOTORS
SATURN L-SERIES

STEERING COLUMN - Parts Cont'd

7-REAR COVER
	Part No.	Price
Saturn 00	21019511	18.00

NOTES FOR: REAR COVER
Order By Color.

········· **STEERING COLUMN** ·········

2-STEERING COLUMN
Saturn 00	21019171	270.00

3-INTERMEDIATE SHAFT
	Part No.	Price
Saturn 00	21019172	70.00

4-LOCK HOUSING

Saturn
Manual Trans 00	26065606	
Auto Trans 00	26061712	

5-LOWER SHROUD
Saturn 00	90585897	17.85

6-UPPER SHROUD
	Part No.	Price
Saturn 00	90585896	17.85

LABOR 24 REAR SUSPENSION 24 LABOR

OPERATION INDEX

BUSHINGS, R&R	10
COIL SPRING, R&R	3
HUB & BEARING ASSY, R&R	4
LOWER CONTROL ARM, R&R	9
REAR TRAILING ARM, R&R	5
STABILIZER BAR, R&R	11
STRUT, R&R	6
SUSPENSION CROSSMEMBER, R&R	7
SUSPENSION, OVERHAUL	1
UPPER CONTROL ARM, R&R	8
WHEEL ALIGNMENT, ALIGN	2

REAR SUSPENSION

········· **SUSPENSION SERVICE** ·········

1-SUSPENSION, OVERHAUL

		Factory Time	Motor Time
Saturn			
One Side	00 B		3.0
Both Sides	00 B		5.8

2-WHEEL ALIGNMENT, ALIGN
Saturn	00 B		1.0

········· **SUSPENSION COMPONENTS** ·········

3-COIL SPRING, R&R

Saturn
One Side	00 C	(0.9)	1.2
Both Sides	00 C	(1.4)	1.6

NOTES FOR: COIL SPRING, R&R
Does Not Include: Wheel Alignment.

4-HUB & BEARING ASSY, R&R

Saturn
One Side			
Disc Brakes	00 B	(0.7)	0.8
Drum Brakes	00 B	(0.5)	0.6
Both Sides			
Disc Brakes	00 B	(1.1)	1.2
Drum Brakes	00 B	(0.8)	1.0

5-REAR TRAILING ARM, R&R

Saturn
One Side	00 B	(1.0)	1.2

6-STRUT, R&R

Saturn
One Side	00 B	(0.9)	1.2
Both Sides	00 B	(1.4)	1.6

NOTES FOR: STRUT, R&R
Does Not Include: Wheel Alignment.

7-SUSPENSION CROSSMEMBER, R&R

Saturn
	00 B	(1.8)	2.3

NOTES FOR: SUSPENSION CROSSMEMBER, R&R
w/ABS, Add	00	(0.1)	0.1

········· **UPPER CONTROL ARM** ·········

8-UPPER CONTROL ARM, R&R

Saturn
One Side	00 B	(2.0)	2.3
Both Sides	00 B	(2.1)	2.6

NOTES FOR: UPPER CONTROL ARM, R&R
To Renew Bushings, Add			
Each	00	(0.1)	0.1

········· **LOWER CONTROL ARM** ·········

9-LOWER CONTROL ARM, R&R

Saturn
One Side			
Disc Brakes	00 B	(1.2)	1.6
Drum Brakes	00 B	(0.9)	1.2
Both Sides			
Disc Brakes	00 B	(1.8)	2.3
Drum Brakes	00 B	(1.5)	1.9

NOTES FOR: LOWER CONTROL ARM, R&R
To Renew Bushings, Add			
Each	00	(0.1)	0.1

········· **STABILIZER BAR** ·········

10-BUSHINGS, R&R

Saturn
One Side	00 C	(0.8)	1.0
Both Sides	00 C	(1.2)	1.6

11-STABILIZER BAR, R&R
Saturn	00 C	(1.0)	1.2

PARTS 24 REAR SUSPENSION 24 PARTS

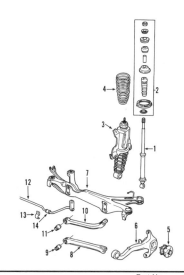

REAR SUSPENSION

········· **SUSPENSION COMPONENTS** ·········

1-SHOCK ABSORBER
	Part No.	Price
Saturn 00	21019032	89.99

2-SHOCK MOUNT
Saturn 00	21018795	54.95

3-SUPPORT

Saturn
Sedan 00	90575755	0.02
Wagon 00	90575757	0.02

4-COIL SPRING

Saturn
Sedan 00	21018797	70.00
Wagon 00	21018796	70.00

5-HUB & BEARING ASSY
	Part No.	Price
Saturn 00	NOT LISTED	

6-KNUCKLE

Saturn
Right 00	21019252	149.95
Left 00	21019253	149.96

7-CROSSMEMBER
Saturn 00	90538987	

········· **UPPER CONTROL ARM** ·········

10-UPPER CONTROL ARM
Saturn 00	09231141	39.95

11-BUSHINGS
Saturn 00	90496686	7.00

········· **LOWER CONTROL ARM** ·········

8-LOWER CONTROL ARM
	Part No.	Price
Saturn 00	09231142	34.95

9-BUSHINGS
Saturn 00	90496686	7.00

········· **STABILIZER BAR** ·········

12-STABILIZER BAR
Saturn 00	21019049	75.00

13-BRACKET
Saturn 00	21019048	6.99

14-BUSHINGS
Saturn 00	90496706	4.00

GENERAL MOTORS 53
SATURN L-SERIES

LABOR — 25 BODY HARDWARE 25 — LABOR

OPERATION INDEX

LATCH, R&R	11
LATCH, R&R	14
LATCH, R&R	2
LATCH, R&R	4
LATCH, R&R	8
LOCK CYLINDER, R&R	12
LOCK CYLINDER, R&R	15
LOCK CYLINDER, R&R	1
LOCK CYLINDER, R&R	5
RELEASE CABLE, R&R	3
STRIKER, R&R	13
STRIKER, R&R	16
STRIKER, R&R	6
STRIKER, R&R	9
WINDOW REGULATOR, R&R	10
WINDOW REGULATOR, R&R	7

INSTRUMENT PANEL

INSTRUMENT PANEL

		Factory Time	Motor Time
1-LOCK CYLINDER, R&R			
Saturn	00 C	(0.4)	0.5
NOTES FOR: LOCK CYLINDER, R&R			
To Code Cylinder, Add	00	(0.2)	0.7

HOOD

HOOD

		Factory Time	Motor Time
2-LATCH, R&R			
Saturn	00 B	(0.4)	0.5

DOOR

		Factory Time	Motor Time
3-RELEASE CABLE, R&R			
Saturn	00 B	(0.6)	0.7

DOOR
FRONT DOOR

		Factory Time	Motor Time
4-LATCH, R&R			
Saturn			
One Side	00 B	(1.0)	1.2
5-LOCK CYLINDER, R&R			
Saturn			
One Side	00 C	(0.7)	0.8
NOTES FOR: LOCK CYLINDER, R&R			
To Code Cylinder, Add	00	(0.2)	0.2
6-STRIKER, R&R			
Saturn			
One	00 C	(0.2)	0.2
7-WINDOW REGULATOR, R&R			
Saturn			
One Side	00 B	(1.1)	1.2

REAR DOOR

		Factory Time	Motor Time
8-LATCH, R&R			
Saturn			
One Side	00 B	(0.8)	0.9
9-STRIKER, R&R			
Saturn			
One	00 C	(0.2)	0.2

		Factory Time	Motor Time
10-WINDOW REGULATOR, R&R			
Saturn			
One Side	00 B	(0.8)	0.9

TRUNK
TRUNK

		Factory Time	Motor Time
11-LATCH, R&R			
Saturn	00 B	(0.2)	0.2
12-LOCK CYLINDER, R&R			
Saturn	00 C	(0.4)	0.5
NOTES FOR: LOCK CYLINDER, R&R			
To Code Cylinder, Add	00	(0.2)	0.2
13-STRIKER, R&R			
Saturn	00 C	(0.2)	0.2

LIFTGATE
LIFTGATE

		Factory Time	Motor Time
14-LATCH, R&R			
Saturn			
5-Door Wagon	00 B	(0.5)	0.6
15-LOCK CYLINDER, R&R			
Saturn			
5-Door Wagon	00 B	(0.4)	0.5
NOTES FOR: LOCK CYLINDER, R&R			
To Code Cylinder, Add	00	(0.2)	0.2
16-STRIKER, R&R			
Saturn			
5-Door Wagon	00 C	(0.3)	0.3

PARTS — 25 BODY HARDWARE 25 — PARTS

PARTS INDEX

LATCH	10
LATCH	13
LATCH	1
LATCH	3
LATCH	7
LIFT CYLINDER	14
LOCK	17
LOCK CYLINDER	11
LOCK CYLINDER	15
LOCK CYLINDER	4
RELEASE CABLE	2
STRIKER	12
STRIKER	16
STRIKER	5
STRIKER	8
WINDOW REGULATOR	6
WINDOW REGULATOR	9

HOOD
HOOD

		Part No.	Price
1-LATCH			
Saturn			
Primary	00	NOT LISTED	
Secondary	00	24403032	6.75
2-RELEASE CABLE			
Saturn	00	90583545	

DOOR
FRONT DOOR

		Part No.	Price
3-LATCH			
Saturn			
Right			
w/o Remote Locks	00	90584736	
w/Remote Locks	00	90584738	
Left			
w/o Remote Locks	00	90584738	
w/Remote Locks	00	90584737	
4-LOCK CYLINDER			
Saturn	00	21019201	19.95
5-STRIKER			
Saturn	00	09134861	0.50
6-WINDOW REGULATOR			
Saturn			
Right	00	09227665	64.95
Left	00	09227666	64.95

REAR DOOR

		Part No.	Price
7-LATCH			
Saturn			
Right			
w/o Remote Locks	00	90585290	
w/Remote Locks	00	90584740	
Left			
w/o Remote Locks	00	90585289	
w/Remote Locks	00	90584739	
8-STRIKER			
Saturn	00	09134861	0.50
9-WINDOW REGULATOR			
Saturn			
Right	00	09227994	64.95
Left	00	24408950	64.95

TRUNK
TRUNK

		Part No.	Price
10-LATCH			
Saturn	00	09152417	34.95
11-LOCK CYLINDER			
Saturn	00	21019051	24.95
12-STRIKER			
Saturn	00	21090085	3.12

LIFTGATE
LIFTGATE

		Part No.	Price
13-LATCH			
Saturn	00	09152058	59.95
14-LIFT CYLINDER			
Wagon	00	90583591	29.95
15-LOCK CYLINDER			
Saturn	00	21019051	24.95
16-STRIKER			
Saturn	00	21097283	

FUEL DOOR
FUEL DOOR

		Part No.	Price
17-LOCK			
Saturn			
w/o Remote Lock	00	09135503	3.00
w/Remote Lock	00	90584440	3.00

DP8V01

Answers to Odd-Numbered Practice Problems

CHAPTER 1

Our Number System

1. Forty-nine   3. Five hundred sixty-one   5. Ten thousand six hundred thirty-four
7. 88   9. 961   11. 7,004

Addition of Whole Numbers

1. 49   3. 220   5. 64,043   7. 908   9. 13,408   11. 389   13. 19,352
15. 33,635   17. 157 in.   19. 96 gal   21. 811 g   23. $19,527

Subtraction of Whole Numbers

1. 41   3. 192   5. 5,859   7. 4,297   9. 110,160   11. 21   13. 10,921
15. $16,717   17. 36 A   19. 323 mi   21. 43 mi   23. 33,468 mi, yes

Multiplication of Whole Numbers

1. 39   3. 567   5. 21,204   7. 41,076   9. 260,373   11. 5,526
13. 333,000   15. 2,250,000   17. $544   19. 540 mi   21. $480
23. $28,560   25. 544 mi   27. 22,800 ohms   29. a. $14,832 b. $2,332

Division of Whole Numbers

1. 12   3. 27   5. 75 r3   7. 12   9. 13 r4   11. 16   13. 290
15. 151 r24   17. 80   19. 80   21. 26 mpg   23. 13 mpg
25. a. 408 quarts per week   b. 68 quarts per day   c. 17 oil changes per day
27. $3 per square foot   29. a. $263   b. $90

Exponents

1. 5^3 3. 62^2 5. 8×8 7. $14 \times 14 \times 14 \times 14$
9. $10 \times 10 \times 10 \times 10 \times 10$ 11. 36   13. 243   15. 529   17. 144
19. 2800   21. 10,000,000   23. 1764 square feet

Order of Operations

1. 18   3. 22   5. 2   7. 16   9. 13   11. 76   13. 1   15. 9   17. $\dfrac{13}{14}$

CHAPTER 2

Reading and Writing Decimal Numbers

1. Three and seven tenths   3. One hundred twenty-five thousandths
5. Sixteen and twenty thousandths   7. Ten and three hundred ten-thousandths
9. 9.2   11. 15.094   13. 0.375   15. 4.0602

Rounding Decimal Numbers

1. 4.5   3. 4.99   5. 6.784   7. 3.001 4

Addition and Subtraction of Decimal Numbers

1. 9.1   3. 220.63   5. 32.983   7. 2300.006 63   9. 683.885 2
11. 10.78   13. 37.1453   15. 64.124 53   17. 786.095 079   19. 21.1
21. 12   23. 69.818   25. 53.4 gallons   27. 3.710 in.   29. 38.1 hours
31. 853.8 g   33. 35.24 mm   35. 2.5 volts   37. 1.428 in.   39. 0.014 in.
41. 0.016 in.   43. $20.73   45. $38.23

Reading a Micrometer

1. 0.699 in.   3. 0.566 in.   5. 0.475 in.   7. 0.102 in.   9. 0.435 in.
11. 7.39 mm   13. 10.64 mm   15. 0.91 mm   17. 10.03 mm   19. 23.60 mm

Multiplication of Decimal Numbers

1. 96.6   3. 0.675 0   5. 82.8   7. 1.44   9. 41.13   11. 0.514 6
13. 7.99   15. 28.578 8   17. 15.761 79   19. 29.293 082   21. 23.971
23. 16.216   25. $1,446.24   27. a. $712   b. 4.2 hr   c. $26.70   d. $112.14
e. $824.14   29. 344.4 miles   31. 1837.44 ft·lb   33. $84.15   35. $22.68
37. $31.14   39. $21.02   41. $196.59

Division of Decimal Numbers

1. 1.6   3. 0.083   5. 2.2   7. 2.7   9. 18   11. 38.2   13. 3.7
15. 26   17. 9.82   19. 45.2   21. 11.6   23. 142.5   25. 6.5   27. 3.19
29. 3.62   31. 11.56   33. $0.23   35. $0.31
37. a. 576 qt   b. $9.12   c. $0.76   39. 19 MPG   41. 23.5 MPG
43. 39.9 MPG   45. 62.4 MPH   47. $15.60

Square Roots

1. 3   3. 10   5. 9   7. 4.359   9. 5.568   11. 35.355   13. 26.230
15. 35.162   17. 36.719

CHAPTER 3

Equivalent Fractions

1. Proper fraction   3. Mixed number   5. Proper fraction   7. $1\frac{1}{2}$ 9. $1\frac{4}{5}$
11. 3   13. $11\frac{1}{4}$ 15. $4\frac{3}{4}$ 17. $1\frac{1}{2}$ 19. $2\frac{3}{5}$ 21. $1\frac{7}{8}$ 23. $\frac{17}{7}$
25. $\frac{37}{20}$ 27. $\frac{9}{1}$ 29. $\frac{13}{1}$ 31. $\frac{769}{64}$ 33. $\frac{87}{11}$ 35. $\frac{141}{8}$ 37. $\frac{2}{4}$
39. $\frac{10}{16}$ 41. $\frac{6}{9}$ 43. $\frac{14}{21}$ 45. $7\frac{24}{64}$ 47. $1\frac{54}{99}$ 49. $\frac{3}{4}$ 51. $\frac{2}{3}$
53. $\frac{2}{3}$ 55. $\frac{1}{4}$ 57. $2\frac{3}{4}$ 59. $7\frac{13}{18}$ 61. $9\frac{11}{15}$ 63. $41\frac{15}{16}$

Common Denominators

1. $\frac{2}{4}, \frac{3}{4}$ 3. $\frac{4}{8}, \frac{5}{8}$ 5. $\frac{2}{12}, \frac{3}{12}$ 7. $\frac{25}{40}, \frac{28}{40}$ 9. $\frac{44}{48}, \frac{27}{48}$ 11. $\frac{100}{120}, \frac{15}{120}, \frac{108}{120}$
13. $\frac{3}{4}, \frac{7}{8}$ 15. $\frac{5}{8}, \frac{2}{3}$ 17. $\frac{1}{3}, \frac{11}{32}$ 19. $\frac{1}{2}, \frac{7}{12}, \frac{3}{5}$

ANSWERS TO ODD-NUMBERED PRACTICE PROBLEMS

Addition of Fractions

1. $\frac{6}{7}$ 3. $1\frac{4}{15}$ 5. $\frac{5}{6}$ 7. $\frac{20}{21}$ 9. $14\frac{5}{6}$ 11. $5\frac{1}{16}$ 13. $19\frac{3}{13}$
15. $12\frac{35}{72}$ 17. $22\frac{23}{30}$ 19. $8\frac{3}{4}$ 21. $15\frac{4}{5}$ 23. $1\frac{1}{8}$ 25. $\frac{15}{16}$ in.

Subtraction of Fractions

1. $\frac{4}{9}$ 3. $\frac{5}{8}$ 5. $6\frac{1}{20}$ 7. $4\frac{1}{2}$ 9. $3\frac{2}{3}$ 11. $2\frac{31}{36}$ 13. $\frac{13}{24}$ 15. $3\frac{31}{36}$
17. $1\frac{1}{4}$ qt 19. $\frac{1}{2}$ in. 21. $\frac{1}{4}$ in. 23. $\frac{15}{16}$ in.

Multiplication of Fractions

1. $\frac{3}{14}$ 3. $\frac{7}{12}$ 5. $\frac{1}{6}$ 7. $2\frac{2}{5}$ 9. 6 11. $1\frac{7}{20}$ 13. $17\frac{3}{5}$ 15. $22\frac{3}{4}$
17. $3\frac{15}{16}$ 19. $\frac{3}{4}$ in. 21. $88\frac{1}{8}$ lb 23. a. $\frac{9}{32}$ in. b. $\frac{27}{64}$ in.
25. a. $\frac{1}{32}$ in. b. $\frac{1}{24}$ in. c. $\frac{7}{64}$ in. 27. a. $\frac{1}{16}$ in. b. $\frac{9}{64}$ in. c. $\frac{15}{64}$ in.

Division of Fractions

1. $\frac{8}{15}$ 3. $4\frac{1}{2}$ 5. $\frac{2}{3}$ 7. 9 9. $1\frac{13}{18}$ 11. $1\frac{7}{11}$ 13. 4 15. $1\frac{91}{225}$
17. 7 (The remaining $\frac{9}{13}$ can't make a whole piece of fuel line, so only the whole number of 7 can be used for the answer.)
19. $15\frac{1}{2}$ in. 21. $1\frac{1}{3}$ 23. $1\frac{3}{4}$ in. 25. $1\frac{29}{32}$ in. 27. $10\frac{1}{2}$ 29. $6\frac{1}{4}$

Converting Fractions to Decimals

1. 0.8   3. 0.7   5. 0.667   7. 6.2   9. 4.182   11. 7.846   13. 0.4375 in.
15. 3.641   17. 15.94   19. 11.403

CHAPTER 4

Equivalent Ratios

1. 40 to 1, 40:1, $\frac{40}{1}$ 3. 4 to 3, 4:3, $\frac{4}{3}$ 5. 528 to 63, 528:63, $\frac{528}{63}$ 7. 2:1, $\frac{2}{1}$
9. 9:10, $\frac{9}{10}$ 11. 4:1, $\frac{4}{1}$ 13. 2:5, $\frac{2}{5}$ 15. 1:8, $\frac{1}{8}$ 17. 256:65, $\frac{256}{65}$
19. 55 MPH   21. 3000 revolutions/min   23. $1.75/gallon   25. $79.16/tire

Writing and Solving Proportions

1. Equal   3. Not equal   5. Equal   7. Not equal   9. Not equal
11. 12   13. 17.333   15. 1.667   17. 25.6   19. 0.156   21. 33
23. 3.574   25. $\frac{3}{2}$ 27. $8\frac{1}{4}$ 29. a. 256 ounces   b. 2 gallons   31. 2.778 lb
33. 79.619 miles

Finding Nearest Fractional Part

1. $\frac{5}{8}$ in. 3. $4\frac{4}{8}$ in. = $4\frac{1}{2}$ in. 5. $1\frac{2}{8}$ in. = $1\frac{1}{4}$ in. 7. $\frac{5}{16}$ in.
9. $8\frac{6}{16}$ in. = $8\frac{3}{8}$ in. 11. $2\frac{13}{16}$ in. 13. $\frac{3}{16}$ in.

CHAPTER 5

Converting Fractions, Decimals, and Percents

1. $\frac{1}{2}$, 0.5 3. $\frac{2}{5}$, 0.4 5. $\frac{18}{25}$, 0.72 7. $\frac{3}{20}$, 0.15 9. $2\frac{39}{50}$, 2.78 11. $3\frac{11}{20}$, 3.55
13. $\frac{117}{250}$, 0.468 15. $\frac{1887}{2000}$, 0.9435 17. $\frac{21}{2500}$, 0.0084 19. $\frac{7}{62,500}$, 0.000 112
21. $\frac{3}{8}$, 0.375 23. $\frac{5}{6}$, 0.833 25. 64% 27. 70% 29. 5% 31. 183.7%
33. 9.8% 35. 0.0051% 37. 47% 39. 98% 41. 80% 43. 71%
45. 48.75% 47. 377.5% 49. 254.5% 51. 0.2% 53. 1.85%

Base, Rate, and Amount

1. Base = 40, Rate = 50%, Amount = 20 3. Base = 400, Rate = 4.75%, Amount = 19
5. Base = 6000, Rate = 0.06%, Amount = 3.6 7. 810 9. 88% 11. 500
13. 8.4 15. 0.06% 17. 180% 19. 21 jobs 21. 11.7 volts
23. 18.2% 25. 10% 27. $5339.81

Discount, Markup, and Sales Tax

1. $89.86 3. $91.79 5. $21.02 7. $99.68 9. $33.73 11. $29.96
13. $89.78 15. $45.78 17. $39.71 19. 54.8% 21. 52.8% 23. 6%
25. 5.5% 27. $95.18 29. $25.59 31. $232.00 33. $124.62
35. $327.10

Technical Applications

1. 117.3 psi, no 3. 130.9 psi, yes 5. Front: 1371.9 lb, Rear: 1318.1 lb
7. Front: 1785.3 lb, Rear: 1460.7 lb 9. 780 CCA 11. 158.4 hp 13. 13.9%
15. 8.1% decrease 17. About 9% 19. About 36%

CHAPTER 6

Metric Units of Measure

1. d 3. a 5. a 7. b 9. c

Metric Prefixes and Conversions

1. 0.145 g 3. 230 m 5. 0.0841 volts 7. 9,700,000 g 9. 5,200 dL
11. 5 L 13. 6.4 Pa 15. 894,000 cm 17. 0.089 dekavolts 19. 870 kL
21. 0.000 000 047 kg

Conversion of Units Using Ratios

1. 0.38 mi 3. 76.41 hp 5. 267.25 mi 7. 3.97 in.^{3} 9. 64.74 cm^3
11. 0.08 qt 13. 3.88 mi 15. 5.74 L 17. 280.66 in.^{3} 19. 38.64 MPH
21. 52.51 ft/second 23. 33.07 MPH 25. 7.35 km/L 27. 29.19 MPG

Conversion of Units Using Tables

1. 38.14 km   3. 527,265 joules   5. 22.47 ft   7. 25.46 ounces
9. 266.7 mm   11. 3.75 m³   13. 35.53 psi   15. 38.28 yards   17. 393.19 cm
19. 12.05 pints   21. 0.08 dm   23. 0.74 q⁺

Temperature Conversions

1. 40°C   3. 100°C   5. 83.89°C   7. −38.33°C   9. 68°F   11. 32°F
13. 188.6°F   15. 21.2°F

CHAPTER 7

Measurement

1. A. $\frac{5}{8}$ in.   B. 1 in.   C. $1\frac{3}{4}$ in.   D. $2\frac{1}{2}$ in.   E. $3\frac{1}{4}$ in.
3. A. $\frac{1}{4}$ in.   B. $\frac{3}{4}$ in.   C. $1\frac{1}{2}$ in.   D. $2\frac{1}{4}$ in.   E. $2\frac{3}{4}$ in.
5. A. 0.2 cm   B. 0.8 cm   C. 1.5 cm   D. 2.1 cm   E. 3.6 cm   F. 5.0 cm
7. 4 in.   9. $\frac{3}{4}$ in.   11. $1\frac{3}{8}$ in.   13. $3\frac{5}{16}$ in.   15. 5.3 cm
17. 0.9 cm   19. 4.2 cm

Perimeter

1. 9.7 in.   3. $8\frac{5}{8}$ in.   5. 21.4 cm   7. 25.6 cm   9. 24.0 cm   11. 43.5 cm
13. 16.5 cm, $6\frac{1}{2}$ in.   15. 12.9 cm, 5.1 in.   17. About 1060 ft

Pythagorean Theorem

1. 12.806 cm   3. 6 in.   5. 5.194 cm   7. 3.010 in.   9. 12.60 cm
11. 38.71 km

Area

1. 20.08 cm²   3. 38.88 cm²   5. 158.29 m²   7. 20,959.15 ft²   9. 7.5 in.²
11. 157.90 in.²   13. 30.44 cm²   15. 17.57 in.²   17. 1.83 in.²   19. 1.57 in.²
21. 1008 cm²   23. 13.85 in.²
25. The round filter is larger (103.9 in.² compared to 78.4 in.²).

Volume

1. 270 in.³   3. 98.91 ft³   5. 407.51 cm³   7. 6104 in.³, 26.4 gal
9. 28,800 in.³, 124.7 gal   11. 521 gal   13. 69.1 in.³, 1.2 qt   15. 0.9 qt
17. 3.1 qt   19. 275 cm³

Angle Measure and Applications

1. 45°   3. 165°   11. $\frac{3}{4}$ 13. 1.7%   15. 3°30′   17. 81°7.5′   19. 42°34.98′
21. 7°23.52′   23. 13.5°   25. 36.45°   27. 60.167°   29. 77.367°
31. 19°42′   33. 59°8′   35. 6°14′   37. 5°44′   39. 0.7°, yes
41. 57.6°, yes   43. 6.225°   45. 55.133°   47. 1.168°   49. 16°50′20.4″
51. 35°12′36″   53. 12°24′0″   55. 13°50′9″   57. 4°3′37″

CHAPTER 8

Signed Numbers

1. 6   3. −5   5. −7.6   7. −60   9. 65   11. $-\dfrac{1}{2}$ 13. −6°F
15. −10°F

Addition and Subtraction of Signed Numbers

1. 13   3. 6   5. −11   7. −18   9. 12   11. −5   13. −1   15. 19
17. −63   19. 2.6   21. −37.8   23. −5.6   25. −57.9   27. $2\dfrac{1}{2}$ 29. $51\dfrac{5}{6}$
31. 8.8   33. −36.69   35. 19°C   37. 14°   39. 1.75°   41. 3.75°

Multiplication and Division of Signed Numbers

1. 42   3. −36   5. 5   7. −1   9. 40   11. 8   13. 4   15. −5.75
17. $-1\dfrac{7}{8}$ 19. $-7\dfrac{3}{5}$ 21. −8°   23. −1.25°

CHAPTER 9

General Engine Measurements

1. 0.004 in.   3. 0.02 mm   5. 0.005 in.   7. 0.03 mm   9. 0.304 in.
11. 0.375 in.   13. 1.5:1   15. 0.448 in.   17. 1.224 cm   19. 1.6:1

Displacement

1. 4.6 cm   3. 1913.2 cm³   5. 411.2 in.³   7. 35.9 in.³   9. 598.3 cm³
11. a. 3962.0 cm³   b. 4.0 L   13. 141.3 in.³   15. 262.4 in.³   17. 2753.4 cm³
19. 260.8 in.³   21. 3932.7 cc, 3.9 L

Compression Ratio

1. 3.78 in.³   3. 0.78 in.³   5. 0.46 in.³   7. 6.69 cm³   9. 0.12 in.³
11. 4.62 in.³   13. 4.89 in.³   15. 9.2:1   17. 11.0:1, no   19. 10.6:1
21. 9.2:1   23. 8.6:1   25. 9.7:1

Changing Compression Ratio

1. a. 9.0:1   b. 9.2:1   3. a. Down   b. 9.2:1   c. 8.5:1   5. 8.8:1   7. 8.0:1

Diesel Applications

1. 17.4:1   3. 17.5:1   5. 18.1:1

CHAPTER 10

Engine Balancing

1. 493 g   3. 510 g   5. 777.5 g   7. 702.5 g
9. a. 385 g   b. 194 g   c. 406 g   d. 579 g   e. 406 g   f. 695.5 g
11. a. 442 g   b. 199 g   c. 432 g   d. 641 g   e. 432 g   f. 752.5 g

ANSWERS TO ODD-NUMBERED PRACTICE PROBLEMS

Torque and Horsepower

1. 124 lb·ft   3. 48 lb·ft   5. 1.5 ft   7. 960 lb   9. 221.1 hp
11. a. 3.2 hp   b. No. The engine may produce more horsepower at a different speed.
13. 65.2 hp, 90.3 hp, 119.3 hp, 163.7 hp, 208.2 hp, 248.5 hp, 261.8 hp

Camshaft Event Timing

1. Intake Duration is the number of degrees through which the crankshaft will turn while the intake valve is open.   3. No. The exhaust and intake valves may never be open at the same moment.   5. 202°   7. 104.5°   9. 204°   11. 107°   13. 198°
15. 0°   17. 212°   19. 8°

CHAPTER 11

Air/Fuel Ratios and Volumetric Efficiency

1. 58.8 lb   3. 20 lb   5. 15:1   7. 30:1   9. 160.6 CFM   11. 143.7 CFM
13. 94.6%   15. 94.5%   17. 105.6%   19. 94.4%

Induction System and Carburetor Sizing

1. 356 CFM   3. 318 CFM, 350 CFM   5. 474.0 CFM   7. 311.9 CFM
9. 245.7 CFM

Indicated Horsepower and Torque

1. 160.4 ihp   3. a. 177.3 ihp   b. 16.9 ihp   5. 149.4 ihp   7. 549.1 ft·lb
9. 236.7 ft·lb

Mechanical Efficiency

1. 79.4%   3. 80.1%   5. 80.9%   7. 82.3%

Altitude Compensation

1. a. 50 hp   b. 11 hp   c. 17 hp   3. a. 165 bhp   b. 182 bhp   c. 179 bhp
5. a. 202 bhp   b. 165 bhp   c. 194 bhp

CHAPTER 12

Gear Ratios

1. 2.00:1, reduction   3. 1.82:1, reduction   5. 3.73:1
7. 41 ring gear teeth, 10 pinion gear teeth   9. 240 lb·ft, 250 RPM
11. 454.5 lb·ft, 742.5 RPM   13. 2.00:1, 340 lb·ft, 1000 RPM
15. 114.0 hp, 1.82:1, 573.3 lb·ft, 1043.9 RPM, 114.0 hp

Overall Drive Ratios

1. 10.05:1, 6.63:1, 4.24:1, 2.73:1, 2.26:1   3. 3.55:1, 2.44:1, 1.70:1, 1:1
5. 1.63:1   7. 1:1   9. a. 2.80:1   b. 821.4 RPM
11. 19.44:1, 174.90 RPM   13. 308 lb·ft   15. 6318 lb·ft

Planetary Gear Ratios

1. 1.67:1   3. 3.33:1   5. 1.46:1   7. 0.47:1   9. 0.33:1   11. 0.70:1

Tire Sizing

1. 205 mm, 20.5 cm, 8.1 in.   3. 235 mm, 23.5 cm, 9.3 in.
5. 173 mm, 17.3 cm, 6.8 in.   7. 177 mm, 17.7 cm, 7.0 in.   9. 24.7 in.
11. 25.7 in.

Speedometer Calibration

1. 13 teeth   3. 21 teeth   5. 20 teeth   7. 27 teeth   9. 26 teeth
11. 19 teeth   13. 24 teeth

CHAPTER 13

Force, Pressure, and Area

1. 8484 lb   3. 282.1 psi   5. 8.1 in.^{2} 7. 785 lb   9. 237.3 psi
11. 0.51 in.^{2} 13. 5299 lb   15. 183 psi   17. 1.38 in.

Braking Systems

1. 4.5:1   3. 4.77:1   5. 112.5 lb   7. 560 lb   9. 173 lb   11. 199 psi
13. 70.2 psi   15. a. 400 lb   b. 905.9 psi   c. 2567 lb   d. 1024 lb

Hydraulic Pumps and Flow Rates

1. 0.79 in.^{3} 3. 7.22 in.^{3} 5. 10.1 GPM   7. 9.6 GPM   9. 13.0 GPM
11. 7.5 GPM   13. 11.7 GPM, 27.4 GPM

Hydraulic Cylinders, Force, and Actuation Speed

1. a. 12.56 in.^{2} b. 9.42 in.^{2} 3. a. 113.0 in.^{2} b. 84.8 in.^{2}
5. a. 16,328 lb   b. 12,246 lb   c. 1.23 in./second   d. 1.63 in./second
7. a. 169,500 lb   b. 127,200 lb   c. 0.43 in./second   d. 0.57 in./second

CHAPTER 14

Current, Voltage, Resistance, and Ohm's Law

1. 0.4 amps or 400 milliamps   3. 3.15 Ω   5. 14.4 volts
7. Yes, the motor draws 20 amps.
9.

Current (Amps)	Potential (Volts)	Resistance (Ohms)
Increased	Increased	No change
No change	**Increased**	Increased
Increased	Increased	No change
Decreased	No change	**Increased**
Decreased	No change	Increased
No change	Decreased	**Decreased**

Electrical Power

1. 1260 watts   3. The saw, 360 watts   5. 8.33 amps
7. 1119 watts or 1.119 kilowatts   9. 1.98 Ω   11. 8.29 volts

Resistor Circuits

1. a. 49 Ω   b. 0.3 amps   c. 3.2 watts   3. a. 28.1 Ω   b. 0.4 amps   c. 5.1 watts
5. a. 1 Ω   b. 12.6 amps   c. 158.8 watts   7. a. 3.5 Ω   b. 3.6 amps   c. 45.6 watts
9. a. 18.7 Ω   b. 0.6 amps   c. 7.7 watts   11. a. 6.3 volts   b. 12.6 volts   c. The voltage drops in a series circuit must add up to the battery voltage.
13. a. 6.3 amps   b. 12.6 amps   c. The individual current values in a parallel circuit must add up to the total current.   15. 12 gauge   17. 10 gauge

CHAPTER 15

Distance, Speed, and Time Relationships

1. 48 MPH   3. 52.2 MPH   5. 52.2 mi
7. No. The actual speed is about 46.9 MPH. The speedometer error is 6.1 MPH.
9. 44.8 MPH   11. b. 12:00 to 12:30, 70 MPH   c. 11:00 to 11:30, 2 MPH   d. 2:00
e. 45.5 MPH

Velocity, Acceleration, and Time Relationships

1. 8.75 ft/second^2 3. −15 ft/second^2 5. 111.69 ft/second   7. 31.4 MPH
9. 11.0 ft/second^2 11. 0.55 G's
13. a.

Time (seconds)	Velocity (MPH)	Velocity (feet per second)
0	5	**7.3**
5.0	47	**68.9**
10.0	63	**92.4**
15.0	71	**104.1**
20.0	71	**104.1**
25.0	68	**99.7**
30.0	70	**102.7**

b.
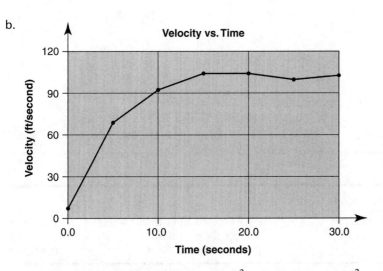

c. From 0 to 5.0 seconds, 12.32 ft/second^2 d. 2.34 ft/second^2 e. 0 ft/second^2
f. The car is traveling at a constant velocity.

15. a.

Time (seconds)	Velocity (MPH)	Velocity (feet per second)
0	60	88.0
0.5	58	85.1
1.0	55	80.7
1.5	49	71.9
2.0	35	51.3
2.5	20	29.3
3.0	11	16.1
3.5	6	8.8
4.0	0	0

b.
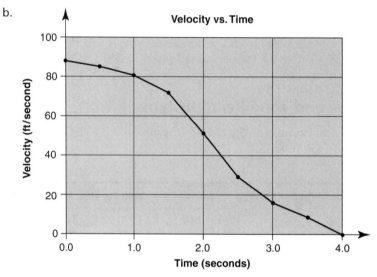

c. From 2.0 to 2.5 seconds, -44 ft/second^2 d. -17.6 ft/second^2 e. -26.4 ft/second^2
f. After coming to a complete stop, the driver began to accelerate once again.

CHAPTER 16

Filling Out a Repair Order

1. 36 minutes   3. 1 hour 18 minutes   5. 8 hours 48 minutes
7. 5 hours 54 minutes   9. 0.5 hours   11. 0.3 hours   13. 5.9 hours
15. $195.00   17. $49.95   19. $149.99   21. 1.9 hours, $110.20
23. 22.5 hours, $1305.00   25. 0.5 hours, $29.00
27.

Parts		Labor	
Air filter	$18.25	Replace air filter	0.2 hours
Muffler	$115.00	Replace muffler	0.5 hours
		Rotate tires	0.4 hours
Parts: $133.25	Labor: $55.00	Tax: $7.33	Total: $195.58

29.

Parts		Labor	
Oil filter	$4.99	Replace oil & filter	0.9 hours
Oil	$11.80		
Fuel filter	$17.99	Replace fuel filter	0.6 hours
Front exhaust manifold	$260.00	Replace front exhaust manifold	4.6 hours
Manifold gasket	$9.25		
Clutch disk	$105.00	Replace clutch disk & flywheel	(9.8 + 0.2) 10.0 hours
Flywheel	$149.99		
Power steering pump	$205.00	Replace power steering pump	1.3 hours
Left window regulator	$64.95	Replace window regulator	1.2 hours

Parts: $817.17   Oil: $11.80   Labor: $1264.80   Tax: $45.59
Total: $2139.36

index

A

acceleration, and velocity, 285
addend, 2
addition
   of angles, 129
   of decimals, 22–24
   of fractions, 49–51
   of signed numbers, 154–157
   of whole numbers, 2
air/fuel ratios, 209
airflow
   actual, 211
   theoretic, 210
alternator, 268
altitude compensation, 216–218
amount, 82
ampere, 265
angle
   right, 117
   units of measure, 126–130
angular
   degree, 126
   minute, 128
   second, 129
annulus, 255
antifreeze
   proportion, 72
   ratio, 69
area, 119 (*See also* force, pressure)
   of a circle, 122
   conversion factors, 104
   and hydraulic systems, 247
   of a rectangle, 120
   of a triangle, 120
aspect ratio, 232
axle ratio (*See also* differential ratio)
   effect on speedometer, 235

B

balance pads
   diagram, 191
   purpose of, 189
base
   of an exponential number, 9
   of a percent problem, 82
   of a triangle, 120
BDC. *See* bottom dead center
bhp. *See* brake horsepower
bobweight, 191
bore, 165
borrowing, 3
bottom dead center (BDC), 165
   and camshaft diagrams, 198
braces, 9
brackets, 9
brake
   caliper, 252
   drum, 23, 253
   pedal, 250
brake horsepower, 196, 215
braking systems
   hydraulic principles, 250
   mechanical advantage of brake pedal, 250
British thermal unit (Btu), 98
burette, 169

C

caliper. *See* brake caliper
camber, 127, 152
camshaft
   diagrams, 197
   event timing, 197–201
   lobe lift, 164
carburetor sizing, 211–212
caster, 129, 161
circumference, 115
clearance volume, 168
combustion chamber volume, 168
common denominators, 47–49
compression ratio, 167
   calculating, 167–174
   changing, 175–176
   of a diesel engine, 176
compression test, 87
connecting rod
   balancing, 189–190
   role in engine balance, 186
contracting speed. *See* ram speed
conversion (*See also* converting)
   factors, 104
   metric prefix, 100
   table, 106
   temperature, 106
converting (*See also* conversion)
   angles, 128–129
   between decimals and percents, 81
   between fractions and percents, 80
   fractions to decimals, 58–59
   metric prefixes, 100–103
   temperatures, 107
   time, 294
   using ratios, 103–105
   using tables, 105–107
cooling system, 53
cost, 84. (*See also* markup; retail)
counterweight, 194
crankshaft
   counterweight, 194
   role in engine balance, 186–188
cross multiplying, 71
current, 265
cylinder
   double-acting hydraulic, 255
   hydraulic, 248, 255
   swept volume, 165
   volume, 125

D

decimal numbers
　addition and subtraction, 22–24
　converting to percents, 81
　division, 28–30
　multiplication, 26–28
　reading and writing, 20–21
　rounding, 21–22
deck height volume, 173
degree (angular), 126
denominator, 11, 43
dependent variable, 282
diameter, 115
diesel
　combustion chambers, 177
　compression ratios, 176
　pistons, 178
difference, 2
differential ratio, 224. (See also axle ratio)
direct proportion, 235
discount, 84–86
displacement. See engine displacement
distance traveled, 284
dividend, 7
division
　of decimals, 28–30
　of fractions, 56–58
　of signed numbers, 157–158
　of whole numbers, 7–9
divisor, 7
double-acting cylinder, 255
drive gear, 222
driven gear, 222
duration
　exhaust, 199
　intake, 199

E

efficiency
　mechanical, 215–216
　volumetric, 210–211
electrical power, 267
electromotive force, 265
electrons, 265
energy
　conversion factors, 104
　units of, 98
engine
　balancing, 186–194
　diesel compression ratio, 176
　displacement, 165–67
　measurements, 163–165
equivalent fractions, 43–47
equivalent ratios, 68–70
exhaust duration, 199
exponents, 9
extension speed. See ram speed

F

factor, 4
flow rate of pumps, 253
force (See also area; pressure)
　conversion factors, 104
　and hydraulic systems, 247
　and torque, 195
　units of, 98
fractions
　addition of, 49–51
　converting to decimals, 58–59
　converting to percents, 80
　division of, 56–58
　equivalent, 43–47
　improper, 44
　lowest terms, 47
　mixed number, 45
　multiplication of, 54–56
　proper, 44
　subtraction of, 51–53
fuel economy
　calculating, 8, 30
　conversions, 106

G

gallons per minute, 253
gas mileage. See fuel economy
gear
　drive, 222
　driven, 222
　overdrive, 225
　planetary. See planetary gear set
　reduction, 225
　ring and pinion, 224
　speedometer, 234
gear ratios, 224
　calculation of, 224
　effect on horsepower, 223, 225
　effect on torque, 223, 225
GPM. See gallons per minute
graph, 282

H

head gasket volume, 170
height of a triangle, 120
horsepower, 196. (See also torque)
　brake, 196
　formula, 196
　indicated, 213–215
　unit of, 98
hydraulic braking systems. (See also braking systems)
　cylinders, 248, 255
　double-acting cylinders, 255
　force, pressure, area relationship, 247
　pumps,
hypotenuse, 117

I

independent variable, 282
indicated horsepower (ihp), 213–215
indicated torque, 213–215
induction system sizing, 211–213
intake
　centerline, 199
　duration, 199
inverse proportion, 234

L

labor charge, 293
LCD. See lowest common denominator
leg of a right triangle, 117
length
　conversion factors, 104
　units of, 98
lever arm, 195
lobe lift, 164
lowest common denominator (LCD), 47
lowest terms. See fractions

M

markup, 84. (See also cost; retail)
master cylinder
　part of braking system, 250
　pressure created by, 251
mean effective pressure (MEP), 213
measurement
　using a micrometer, 24
　using inches, 113
　using centimeters, 114
　using millimeters, 114
mechanical advantage, 250
mechanical efficiency, 215–216
MEP. See mean effective pressure
metric
　conversions, 100–103
　prefixes, 100
　staircase, 102
　units of measure, 98–100
micrometer
　English, 24
　metric, 26
minuend, 3
minute (angular), 128
mixed number. See fractions
multiplication
　of decimals, 26–28
　of fractions, 54–56
　of signed numbers, 157–158
　of whole numbers, 4–6

N

nearest fractional part, 74–75
negative numbers. See signed numbers
Newton, 98

number line, 152, 153
number system
    decimal, 20
    whole, 1
numerator, 11, 43

O

octane rating, 167
odometer, 3
Ohm's law, 266–267
ohms, 265
order of operations, 9–11
out-of-round, 164
overall drive ratios, 226–228

P

parallel circuit, 271
parentheses, 9
Pascal, 98
percents, 79
    base, rate, and amount, 82
    converting to decimals, 81
    converting to fractions, 80
    as a ratio, 80
perimeter, 114
    of a figure, 114–117
pi, 115
piston pump
    displacement, 253
    flow rating, 255
piston relief volume, 174
place values, 1
planetary gear set
    carrier, 229
    components, 228–229
    ratio, 229–231
    table, 231
potential, 265
power
    conversion factors, 104
    electrical, 267
    of a number, 9
    units of, 98
pressure. (See also area; force)
    conversion factors, 104
    and hydraulic systems, 247
    mean effective, 213
    units of, 98
product, 4
proportions
    direct, 235
    inverse, 234
    solving, 71–74
protractor, 127
    writing, 71
pump
    displacement, 253
    flow rating, 255
    types of hydraulic, 253–254
Pythagorean theorem, 117–119

Q

quotient, 7

R

radial tire, 232
radius, 115
ram
    hydraulic cylinder, 255
    speed, 258
rate
    of a percent problem, 82
    unit, 70
ratios
    air/fuel, 209
    antifreeze and water, 69
    axle (effect on speedometer), 234
    conversions using, 103–105
    equivalent, 68–70
    fractional, 69
    gear. See gear ratios
    gasoline to oil, 68
    overall drive, 226–228
    planetary gear. See planetary gear set
    rocker arm, 165
    unit rates, 70
reading micrometers. See micrometer
retail, 85. (See also cost; markup)
reciprocal, 56
reciprocating
    assembly, 187
    weight, 190
resistance, 265
    of parallel circuits, 271
    of series circuits, 269
    of a spark plug wire, 6
resistor circuits
    parallel, 271
    series, 269
ring and pinion gears, 225
ring gear (planetary component), 229
repair orders, 293–297
rocker arm ratio, 165
rotating
    assembly, 187
    weight, 190
rounding, 21–22
ruler
    inch, 113
    metric, 114

S

sales tax, 86, 293
schematic symbols, 272
second (angular), 129
series circuit, 269
signed numbers
    addition and subtraction, 154–157
    multiplication and division, 157–158
    value, 152

slope, 282, 287
solenoid, 276
speed
    average, 280
    instantaneous, 280
speedometer
    calibration, 233–236
    gears, 234
    showing instantaneous speed, 281
sphere, 126
square root, 30
starter, 275
stroke, 165
subtraction
    of angles, 129
    of decimals, 22
    of fractions, 51–53
    of signed numbers, 154–157
    of whole numbers, 2–4
subtrahend, 3
sum, 2
sun gear, 229
swept volume, 165

T

taper of a cylinder, 163
TDC. See top dead center
temperature conversions, 107
theoretic airflow, 210
time
    allowed for repairs, 293
    converting hours and minutes, 293–294
    and speed, distance, 280–281
    and velocity, acceleration, 285
tire
    aspect ratio. See aspect ratio
    calculating diameter, 233
    effect on speedometer, 234
    sidewall height, 232
    width, 232
top dead center (TDC), 165
    and camshaft diagrams, 198
torque, 194. (See also horsepower)
    conversion factors, 104
    indicated, 213–215
    units of, 98
transmission
    countershaft, 227
    gear ratios, 226–228
    input shaft, 227
    output shaft, 227
triangle, area of right, 117
turbocharger, 88

U

unit rate, 70

V

valve overlap, 199
variable
   dependent, 282
   independent, 282
velocity
   and acceleration, 285
   change in, 287
   conversions, 106
volt, 265
volume, 123
   clearance, 168
   conversion factors, 104
   of a cylinder, 125
   deck height, 173
   head gasket, 170
   piston relief, 173
   of a rectangular box, 124
   of a sphere, 126
   units of, 98
volumetric efficiency, 210–211

W

Watt, 98, 267
weight
   conversion factors, 104
   units of, 98
weight distribution, 89
wheel cylinder, 250, 253
whole number
   addition, 2
   division, 7–9
   multiplication, 4–6
   reading and writing, 1
   subtraction, 2–4
winch, 195
wire gauge table, 273